Advances in Intelligent Systems and Computing 209

Editor-in-Chief

Prof. Janusz Kacprzyk
Systems Research Institute
Polish Academy of Sciences
ul. Newelska 6
01-447 Warsaw
Poland
E-mail: kacprzyk@ibspan.waw.pl

For further volumes:
http://www.springer.com/series/11156

Phayung Meesad, Herwig Unger,
and Sirapat Boonkrong (Eds.)

The 9th International Conference on Computing and Information Technology ($IC^2IT2013$)

May 9th–10th, 2013 King Mongkut's
University of Technology North Bangkok,
Bangkok, Thailand

Springer

Editors
Phayung Meesad
Information Technology
King Mongkut's
University of Technology North Bangkok
Bangkok
Thailand

Sirapat Boonkrong
Information Technology
King Mongkut's
University of Technology North Bangkok
Bangkok
Thailand

Herwig Unger
Communication Networks
FernUniversität in Hagen
Hagen
Germany

ISSN 2194-5357 ISSN 2194-5365 (electronic)
ISBN 978-3-642-37370-1 ISBN 978-3-642-37371-8 (eBook)
DOI 10.1007/978-3-642-37371-8
Springer Heidelberg New York Dordrecht London

Library of Congress Control Number: 2013934473

Printed on acid-free paper

Springer is part of Springer Science+Business Media (www.springer.com)

Preface

This volume contains the papers of the 9th International Conference on Computing and Information Technology (IC^2IT 2013) held at King Mongkut's University of Technology North Bangkok (KMUTNB), Bangkok, Thailand, on May 9th-10th, 2013. Traditionally, it is organised in conjunction with the National Conference on Computing and Information Technology, one of the leading Thai national events in the area of Computer Science and Engineering.

For the first time, the conference has been structured into 3 main tracks on Data Networks/Communication, Data Mining/Machine Learning, and Human Interfaces/Image processing. This year, the international program committee received exactly 100 submissions from authors of 21 countries on 5 continents. Each submission was reviewed by at least 2, mostly 3 members of the program committee to avoid contradictory results. On these judgments, the committee decided to accept 29 papers for oral presentation and inclusion in the conference proceedings. Moreover, this is the first time we published within the Springer series on Advances in Intelligent and Soft Computing. In addition, four internationally well-known scientists have been invited and accepted to give keynote talks to our participants.

A special thanks is given to KMUTNB President, Professor Dr. Teeravuti Boonyasopon for his support of our conference from the first year on, and for providing us with a lot of resources from KMUTNB. We hope that IC^2IT again provides great opportunities for academic staff, students and researchers to present their work. IC^2IT is also a platform for exchange of knowledge in the field of computer and information technology and shall inspire researchers to generate new ideas and findings and meet partners for future collaboration. We also hope that our participants use this occasion to learn more about Thailand and its beautiful scenery, people, culture and visit its famous historic sights before or after the conference.

We would also like to thank all authors for their submissions and the members of the program committee for their great work and valuable time. A lot of technical and organizational work has been done by the staff of the Information Technology Faculty at KMUTNB. A very special and warm thank you is given to our web masters: Ms. Kanchana Viriyapant, Mr. Jeerasak Numpradit, and Mr. Armornsak Armornthananun. Without the meticulous work of Ms. Watchareewan Jitsakul and Ms. Thitinan Ngamsanguan the proceedings could not have been completed in the needed form at the right time.

After so much preparation, all of the organisers of course hope and wish that IC^2IT 2013 will again be a successful event and will be remembered by the participants for a long time.

February 15th, 2013
Bangkok

On behalf of all organizers
Phayung Meesad
Herwig Unger
Sirapat Boonkrong

Organization

Program Committee

T. Bernard	Syscom CReSTIC, France
W. Bodhisuwan	Kasetsart University, Thailand
H. Cam Ha	Ha Noi University of Education, Vietnam
M. Caspar	Chemnitz University of Technology, Germany
T. Chintakovid	KMUTNB, Thailand
H. K. Dai	Oklahoma State University, USA
D. Delen	Oklahoma State University, USA
N. Ditcharoen	Ubon Ratchathani University, Thailand
T. Eggendorfer	German Police University, Germany
R. Gumzej	University of Maribor, Slovenia
M. Hagan	Oklahoma State University, USA
C. Haruechaiyasak	NECTEC, Thailand
S. Hengpraprohm	Nakhon Pathom Rajabhat University, Thailand
K. Hengproprohm	Nakhon Pathom Rajabhat University, Thailand
U. Inyaem	RMUTT, Thailand
T. Joochim	Ubon Ratchathani University, Thailand
A. Kongthon	NECTEC, Thailand
S. Krootjohn	KMUTNB, Thailand
M. Kubek	FernUniversität in Hagen, Germany
S. Kukanok	Mahasarakham Rajabhat University, Thailand
J. Laokietkul	Chandrakasem Rajabhat University, Thailand
B. Lent	HERMES GROUP/ ISB, Germany
K. Limtanyakul	KMUTNB, Thailand
A. Mahaweerawat	Ubon Ratchathani University, Thailand
A. Mikler	University of North Texas, USA
A. Mingkhwan	KMUTNB, Thailand

C. Namman	Ubon Ratchathani University, Thailand
C. Netramai	KMUTNB, Thailand
K. Nimkerdphol	RMUTT, Thailand
S. Nitsuwat	KMUTNB, Thailand
S. Nuanmeesri	Suan Sunanta Rajabhat University, Thailand
P. P. Na Sakolnakorn	Mahasarakham Rajabhat University, Thailand
N. Porrawatpreyakorn	KMUTNB, Thailand
P. Prathombutr	NECTEC, Thailand
A. Preechayasomboon	TOT, Thailand
P. Saengsiri	TISTR, Thailand
P. Sanguansat	Panyapiwat Institute of Management, Thailand
S. Smanchat	KMUTNB, Thailand
M. Sodanil	KMUTNB, Thailand
S. Sodsee	KMUTNB, Thailand
B. Soiraya	Nakhon Pathom Rajabhat University, Thailand
T. Srikhacha	TOT, Thailand
W. Sriurai	Ubon Ratchathani University, Thailand
P. Sukjit	FernUniversität in Hagen, Germany
W. Tang	City University of Hong Kong, Hong Kong
T. Tilli	Ingenieurbüro Tilli, Germany
D. H. Tran	Ha Noi University of Education, Vietnam
K. Treeprapin	Ubon Ratchathani University, Thailand
H. M. Tsai	National Taiwan University, Taiwan
D. Tutsch	University of Wuppertal, Germany
N. Utakrit	KMUTNB, Thailand
N. Wisitpongphan	KMUTNB, Thailand
K. Woraratpanya	KMUTL, Thailand
C. Yawut	KMUTNB, Thailand

In Cooperation with

King Mongkut's University of Technology North Bangkok (KMUTNB)
FernUniversitaet in Hagen, Germany (FernUni)
Chemnitz University, Germany (CUT)
Oklahoma State University, USA (OSU)
Edith Cowan University, Western Australia (ECU)
Monash University, Australia
Hanoi National University of Education, Vietnam (HNUE)
Mahasarakham University (MSU)
Ubon Ratchathani University (UBU)
Kanchanaburi Rajabhat University (KRU)
Nakhon Pathom Rajabhat University (NPRU)
Mahasarakham Rajabhat University (RMU)
Rajamangala University of Technology Lanna (RMUTL)
Rajamangala University of Technology Krungthep (RMUTK)
Rajamangala University of Technology Thanyaburi (RMUTT)
Council of IT Deans of Thailand (CITT)

IC^{2}IT 2013 Organizing Committee

Conference Chair	Phayung Meesad, KMUTNB, Thailand
Technical Program Committee Chair	Herwig Unger, FernUni, Germany
Secretary and Publicity Chair	Sirapat Boonkrong, KMUTNB, Thailand

Contents

Bipolarity in Judgments and Assessments: Towards More Realistic and Implementable Human Centric Systems

Janusz Kacprzyk

Fellow of IEEE, IFSA
Full Member, Polish Academy of Sciences
Foreign Member, Spanish Royal Academy of Economic and Financial Sciences (RACEF)
Systems Research Institute Polish Academy of Sciences
kacprzyk@ibspan.waw.pl

Abstract. We are concerned with the conceptualization, analysis and design of human centric systems. Such systems are meant, roughly speaking, as those in which a human being is a relevant (if not principal) element of a computer based system, and – by obvious reasons – there exists an inherent communication and articulation gap between the human and the computer implied first of all by different languages employed, i.e. strings of bits and natural language. Some human-computer interface (HCI) should therefore be employed to bridge that gap. Its very essence and purpose boils down to the use of most human consistent tools, techniques and solutions assuming that it is easier to change the machine than the adult human being. Obviously, there is a multitude of possible options in this respect and in our talk we consider one that is related to a proper representation and processing of human judgments and assessments that are crucial while considering any human – computer interaction. In this context we consider a known fact that when a human being is requested to provide a judgment or assessment concerning an option or course of action, he or she very often tends to to provide them in a bipolar version. This is meant in the talk in the sense of providing testimonies concerning separately positive and negative aspects (pros and cons), mandatory and optional conditions to be fulfilled, etc. To start with, we review two types of scales employed to quantified such type of bipolar testimonies. The first is the bipolar univariate scale, in which there is a neutral (0) point, and the negative part (0-1], related to a negative testimony, and a positive part (0,1], related to a positive testimony. The second is the unipolar bivariate scale in which two separate scales are employed, both with values in [0,1], expressing separately the positive and negative testimonies. The main problem is how to aggregate the positive and negative testimonies related to an option in question. We present two basic approaches, one that is decision theoretic and is based on some special multicriteria decision making scalarizing functions, and one which is logical and is based on multivalued logic with properly chosen definitions of logical operations. We present applications of these approaches to database querying and information retrieval, and to a multicriteria choice of design options of computer systems. We advocate such a bipolar setting and outline some possible future direction.

P. Meesad et al. (Eds.): *IC²IT2013*, AISC 209, pp. 1–2.
DOI: 10.1007/978-3-642-37371-8_1

Curriculumn Vitae: Janusz Kacprzyk graduated from the Department of Electronic, Warsaw University of Technology in Warsaw, Poland with M.Sc. in automatic control, his Ph.D. in systems analysis and D.Sc. ("habilitation") in computer science from the Polish Academy of Sciences.

He is Professor of Computer Science at the Systems Research Institute, Polish Academy of Sciences, Professor of Computerized Management Systems at WIT – Warsaw School of Information Technology, and Professor of Automatic Control at PIAP – Industrial Institute of Automation and Measurements, in Warsaw, Poland, and Department of Electrical and Computer Engineering, Cracow University of Technology, in Cracow, Poland.

He is Honorary Foreign Professor at the Department of Mathematics, Yli Normal University, Xinjiang, China, and Visiting Scientist at the RIKEN Brain Research Institute in Tokyo, Japan. He is Full Member of the Polish Academy of Sciences and Foreign Member of the Spanish Royal Academy of Economic and Financial Sciences (RACEF). He is Fellow of IEEE and of IFSA.

He was a frequent visiting professor in the USA, Italy, UK, Mexico and China. His main research interests include the use of computational intelligence, notably fuzzy logic, in decisions, optimization, control, data analysis and data mining, with applications in databases, ICT, mobile robotics, etc.

He is the author of 5 books, (co)editor of 60 volumes, (co)author of ca. 400 papers. He is the editor in chief of 5 book series at Springer, and of 2 journals, and a member of editorial boards of more than 40 journals. He is a member of Award Committee of IEEE CIS, a member of Adcom (Administrative Committee) of IEEE CIS, and a Distinguished Lecturer of IEEE CIS.

He received many awards, notably: The 2006 IEEE CIS Pioneer Award in Fuzzy Systems, The 2006 Sixth Kaufmann Prize and Gold Medal for pioneering works on soft computing in economics and management, and The 2007 Pioneer Award of the Silicon Valley Section of IEEE CIS for contribution in granular computing and computing in words, and Award of the 2010 Polish Neural Network Society for exceptional contributions to the Polish computational intelligence community. Currently he is President of the Polish Operational and Systems Research Society and Past President of IFSA (International Fuzzy Systems Association).

Critical Issues and Information Security and Managing Risk

Mark Weiser

Associate Dean and Fleming Professor of Technology Management
Director, Center for Telecommunications and Network Security
William Spears School of Business
Oklahoma State University, Stillwater, OK, USA
weiser@oksate.edu

Abstract. Threat vectors against information systems are constantly changing and increasing in both diversity and frequency. This talk will review the latest threats to global information assets and mechanisms to assess risk exposure and mitigation approaches. Using examples from academia, industry, personal experience, and audience members; a spotlight will be cast on the major vulnerabilities that pervade our daily lives.

Appropriate access to most information technology resources inherently requires some risk. Assessing, eliminating, mitigating, and accepting risk then become functions that are necessarily performed by both individuals and organizations. Just as the threats themselves are misunderstood, so too are each of these four risk management elements often mismanaged. We'll explore structures to address each element, common theoretical and practical errors in application, and how these gaps might be closed by a different approach or through future research.

Finally, we'll review how the very actions that expose individuals and companies to significant risk may be exploited to thwart and prosecute criminals, by looking at recent approaches in digital forensics.

Curriculum Vitae: Mark Weiser is the Fleming Professor of Information Technology Management and Professor in Management Science and Information Systems and serves as Associate Dean for Undergraduate Programs, Spears School of Business at Oklahoma State University, Stillwater, USA since 2006. He is also the director of center for Telecommunications and Network Security since 2003. In addition, he is in charge of the Master of Science in Telecommunications Management program. His research interests are in the area of Information Assurance & Forensics, Applied telecommunications and applications, Technology to support education and training, and Organizational Memory.

P. Meesad et al. (Eds.): *IC²IT2013*, AISC 209, p. 3.
DOI: 10.1007/978-3-642-37371-8_2

Visualising the Data: Now I Understand

Ken Fowle

Head of School, Computer and Security Science
Edith Cowan University (ECU), Western Australian
k.fowle@ecu.edu.au

Abstract. Visualising information is increasing in situations where complex events are being presented to people who often have no understanding of what has/could have happened, procedures, methodologies or science. Computer Graphics (CG) can visually present scenarios based on scientific methodologies as well as depicting the perception of a witness to show what may have occurred. But more importantly CG can illustrate "what if…" questions and explore the inconsistencies and discrepancies within evidence and expert witness testimony. Therefore representing an important development in forensic graphics that are unparalleled due to its ability to assimilate and analyse data. However it is very important that when we use "science" to determine the validity of evidence or information that is done in a manner that is acceptable to the scientific community.

Curriculumn Vitae: Ken Fowle is currently the Head of School, Computer and Security Science, at Edith Cowan University (ECU), Western Australian and a Adjunct Associate Professor at the University of Western Australia, Centre of Forensic Science (CFS). Prior to moving over to academia in 2011, Dr. Fowle was employed by the Department of Mines and Petroleum in the Investigation Branch.

His research interest is in the use of 3D laser scanners in incidents and accidents and the use of visualisation as a tool for law enforcement and security. He works closely with the WA Police Sevices Forensic Surveying Branch and other national and international law enforcement agencies.

Dr. Fowle's interest in visualisation and accident reconstruct started back in 1996, when seconded to the departments Mine Safety Branch to assist with developing computer applications for mining accident and incidents. This interest was further enhanced in 1999 when he was seconded to Central Tafe to establish a research and development group specifically for developing computer graphics for the resource sectors of Western Australia. During his time at Central Tafe, Dr. Fowle undertook a PhD with the University of Nottingham (UK) and was conferred in 2003.

P. Meesad et al. (Eds.): *IC²IT2013*, AISC 209, pp. 5–6.
DOI: 10.1007/978-3-642-37371-8_3

In 2003 Dr. Fowle returned to the Department of Mines and Petroleum where he continued his research into visualisation and won funding from the Western Australian Government, to continue research in the use of 3D environments for accident reconstruction.

Dr. Fowle is past president of the Australian and New Zealand Forensic Science Society (WA) and is still an active committee member, a member of the International Association for Forensic Survey and Metrology, American Society for Industrial Security, Australian Computer Society and the Australian Law Enforcement Forensic Surveying Working Group.

Improved Computational Intelligence through High Performance Distributed Computing

David Abramson

Director of Research Computing Centre
The University of Queensland, Australia
david.abramson@monash.edu

Abstract. Modern computational and business intelligence techniques are increasingly used for product design and complex decision making. However, as the problems they solve become more difficult, the need for high performance computing becomes increasingly important. Modern Cloud and Grid platforms provide an ideal base for supporting this work, but typically lack software support. Over the past 20 years we have developed a computational framework, called Nimrod that allows users to pose complex questions underpinned by simulation technologies. Nimrod allows users to perform what-if analysis across an enormous number of options. Nimrod also supports experimental design techniques and provides automatic optimisation algorithms (e.g. genetic algorithms) that search through design options. Nimrod has been used globally to across a wide range of application in science, environmental modelling, engineering and business.

In this talk I will describe the Nimrod framework, and show examples of how it has supported a range of scientific and engineering questions. I will show how Clouds and Grids support these case studies, and outline our continued research in the area.

Curriculum Vitae: David Abramson is currently a Professor and the Director of the Centre for Research Computing at the University of Queensland, Brisbane, Australia. Prior to that, he was the Director of the Monash e-Education Centre, Science director of the Monash e-Research Centre and a Professor of Computer Science in the Faculty of Information Technology at Monash University, Australia. He has also held senior appointments at Griffith University, CSIRO, and RMIT. He is a fellow of the Association for Computing Machinery (ACM), the Australian Computer Society and the Academy of Science and Technological Engineering (ATSE), and a Senior Member of the IEEE.

P. Meesad et al. (Eds.): *IC²IT2013*, AISC 209, pp. 7–8.
DOI: 10.1007/978-3-642-37371-8_4

Abramson has served on committees for many conferences and workshops, and has published over 200 papers and technical documents. He has given seminars and received awards around Australia and internationally and has received over $8 million in research funding. He also has a keen interest in R&D commercialization and some of his research tools are available commercially.

Abramson's current interests are in high performance computer systems design and software engineering tools for programming parallel and distributed supercomputers.

Lying in Group-Communication in El-Farol-Games Patterns of Successful Manipulation

Frank Großgasteiger and Coskun Akinalp

Fernuniversität in Hagen,
Fakultät für Mathematik und Informatik
Universitätsstraße 27, PRG
D-58084 Hagen, Germany
frank@grossgasteiger.de, coskun@akinalp.com

Abstract. The El-Farol-Bar-Problem is a well-established tool for behavior analysis. Recent studies focused on group communication in minority games of this kind. We introduce the possibility of lying to the behavior of agents in communicating groups. We found a successful strategy for lying, if the group is composed by specific characters.

Keywords: El-Farol, Group-Communication, Limbic, Lying, Manipulation, Minority Game, Psychology.

1 Introduction and Motivation

Group communication may be viewed as a tool to accomplish a common task and therefore a way to identify the best decision for all. In this case the efficiency of decision-making tends to increase as the complexity of the task increases [1].

On the other hand group communication may be used to gather information (passive) and to influence others (active) to accomplish individual success. In situations of competition it is impossible to achieve individual success for all group members. Situations like these are modeled in minority games, e.g. the El-Farol-Bar-Problem by Brian Arthur [2]:

"*N* people decide independently each week whether to go to a bar that offers entertainment on a certain night. For correctness, let us set *N* at 100. Space is limited, and the evening is enjoyable if things are not too crowded specifically, if fewer than 60 percent of the possible 100 are present. There is no sure way to tell the numbers coming in advance; therefore a person or an agent goes (deems it worth going) if he expects fewer than 60 to show up or stays home if he expects more than 60 to go." [3]

The original El-Farol-Bar-Problem has no kind of group communication. Each agent had to decide individually and without any knowledge about the decisions and motivations of the others.

When introducing group communication in El-Farol, the population of players can be understood as a social network. Within this network the individual decisions are submitted to the peers of an agent, which may their decision subsequently influenced

P. Meesad et al. (Eds.): *IC²IT2013*, AISC 209, pp. 9–18.
DOI: 10.1007/978-3-642-37371-8_5

by the knowledge about the others. The ratio of agents changing their decisions tends to increase, when the communication between agents is increased. Also, communication increases the success of the group itself proportionally [4].

Daniel Epstein replaced in his work the information-basis for individual decision in El-Farol with the input of peers in randomly created social networks. The group communication is the only phase of decision making for him. He concluded, even with this kind of limited information, "(…) each agent was able to select an action that would bring good result for itself and for other agents as well." [5].

We will enhance the agents in El-Farol with the possibility of communicating their decision to each other and eventually alter their decision as a result of communication. They will be able to lie within the group-communication-process. The lying may be directly caused by a character of an agent or indirectly a strategic act of manipulation. We will analyze the effects and results for the individual agents and for the group(s). Is lying a viable strategy to achieve individual success in competing groups, e.g. in minority games? Which circumstances are advantageous for lying? Is it in general good to change a decision after submitting it to others?

The decision about lying should be individual and according to the current situation of the game. Therefore we need a psychological model to create different characteristics and algorithms for the agents.

We decided to base our work on the psychological theory of limbic characteristics for human behavior. This concept, described by Häusel [6], defines human behavior as a result of three basic instructions: balance, dominance and stimulant. He further forms eight stereotypes with different emphases on these desires (see Table 1).

Table 1. Limbic types as described by Häusel [6]. Each instruction may be activated (1) or deactivated (0). The binary representation of instructions allows eight different types.

No.	Name	Balance	Dominance	Stimulant
0	Apathetic person	0	0	0
1	Hedonist	0	0	1
2	Technocrat	0	1	0
3	Entrepreneur	0	1	1
4	Harmonizer	1	0	0
5	Epicure	1	0	1
6	Stress-Type	1	1	0
7	Eccentric	1	1	1

There is already a complete simulation framework for El-Farol based on limbic characteristics. This includes established algorithms for individual decision-making for the limbic types [7].

In section 2 of this paper we extend this framework with algorithms for group communication and eventually lying. These algorithms are also based on limbic characteristics and define different strategies for the limbic types. This way we obtain a complete and well-founded simulation framework for group communication in El-Farol-Bar-Problems.

In section 3 we consecutively test our framework with different configurations and discuss the results and findings. First we run simulations within the complete framework without lying applied by the agents. Subsequently we repeat these simulations, but then with lying applied. We compare the results to look for significant effects of lying. Are there configuration patterns favorable for lying? Is there an efficient strategy or pattern that includes lying?

Finally in section 4 we make conclusions and suggest future studies and researches based on the established simulation framework.

2 Extending the Simulation Framework

Taking the established framework for limbic types in El-Farol-Bar-Problems [7], we extend it by implementing group communication and possible lying. First we define the new decision making algorithm (2.1). Then we describe the type-specific algorithms for lying (2.2) and eventual decision-altering, called "reflection" (2.3).

2.1 Decision-Making with Group Communication and Reflection

The population of players (the agents) is distributed into one or more groups. Every agent must be a member of only one group. Each group is understood as a social network. This means each agent is linked with every other agent in the group. Linked agents are able to communicate with each other.

The extended algorithm for decision-making is done every round for every agent in the game. The respective next step of the algorithm starts only, when the previous step is finished for all agents in the game.

1. Individual decision-making based on the history of the game as described by its limbic type [7].
2. Polling the individual decisions of its group-members. A polled agent may decide to lie about its true decision. If so, they submit the inverted version of their decision. The algorithm to decide about lying is based on the limbic type of the agent (described in section 2.2).
3. Viewing the results of the poll and eventually altering the decision. This is called "reflection". The algorithm for reflection of an agent is also based on its limbic type (described in section 2.3).

For example: An agent decides individually to visit the bar. It polls its group members for their respective individual decision. As a result, it will get percentages of the decisions of the group members. X percent submit that they will go to the bar, Y percent submit that they will stay home. The agent now reflects its own decision against these percentages. It may eventually decide to alter its decision. It will try to stay or become part of the minority. But this is (mostly) the exact consideration of the other agents, too.

This dilemma is solved in a different way by each limbic type. Indeed an agent does not know about the limbic type or characteristics of its other group members. It only knows their decision, as they submitted it during polling.

2.2 When to Lie or Not to Lie

When an agent has made the individual decision to visit the bar and is polled by the other players of its group, it may either decide to tell the truth (that it will visit the bar) or to "lie" (that it will not visit the bar) and vice versa.

In short, if an agent decides to lie, it will report the inverted value of its individual decision. The decision to lie is based on the limbic type of each agent again. It is made individually in every round of a game as described in the table below. The algorithms are based on the description of the limbic types by Häusel [6 S. 99-103].

Table 2. Type-specific algorithms for each limbic type, to lie or not to lie when polled about its individual decision

Type	Decision to lie - algorithm
0	Is not driven by any limbic instructions. No curiosity. No fear. Totally unpredictable. Decides randomly to lie or not.
1	As a hedonist character follows only its own stimulant desire. Has a high inner stability und is not very fearful. Does not strive for power and is not worried a lot. Always tells the truth.
2	As a totally dominant type it is unethical, tries to win at all costs and to prove its superior methods. So this type tries to manipulate strategically: It will lie if it lost the last round and its current decision is the same as in the round before. This way it tries to manipulate the others against its own decision to become minority itself. This is complementary to its reflection logic (see section 2.3).
3	Not only dominant, but also stimulant. This type tries to balance both instructions for success. It considers lying, if it lost the last round. If so, its decision to lie is based on the risk tolerance configured for the stimulant instruction in the simulation framework [6]. In other words, it listens to its "stomach".
4	The fearful type does not want to anger its group members and has no "claim to power". This purely balanced type will always tell the truth.
5	The combination of stimulant and balanced characteristics makes this type likable to others. Has some good ideas, but does not dare to enforce those. Therefore it will always tell the truth.
6	The stress-type will always lie, if it lost the last round. Representing stressful counter-actions.
7	This type is driven by all limbic instruction. Has good ideas, but is choleric in times and driven by its erratic mood. Therefore it is as unpredictable as type 0. Decides randomly to lie or not.

There are basically three types of algorithms:

- Random lying.
- Always lie/always tell the truth.
- Analyze the situation and try to determine the more advantageous action.

The types 0 and 7 are totally unpredictable and will always decide randomly. Types 1, 4 and 5 will always tell the truth, but caused by different motivations. Types 2, 3 and 6 will try to implement strategic lying in some kind.

This will probably make the composition of a group a big influence by itself. The distribution of limbic types among the agents is expected to have a significant effect on the success of lying.

An important feature of the simulation framework: The ability of lying may be deactivated for the agents. This enables the possibility to compare identical configurations with the only difference in the implementation of lying.

2.3 Algorithms for Reflecting the Individual Decision

The reflection starts, when all agents have finished polling their group members. No agent will submit a reflected decision when polled. So this phase is clearly differentiated from the others.

The result of reflection may be to keep the individual decision or to invert it. This means, going to the bar when the former decision was to stay home and vice versa. Parameter for this decision is the complete polling result: The submitted individual decision of the group members. Within the reflection, the submitted decisions are not doubted, but always taken as the truth.

Each type (expect type 0) will try to reflect and alter its decision for its own individual success. There is no kind of consciousness for the success of the group as a whole. These algorithms are again formed after the description of the limbic types by Häusel [6 S. 99-103, 199-202].

Table 3. Reflection algorithms, specific for each limbic type. This happens after an agent has polled its group members about their decision.

Type	Reflection algorithm
0	Like in lying, it is unpredictable and decides randomly.
1	As a "player" it likes to risk and speculate. Its reflection is based on the configured risk tolerance for the stimulant instruction in the simulation framework [7]. According to the risk it decides to alter its decision to the one of majority of its group. This is against the logic of a minority game, but it speculated on enough group members altering their decision, because of the poll.
2	Does not trust in its group members, but believes only in its calculations. It will never change its decision, because of the others. Indeed, it will try to manipulate if possible (see for type 2 in section 2.2).

Table 3. (*continued*)

3	Based on the configured risk tolerance for the stimulant instruction in the simulation framework [7]. According to the risk, it will eventually consider to alter its individual decision to the decision of the majority of its group, If it decides to consider, it will take a look at the visitor-quota of its group. If this quota is lower than the threshold to win the game, it will alter its decision to visits the bar, regardless it previous decision. Otherwise it will alter its decision to not visit the bar.
4	Tends to avoid risks and tries to play safe. So, it will take the obvious route and alter its decision to the minority. It takes a look at the visitor-quota in its group. If this quota is lower than the threshold to win the game, it will alter its decision to visit the bar, regardless its previous decision. Otherwise it will alter its decision to not visit the bar.
5	Will always stick with the majority of its group, regardless the chances of winning the game. When the majority of its group decides to visit the bar, it will alter its decision to that. When the majority of its group decides to not visit the bar, it will follow that either. The agent may have the idea that this is not a clever behaviour for a minority game, but as in lying, it does not dare to enforce its own way consequentially.
6	Full of stress, it tends to immediate counter-actions. If the previous game was won, it will keep its individual decision. If not, it acts like type 4.
7	When the majority of its group decides to visit the bar, it will alter its decision to not visit the bar. When the majority of its group decides to not visit the bar, it will alter its decision to visit the bar.

3 Results and Discussions

With the simulation framework established, we translate our questions into valid configurations. We define the scope of simulations in this paper (3.1) and determine the settings and configurations needed for those (3.2). Finally we look at the results and discuss the findings (3.3).

3.1 Introduction

The extended simulation framework was implemented in a configurable software-client. It allows the configuration of one or more groups of agents, composed of different limbic types. These groups play a defined number of games of the El-Farol-Bar-Problem. A game is divided into one or more rounds, where each round represents the decision-making. The minority (winners) of a round is defined by a configurable threshold of visitors.

The software allows analyzing the simulation results in different views by behavior, success, visitor-composition by round, etc. Finally the results are also exported as a structured text file for further analysis.

We focus on validating our simulation framework and the questions stated in the introduction. For this first time we limit ourselves to the study of one group, which is composed evenly of all limbic types. During the first run of simulations its members will be forced to say the truth during reflection. During the second run of simulations its members will be allowed to lie, so we may compare the differences in behavior and success of its members.

Both runs of simulation will increase the size of the group, if necessary to provide significant results, but with keeping the even composition of limbic types.

3.2 Environmental Settings of the Simulation

We start with one group of 400 players with 50 players of each limbic type. The specific settings of the limbic instructions were taken from the previous study and are explained there [7]:

Table 4. Configuration of the limbic instructions

Balance	
Visiting orientation	35%
Weighted value for other limbic instructions	60%
Change strategy after X rounds lost	3 rounds
Dominance	
Use prediction function for at least X rounds	3 rounds
Weighted value for other stimulant instructions	65%
Stimulant	
Chance of repeating a decision in the next round	45%
Risk tolerance	60%

With these settings, we run four simulations with configurations as shown in table 5. The first two configurations (A and B) test the reflection without the effect of lying. We increase the count of players in the second configuration, to confirm our results against a larger population. We basically repeat these configuration in C and D, but then with possible lying during reflection.

Table 5. Sets of configuration for the simulations in this paper

Config.	Players per limbic type	Games	Rounds/Game	Lying allowed?
A	50	100	500	no
B	100	100	500	no
C	50	100	500	yes
D	100	100	500	yes

3.3 Results

Reflection Only, No Lying Applied. The first simulation shows a reliable and significant pattern. An average of 30% of all players changes its opinion per round. It never exceeds the range of 20%-40%. Interestingly there is always a stable ratio of players that wins and has kept its decision during reflection. The ratio of those who win and has changed its decision previously during reflection was highly erratic in opposite to that.

The highest cluster of visitor-counts is around 50% of the population. The second highest cluster is nearly above 60%, missing the threshold of the visitors to win. So the group interaction leads to a maximizing the count of successful agents within the group. The average percentage of visitors never drops below 40% and also never exceeds 80%. The even composition of the group and therefore the even distribution of reflection-algorithms seem to balance each other out.

The aggregated ranking and success of the limbic types is nearly invariant through all games. Type 7 is by far the most successful one. This was expectable to some degree, since its reflection algorithm alters its decisions always against the majority. Obviously a good move in minority games in general.

Limbic type 2 is the by far the most unsuccessful one. It never changes its decision during reflection, making it predictable and inflexible.

The mainly stimulant types 1 and 5 are ranked below average. Their behavior to stick with the majority of a decision does not pay off within a group of this composition. Types 4 and 6 both accomplish to balance wins and losses, with type 4 being a bit more consistent.

Types 0 and 3 achieve both aggregated results above average. For type 0 this is surprising, as its random behavior would be expected creating purely average results. The result of type 3 is n total opposite to the other dominant type 2. The addition of stimulant instruction to the dominant behavior drastically increases its success.

These results scale up consistently in the simulation of configuration B. It shows similar patterns as in the previous simulation. This is in opposite to Redmondinos work [4]. The ratio of changed decisions does not increased with the count of links between the players. Reason for that may be the different communication protocol and the more complex characteristics of the players.

Both simulations showed, it is not per se good to change decision in minority games. But with a clever algorithm (e.g. type 7) and other group-members with a complementing “not-clever” algorithm (e.g. type 2*)*, it may become a huge factor of success.

Reflection with Lying Applied. In average between 20% and 40% of the agents apply lying per round. This is a very consistent pattern and nearly never exceeds this range. Up to 80% of the lies are applied successful, meaning an agent who lied won that round. The total number of liars that win a round is higher most of the times as the number of liars that lose a round.

In total, around 50% of the winners of a round are liars. Compared to the total amount of liars within the population (see above), the minority of liars is over-represented as winners. Therefore it is a viable strategy in this type of group composition.

Surprisingly the aggregated results for the limbic types do not change a lot compared to the simulations without applied lying. Despite a bit more variability, type 7 is still the most successful, type 2 still the most unsuccessful. The aggregated ranking of the limbic types by success over all games is the same.

Again the results scaled up consistently in configuration D, confirming the patterns observed. The ranking in success of the limbic types did not change significantly to these in the previous simulations.

There is no significant effect of lying in the resulting success of limbic types in this even composed configuration. Still, there is a significant influence of lying on the behavior of the agents during the simulation. The limbic algorithms balanced each other out in the end. So if all agents apply lying with their own respective strategies, they are able to retain their level of success. Lying does not better their results, but ensures that the distribution of success between the types stays persistent. This defines successful lying in our simulations.

4 Conclusions and Future Work

We extended the simulation framework of limbic types in the El-Farol-Bar-Problem by introducing group communication and the ability of lying by the agents. The group communication is implemented by polling the individual decisions of its group members by a player. The player looks at the results afterwards. Eventually it alters its decision to visit the bar according to the decisions of others. Additionally the players may lie about their own individual decision to each other during polling.

The developed software was used to test our simulation framework and reviewed the application on configuration with and without applied lying to compare the effects. The effects observed were significant and reproducible. The ratio of players changing their decision was within a stable range. There was no overall advantage of changing decisions by reflection. But with the right strategy and group composition the limbic type 7 could outperform the others by far.

If lying is allowed during communication, *the final results did not change* significantly. The ratio of changed decisions was nearly the same, as the ranking of the success of limbic types was the same. The direct effects of lying during communication and reflection were balanced out due to the even composition of the group. But to succeed and retain success, *an agent has to adapt to the lying* situation and eventually apply it, to retain its level of success.

Further studies will observe different group compositions and the possibility for competing groups with its members playing for success of their group, even if this results in losing individually. This collective style of player may require new algorithms and strategies.

References

1. Hirokawa, R.Y.: The role of communictation in group decision-making efficacy. Small Group Research 21, 190–204 (1990)
2. Whitehead, D.: The El Farol Bar Problem Revisited: Reinforcment Learning in a Potential Game. University of Edinburgh (2008)
3. `http://www.econ.ed.ac.uk/papers/The%20El%20Farol%20Bar%20Problem%20Revisited.pdf` (last visited: December 15, 2012)
4. Arthur, W.B.: Bounded Rationality and Inductive Behavior (the El Farol Problem). American Economic Review 84, 406–411 (1994)
5. Remondino, M.: Introducing Communication in a Minority Game: an Agent Based Simulation. University of Turin, Italy (2004)
6. Epstein, D., Bazzan, A.L.C.: Decision-making using random boolean networks in the El Farol Bar Problem. Instituto de Informática, UFRGS, P.Alegre, RS, Brazil (2011)
7. Häusel, H.: Think Limbic!, vol. 3. Rudolf Haufe Verlag, Auflage (2003) 3-448-05661-8
8. Akinalp, C., Unger, H.: The limbic characteristic and el-farol games. In: Proc. of the 11th International Conference on Innovative Internet Community Services, I2CS, pp. 190–195 (June 2010)
9. Grabisch, M., Rusinowska, A.: A model of influence in a social network. Université Paris I PanthéonSorbonne (2010)

Identifying Limbic Characteristics on Twitter

Christine Klotz and Coskun Akinalp

Fernuniversität in Hagen,
Fakultät für Mathematik und Informatik
Universitätsstraße 27, PRG
D-58084 Hagen, Germany
c._klotz@web.de, coskun@akinalp.com

Abstract. An adaptive, intelligent system requires certain knowledge of its users. Patterns of behavior, preferences, and motives for decision making must be readily identifiable for the computer to react to. So far, typifying of users needed either a huge collection of empirical data via questionnaires or special hardware to track the user's behavior. We succeeded to categorize users by analyzing only a small amount of the data trace a user leaves while using the online social network (OSN) Twitter. Our approach can be adapted to other platforms easily. Thus, human behavior is made understandable for computer systems and will help to improve the engineering of human-computer-interactions.

Keywords: Limbic Characteristics, Twitter, Classification, Social Network.

1 Introduction

The behavior of people bears a lot of their personality. It gives evidence about their way of thinking, feeling, and deciding. An adaptive, intelligent system requires that knowledge of its users to be able to show specific reactions to each different user. Therefore, the automatic identification of different personalities is a challenge. So far, typifying of users needed either a huge collection of empirical data via questionnaires or special hardware to track the user's behavior. We succeeded to categorize users by analyzing only a small amount of the data trace a user leaves. Based on the model of limbic characteristics by Häusel [1][2], we deduced six hypotheses about the behavior of Twitter users conveying the specific limbic characteristics. The accuracy of Häusel's approach for predicting human behavior was proofed by empiric validation with more than 60,000 consumers as subjects [2]. Akinalp showed how this approach can be adapted for decision making processes in economic systems with limited resources [3]. By analyzing the huge mass of data a user produces while using online services the inference from behavior to characteristics can be done automatically by the system. That is the achievement of our work.

Related work on social networks presented methods for measuring social influence [4] and message propagation behaviors [5] in interpersonal relationships, but does not consider the personality of humans. The connection between Twitter users and

P. Meesad et al. (Eds.): *IC²IT2013*, AISC 209, pp. 19–27.
DOI: 10.1007/978-3-642-37371-8_6

personality traits based on the psychological model called "The Big Five" has been examined in [6]. In our work, we quantify user behavior using basic functionality of the platform. Thus, human behavior is made understandable for the system and will help to improve the engineering of human-computer-interactions.

2 Fundamentals of Limbic Type Identification

2.1 Functionality of Twitter and Its Potential for Data Analysis

One of the most popular online platforms has been chosen as the subject of our work because of the multiple possibilities for data mining. Quantitative studies about the user's personnel data, e.g. their origin, and their activities on the platform can be found in [7]. In [8], subject matter of analysis was mainly the network topology and the retweet tree, and a ranking of users as well as of topics.

Data for analysis can be sampled by querying the Twitter Search API [9] with random generated user-IDs. Not only the user's activities are documented but also information related to emotions, interests, personal opinions and so forth, conveyed by functions e.g. replies, retweets, hashtags. This opens up possibilities for several methods: text and sentiment analyzes, analyzes of the network structure as well as of personal data. We categorize the subject matter of a data analyzes on Twitter as follows:

1. Private user's profile data and user settings
2. Data, resulting from online activities of the user (connected to Twitter)
3. The user's position within the social network
4. Content analyzes of the user's tweets
5. Formal analytic interpretation of the user's tweets

All five categories include a great range of variables. In addition to parameters that can be directly extracted, measurements were proposed, e.g. the klout score for influence [10], which combine multiple parameters.

2.2 The Approach of Limbic Characteristics and Limbic Types

As an approach to model human characteristics, Häusel states in [1] that every person is influenced by three key motivational and emotional systems. The different magnitudes of influence of one person compared to another form the individual character and determine in this way behavior and decisions.

These three systems, called limbic instructions, are the following:

- **Stimulant:** The impulse responsible for curiosity, inventiveness, restlessness, communicativeness and the pursuit for amusement.
- **Dominance:** Influences human beings to fight, to strive for power, autonomy, leadership and to overtake responsibility.
- **Balance:** The wish for security, stability, home and social integration.

By drawing the limbic instruction as poles and interpreting the distances as rates of influence, seven different limbic types can be described as resulting character categories. Each limbic type is primarily influenced by one limbic instruction. Apart from the Hedonist- and the Performer-type, which have only the stimulant respectively the dominant instruction as motivation, a second instruction provides for a different impulse [2].

3 Classification of Limbic Types on Twitter

3.1 Detecting Specific Behavior on Twitter

Every influence leads to a specific behavior. Vice versa, a specific behavior can be a hint for the characteristics of the actor. By analyzing the behavior of a user/actor on Twitter, or more precisely by analyzing the data trace he leaves, he can be classified. Therefore as a first step, it is necessary to transfer the limbic characteristics into functionality of the Twitter platform.

In the following, the parameters we used for our experiment are described and linked with six hypotheses about the influence of limbic instructions. Table 1 summarizes our hypotheses. Col. 2 of the table names the variables and col. 3 gives the height of the value used as indicator for the limbic instruction in col. 4. In that way, limbic characteristics are identifiable in a data sample.

Table 1. Hypotheses about data values in relation to limbic instructions

Hyp.	Variable(s)	Value(s)	Limbic Instruction(s)
1	All	Mainly extreme values	Stimulant
2	All	Mainly moderate values	Balance - Dominance
3	*tweets*	High value	Stimulant
4	*followers*	High value	Dominance
5	*followings*	High value	Balance
6	*favorites*	High value	Balance - Dominance
-	*Lists*	Indicator for participation in social network	None particular

First, two opposed behavior patterns can be worked out from the model of limbic characteristics in relation to the Twitter platform: The obsessive usage of some functions attended by disregard of the full range of functionality (Hyp. 1). These persons are mainly instructed by stimulant searching for novelty, amusement, and self-expression. If they like something they can hardly restraint.

The second behavior pattern specifies persons, who act more purposefully because their motive for doing something is mainly not based on joy but on plan and worth (Hyp. 2). Aim of their work is to build up trusting relationships and stable, functional structures. To achieve this aim, they use every function restraint but effectively.

Hypothesis 1: Mainly extreme parameter values indicate a high stimulant instruction.

Hypothesis 2: Mainly moderate parameter values indicate dominance and balance instructions.

With regards to the blogging aspect the number of posted tweets is surely a strong indicator for the expressivity of a person. Expressivity is connected with the stimulant instruction in the model of limbic characteristics, so a high number of posted tweets results from a stimulant instruction, or vice versa:

Hypothesis 3: Users with a strong stimulant instruction post a high number of tweets.

In contrast to other social network platforms, the relationship between two persons is modeled as a unidirectional link. The incoming link is named a *follower*, the outgoing link a *following*. That means that one user can follow another ones statements without being interesting for the opponent. In particular, by investigating these relationships it is possible to identify users who are opinion formers with a strong dominance instruction and those who are influenced by the balance instruction searching for guidance and social integration.

Hypothesis 4: Users with a strong dominance instruction have got a high number of followers.

Hypothesis 5: Users with a strong balance instruction have got a high number of followings.

As a third indicator of participation and presence within the social network the number of lists to which a user is subscribed is a matter of data analysis. This parameter is not yet linked with a specific limbic instruction because the motivation to subscribe for lists can be various. In combination with other network indicators it is nevertheless a useful information about network activities.

Another interesting aspect in people's behavior is the tendency either to marshal things or not. The capability of organization, discipline, and efficiency is assigned to the mixture of dominance and balance in the model of limbic characteristics. As a fifth variable for data exploration the bookmark function - named *favorites* function on Twitter - is used to detect the way in which users act purposefully while using the platform services.

Hypothesis 6: Users with a mixture of dominance and balance instructions have got a higher number of favorites.

To verify those hypotheses stated, we applied a cluster analysis with both, the k-means-algorithm and the hierarchical complete-linkage-algorithm. We used the parameter given in Table 1 as characteristics for the objects to group. The idea behind a cluster analysis is that similar objects form homogenous groups and detecting them will give evidence about the number of behavior categories. As the approach of limbic characteristics names seven limbic types, seven resulting clusters with the characteristics as described will prove the assumptions.

3.2 Detecting the Limbic Type

Cluster analysis of a data sample helps to discover significant structures by searching for homogenous groups. Thereby, clusters have to be homogenous concerning similar objects but not even in size and expansion. As a specific medium attracts mainly a specific group of users, a platform like Twitter has probably a lower average age than the complete population. For example, 87 percent of the Europeans between 16 and 24 years use online communication platforms like chats, (micro) blogs and messengers whereas only 3 percent older than 65 years use these services [11]. In terms of limbic characteristics it is most likely that some of the limbic types will be over-represented whereas others will sparsely participate.

The limbic model names seven types, so we expected seven groups to appear. For analysis a clustering with the k-means-algorithm was applied. The determination of the number of clusters was based on a data sample with 1400 entities and the five variables *tweets*, *followers*, *followings*, *favorites* and *lists*. A scree-diagram relating to the optimization criteria showed good solutions for three and seven clusters. In conformance with the scree-diagram a plot of the silhouettes showed most homogenous clusters for three, seven and sixteen groups [12]. The emergence of a homogenous classification with seven groups proofed our assumption with seven identifiable limbic types.

A look at the seven clusters showed the closest similarity between the characteristic mean values for each cluster (Table 4) and the characteristic behavior of the seven limbic types as described in hypotheses above. In the following, we exemplify the interpretation for a resulting cluster. The complete process can be found in [12].

A cluster is specified by its center point and its dispersion as indicator for the homogeneity. To characterize the clusters, the mean value and the standard deviation for all variables were calculated. As additional indicator for a particularly representative value, the F_l-criteria [13] was calculated. The F_l-criteria is a relative deviation and is defined as $F_l = s^2_{lp} / s^2_l$. The numerator is the variance for variable l in the examined group p whereas the denominator is the variance for the same variable in all groups.

Table 2. Characterization of resulting clusters

CLUSTER 1	VARIABLE(S)	VALUE(S)	F_L	HYP.	LIMBIC INSTRUCTION
	all	Mainly moderate values		2	Balance - Dominance
	tweets	High value		3	Stimulant
	favorites	Medium value			
	lists	Medium value	*		
	followers	Medium value	*		
	followings	High value	*	5	Balance

Table 3. Mapping of resulting clusters with limbic types

CLUSTER	FORMATIVE INSTRUCTION	OTHER INFLUENCES	LIMBIC TYPE
Cluster 1	Balance	Dominance + Stimulant	Traditionalist

All seven clusters were examined and their variable values compared with the hypotheses. Table 2 shows the characterization for a cluster. The table is drawn on the lines of table 1 with the names of the variables in col. 2 and the height of the value used as indicator for the limbic instruction in col. 3. The * in col. 4 marks particular representative values with $F_l \leq 0,5$ [13]. Col. 5 names the pointed hypotheses and col. 6 the limbic instruction.

To map the resulting clusters with limbic types, three matters were considered: Is a value representative for the cluster? How many hypotheses indicate a limbic instruction? And which behavior pattern (in regards to Hyp. 1 and Hyp. 2) is relevant? In such a way, a formative limbic instruction could be quantified for each cluster. As a result, a mapping table like Table 3 names the cluster (col. 1) with a limbic type (col. 4). Col. 2 gives the formative instruction and col. 3 shows other influences as second impulse (see the approach of limbic characteristics in section 3.1). In our experiment, all seven resulting clusters matched the seven different limbic types.

Our suggested way for the identification of limbic types can be automated after adapting the shown interpretation process to a specific platform. Therefore, the platform functionality must be examined and hypotheses between limbic instruction and the assumed behavior specified. The heights of valuable parameter values serve as indicators for the limbic instructions. Using this method, the system will learn to identify a user's personality and be able to show specific reaction to each different user.

4 Identified Limbic Type on Twitter

All seven limbic types could be identified by clustering the data sample and interpretation of the resulting cluster as developed in chapter 3. The mean values and standard deviations are given for all clusters in Table 4. The mean values represent center points. The standard deviation is used as indicator for size and homogeneity of the cluster. A high s-value shows unequal distribution of a parameter value within a cluster, and therefore an unequal behavior of objects within a cluster. Especially s-values higher than the mean indicate a few objects as outliers which do not represent the cluster (concerning this variable) by showing extreme behavior whereas the majority behaves in a nearly similar way.

With regards to our six hypotheses described above, we identified and named the clusters as follows:

Hedonist: The group of six objects shows extreme values in all five parameters that indicate - referring to Hyp. 1 - a basic stimulant instruction. The peak value for posting tweets points to an extremely expressive and extroverted person influenced by stimulant (see Hyp. 2). High self-centered attitudes result in low values for participation within the social network.

Adventurer: This group also shows extreme values in all five parameters indicating a basic stimulant instruction. In contrast to the Hedonist cluster, extremely high values can be noticed for all three networking parameters (follower, followings and lists), whereas the number of tweets is very low. The two objects in this cluster have a high presence despite of a low messaging rate and have most likely a prominent personality.

Table 4. Mean values and standard deviations for all seven clusters

CLUSTER	X(TWEETS)	S(TWEETS)	X(FAV.)	S(FAV.)	X(LISTS)	S(LISTS)	N
Hedonist	198187,8	36919,7	970,8	2367,3	162,7	217,7	6
Adventurer	4585,0	432,7	0,0	0,0	148032,0	34345,6	2
Performer	9830,1	15435,2	125,1	259,8	42624,0	22579,5	28
Disciplined	25935,0	28046,0	7314,4	3537,7	1145,2	1570,9	9
Traditionalist	35433,0	19244,0	308,3	619,6	1264,2	3542,0	75
Harmonizer	15219,0	11540,2	452,3	986,4	9331,8	6485,3	8
Open-minded	2147,9	3623,6	46,0	230,9	467,7	1997,8	1269

CLUSTER	X(FOLLOWERS)	S(FOLLOWERS)	X(FOLLOWINGS)	S(FOLLOWINGS)	N
Hedonist	4852,3	5928,0	3868,8	3877,8	6
Adventurer	20576908,0	4120585,6	337785,5	477556,6	2
Performer	5370149,4	2556630,1	1042,3	2725,5	28
Disciplined	134812,2	200693,6	2751,6	3880,9	9
Traditionalist	127508,6	359369,1	7273,0	14717,0	75
Harmonizer	1105861,8	1146033,1	254322,4	85526,9	8
Open-minded	53128,8	262906,1	1776,1	6677,4	1269

Performer: The Performer group is characterized by a combination of high values for follower and lists and a very low number of followings. This indicates an opinion leader with power and presence within the social network in regards to hypotheses 4 and 5. A medium value for favorites also distinguishes the Performer from the Adventurer and shows a more organizational attitude.

Disciplined: The second group with a significant low number of followings can be identified as disciplined people mainly ruled by the limbic dominant instruction. In contrast to the performer a much higher value for the bookmark function favorites can be interpreted as efficient and disciplined in reference to Hyp. 6. Predominantly mean values back dominant and balance influences (see Hyp. 2).

Traditionalist: Similar parameter values within this cluster to the Disciplined cluster point to a neighboring limbic type. Predominantly mean values also indicate dominant and balance influences regarding Hyp. 2. Comparatively higher values for the numbers of followings and tweets show more social interaction and search for guidance and distinguish the Traditionalist group from the Disciplined.

Harmonizer: Similar parameter values within this cluster to the Traditionalist group point to another neighboring limbic type. All three networking parameters are above average in contrast to the Disciplined cluster but, compared to all clusters, not outstanding, which indicates a moderate and communicative attitude.

Open-Minded: The biggest cluster with about ninety percent of the entities can be identified as open-minded people who represent the prototype of a Twitter user. These people like to express themselves, get in contact with other people, and participate within a social network. As Twitter is a very modern form of communication they need to be curious. On the other hand, the moderate values show that these people do not have an outstandingly expressive but a common personality.

To test the stability of the classification, the complete-linkage-algorithm was applied and both cluster solutions compared [12]. That classification was not entirely but very nearly similar, so the described classification can be seen as stable and reproducible and is therefore highly induced by the data. As expected, the numbers of objects forming a cluster vary. The target group of a microblogging and social network service is surely a talkative, open-minded person with an affinity for modern technology. Probably, other platforms will show different distributions of the limbic types. With the adaptation of our hypotheses to the specific platform and selection of the correct parameters the system will be able to identify the user's limbic characteristics automatically in the future.

5 Conclusions and Future Work

Our experiment proves the suggested way of identifying limbic types with a small amount of parameters and is therefore an achievement for automated categorization of system users. For our data analysis we used parameters, which refer to the user's position within the social network or his online activities on Twitter. As described in section 2.1, the subject-matter of data analysis can be more various. Future research will inspect user tweets for keywords, which refer to limbic instructions. In the same way formal text parameters may lead to limbic characteristics, as first surveys indicate [12]. Another approach focuses on how to identify limbic types only by the network structure, and which types form homogenous or mixed groups.

References

1. Häusel, H.G.: Think Limbic. Haufe Gruppe, Freiburg (2008)
2. Häusel, H.G.: Die wissenschaftliche Fundierung des Limbic-Ansatzes. Citing web resource (March 2011), `http://www.nymphenburg.de/tl_files/pdf/LimbicScience110220.pdf`
3. Akinalp, C.: The Limbic Characteristic and El-Farol Games. GI-Edition Lecture Notes. In: 10th International Conference on Innovative Internet Community Systems (2010)
4. Cha, M., Haddadi, H., Benevenuto, F., Gummadi, K.P.: The million follower fallacy. In: Proceedings of International AAAI Conference on Weblogs and Social Media, ICWSM 2010 (2010)
5. Achananuparp, P., Lim, E.P., Jiang, J., Hoang, T.A.: Who is Retweeting the Tweeters Modeling, Originating, and Promoting Behaviors in the Twitter Network. ACM Transactions on Management Information Systems 3(3) (October 2012)
6. Quercia, D., Kosinski, M., Stillwell, D., Crowcroft, J.: Predicting personality with twitter. In: Proceedings of IEEE SocialCom. (2011)
7. Krishnamurthy, B., Gill, P., Arlitt, M.: A few chirps about Twitter. In: ACM SIGCOMM WOSN, Proceedings of the First Workshop on Online Social Networks (2008)

8. Kwak, H., Lee, C., Park, H., Moon, S.: What is Twitter a social network or a news media. In: Proceedings of the 19th International Conference on World Wide Web (2010)
9. Twitter Developer Page. The Twitter REST API (December 2012), https://dev.twitter.com/docs/api
10. Klout Score (December 2012), http://klout.com/corp/klout_score
11. Statistisches Bundesamt.Wirtschaftsrechnungen, Private Haushalte in der Informationsgesellschaft – Nutzung von Informations- und Kommunikationstechnologien, Wiesbaden (2011)
12. Klotz, C.: Textanalyse Wortanalyse und Bestimmung der limbischen Charaktere. Masterthesis, FernUniversität Hagen (2012)
13. Rasch, D., Kubinger K.D.: Statistik für das Psychologiestudium, Elsevier, Spektrum Akademischer Verlag, München (2006)

Mobile Agent Administrator for Computers over Network (MAACN): Using the Concept of Keyboard Buffer

Rajesh Dontham[1], Srinivas R. Mangalwede[2], Subodh Ingleshwar[3], Manoj A. Patil[3], and Arun W. Kumar[4]

[1] Dept. of IT, RIT, Sakharale, Walwa, Sangli, Maharashtra, India
rajesh.dontham@ritindia.edu
[2] Dept. of CSE, GIT, Belgaum, Karnataka, India
mangalwede@yahoo.com
[3] RIT, Sakharale, Walwa, Sangli, Maharashtra, India
{subodh.ingleshwar,manoj.patil}@ritindia.edu
[4] TKIET, Warnanagar, Panhala, Kolhapur, Maharashtra, India
arunkumar@tkietwarna.org

Abstract. This paper proposes an idea of using the concept of Keyboard Buffer for performing Network Administrator and Computer System Administrator activities by passing the text commands to computers for performing system functions without the need for sharing or transferring the script files or the need for the administrators to personally visit the computer systems over the network. This paper also describes a system that puts the proposed idea into practice by using Java Aglets: - Mobile Agent Technology as a platform. The proposed idea is an extension to the system "The Agent Admin: Administration of Remote nodes on huge Networks using Mobile Agent Technology" that performs few administration activities like retrieving systems information, list of softwares installed, list of active applications and software installation by sharing the files over the network. The performance of "The Agent Admin" was tested, analyzed and enhanced with the Commander Agent.

Keywords: Keyboard Buffer, Mobile Agent, Java Aglets, Network Administrator, System Administrator, System functions, Commander Agent.

1 Introduction

The present day organizations/institutions can never be thought of without computers and they connected by communication links forming a computer network: for managing these computers, resource sharing, availability of data, distribution of workloads and centralization of applications etc. The networks maintained by the organizations/institutions today contain hundreds to thousands of computers and are getting larger and larger driven by growing requirements. Administration of such a sizeable hosts and their software's with changing in needs poses a new challenge for both Network Administrators and System Administrators. A brief list of System Administrator activities are as follows:

P. Meesad et al. (Eds.): *IC²IT2013*, AISC 209, pp. 29–40.
DOI: 10.1007/978-3-642-37371-8_7

1.1 System Administrator Activities

A system administrator is a person employed to maintain and operate computer system(s) and/or network. The responsibilities of a system administrator and network administrator often overlap; however, the system administrator's main concern is about the computer systems and less on the network.

The system administrator activities specific to computer systems may include: Installing, supporting, updating, and maintaining servers/computer systems hardware, software deployment, assigning names and addresses to each computer and device on the network, availability check, resource sharing, directory and other services, troubleshooting, performance tuning, ensuring system efficiency, backup, recovery and archives, security measures, performing the commands required to share, remove, and restrict resources etc, scripting or light programming, project management for systems-related projects, supervising or training computer operators, and being the consultant for computer problems beyond the knowledge of technical support staff.

Other low level administrative tasks considered in this paper are: shutting down the entire/selected range of remote systems at the end of working hours which helps in power management, changing the IP address of a computer, rebooting the systems when required, retrieving the current details of remote systems and extending it to E-Learning and Student/Staff Training and basic demonstrations like opening a website/a Moodle server and viewing video lectures, conducting mock tests, displaying notices/images and LAN booting of systems without inputting any data at the user end by the user.

This paper address most of the System Administrator's activities mentioned above that are carried out traditionally by manually/personally visiting each system. In the proposed approach, by developing the system using mobile agent technology and the concept of Keyboard Buffer, we are eliminating the need for manually visiting the systems over the network which is proven to be a burden for administrator's for a sizeable collection of computers.

1.2 Mobile Agents

A Mobile Agent is a specific form of mobile code that consists of the program code and the program execution state (the current values of variables, next instruction to be executed, etc.) with the unique ability of transporting themselves (state and code) from one system in a network to another. A mobile agent is not bound to the system where it begins execution. The agents have certain attributes associated with them that distinguish them from the standard programs. They are categorized into two types:

Mandatory Properties. The mandatory properties provide weak notion of agents. The mandatory properties of an agent are as follows:

Autonomy: An agent operates without the direct intervention of humans or others, and has some kind of control over their actions and internal state.

Reactivity: An agent perceives their environment, and responds in a timely fashion to changes that occur in it.

Proactive: An agent exhibits goal-directed behavior by taking the initiative.

Adaptive: An agent is capable of adjusting itself to user/environment using either rules or some form of preferences.

Temporal continuity: An agent is continuously running processes (either running active in the foreground or sleeping/passive in the background).

Goal oriented: An agent is capable of handling a task to meet its desired goal.

Believable: An agent created/dispatched should be trust worthy.

Orthogonal Properties. The orthogonal properties provide strong notion of the agents. The orthogonal properties of agents are as follows:

Mobility: An agent is capable of roaming around in an electronic network.

Collaborative: An agent should be capable of computing the desired tasks of the user/process by cooperating with other agents; sometimes it may refuse to execute certain tasks.

Learning: An agent can learn the environment factors, user preferences, etc., and develop certain degree of reasoning to take intelligent decisions/actions. Learning mechanisms could be: reinforcement learning; neural networks; user training by examples.

Communicative: An agent interacts with other agents and (possibly) humans via some kind of agent communication language.

Inferential capability: An agent is able to share a set of knowledge in order to achieve a specific goal.

Rationality: Agents embody (give visible form to idea) the capacity to analyze and solve a problem in a rational (ability to think logically) manner [1].

This paper is organized as follows: Section 2 discusses the related work for performing system administration activities over network using mobile agent technology; in section 3, we describe the proposed system; Section 4 demonstrates the proposed idea of using Keyboard Buffer by the agents; Section 5 concludes this paper and in Section 6, we present the future work.

2 Related Work

In this section, we first discuss the related work done for software deployment. The software deployment, an activity of system administrator, is an important and complex procedure that involves the release, installation, adaptation, reconfiguration, update, activation, deactivation, retirement and removal of software. The following are some of the existing technologies that support various aspects of the software deployment process: Installation Tools, Package Managers, Application Management Systems, Disk image-based deployment, Behavior-based deployment (Zju Software Delivery Toolkit), Package-based deployments (XcelleNety using Prism Deploy), Novel Software Deployment System, EMCO Remote Installer, Windows Remote

Desktop Connection, TeamViewer [2-7]. Each of the tool/system mentioned has its own disadvantages which are discussed in [10].

In [8], the authors describe a system that performs software installation on a heterogeneous network using mobile agents which is closest to our own work for software installation. This system is suited only for installation of mandatory software. Individual nodes over LAN cannot be handled. Two separate Mobile agents are used. Creating N clones of AgentController for N entries in Inter tables and M Agents for M entries in Intra table increases network traffic drastically and doesn't perform any other administration activities.

The second part of this section discusses the related work done in performing System Administration activities using mobile agent technology.

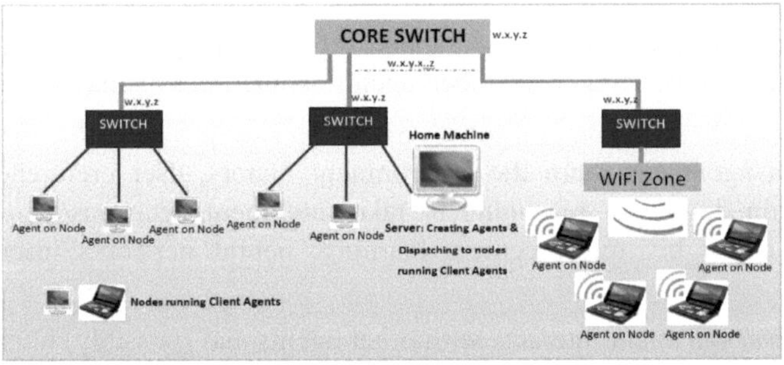

Fig. 1. The Topology view of "The Agent Admin" system [10]

In [9], the authors develop the "ABSA: An Agent-Based Tool for System Administration" to automate the System Administration activities. This system seems to be complex as it involves several agents-to-agents communication where agent to agent communication itself is complex. The authors aim for achieving scalability gets into trouble in some instances with increase in number of hosts as the architecture uses synchronized circular shared buffer and First In First Out (FIFO) for receiving requests from internet also limiting the number of requests to handle at a time. The system uses Client-Server architecture where client nodes request the manager node for performing administration requests and the system performs only few administration activities and each activity needing a new agent. The scheduling of agents in this system is motivating us in incorporating scheduling in to our system.

"The Agent Admin: Administration of Remote nodes on huge Networks using Mobile Agent Technology" [10] is used as the base system on which the proposed idea of using Keyboard Buffer will be experimented and also extending its capabilities to perform most of the System Administration activities as ever before. This system uses script files similar to "NOVEL SOFTWARE DEPLOYMENT SYSTEM" [4] for executing sequence of commands once the files are made available by the home machine running the server agent which involves additional processing both at server agent and dispatched agent. The proposed system provides an alternate by creating a Commander Agent that carries only the commands to be executed which will be discussed in the Section 3.

3 Proposed System

The limitations of existing systems and the interesting properties; the mobility and the autonomy of mobile agents strongly motivated to continue with this work of upgrading "The Agent Admin". The Figure 1 shows the topology of "The Agent Admin" running the agents. The home machine is the system that contains the necessary files and creates the agents that dispatch to remote nodes to perform the tasks.

The Figure 2 shows the Graphical User Interface (GUI) of "The Agent Admin" with options available to perform system administrative tasks where an individual system or a group of systems can be selected. The Server Agent creates agents equal to number of computers selected and dispatches the agents to their destinations; simultaneously making the required files available over the network. The dispatching agents do not carry any files along with them as they are made available by the server. Once the agents successfully dispatch to their destinations, they perform the prescribed tasks. Though the files are not carried along with the dispatching agents nor do they keep a copy at the destination, there is an increase in the network traffic when the dispatched agents start accessing them according to their current requirement.

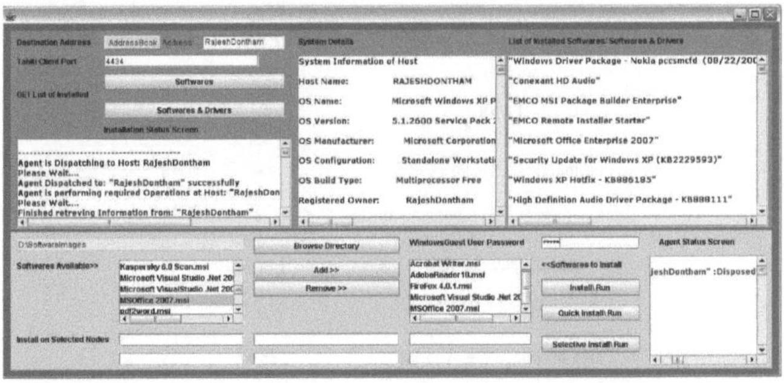

Fig. 2. GUI of "The Agent Admin" system [10]

The system has been tested in a computer laboratory consisting of 21 systems with identical hardware and software configurations. One system acts as a Home Machine and the remaining 20 computers as clients. The hardware and software configurations of the systems are given in Table 1. Several system commands and audio-video files were executed neither by visiting the computers manually nor the need for end user to input any kind of data. The performance results were recorded and shown in Figure 3.

The systems were connected over broadcast LAN via Layer 2 switch. The results are promising over the traditional approach of performing the same by visiting the systems manually. The performance of "The Agent Admin" is expected to increase drastically when executed over a point to point star topology LAN and with configured server system.

Table 1. Hardware and Software configurations

Hardware	
Processor	Intel Core I5 3.1GHz
RAM	3.1 GB
Hard Disk	500 GB
Monitor	Standard Color
Keyboard & Mouse	Standard
Network Interface Card	Realtek PCIe GBE Family Controller with Realtek Ethernet Controller All-In-One Windows Driver
Software	
Operating System	Windows XP-SP3
Backend	Aglets-2.0.2 (Aglets Software Development Kit) and JDK1.5.0

Number of Computers		**10**	**20**
TASK	File Size	Time (Sec)	Time (Sec)
Command "shutdown -s -t 0"	--	14	32
Open a website "www.google.com"	--	19	33
Install VLC 0.8.msi	13.3 MB	36	63
Play a song	04.2 MB	19	33
Open an Image file "123.jpg"	25 KB	19	33
Command "Taskkill /f /im app.exe"	--	20	36
Total Computer IP Range; **From:** 172.22.7.111 **To:** 172.22.7.130 Total No. of Computers in the Range= 20 Type of LAN: Broadcast with Layer 2 Switch			

Fig. 3. Performance results of "The Agent Admin" System

The systems in an organization can be shut down after their working hours to save electric power; Softwares can be installed, Facilitators can demonstrate basic usage of computers and also play audio/ video lectures without the required files being copied/ carried and even without the need of end user to input any data/ instructions. The results are encouraging and may prove effective in distributed computing applications such as e-Learning.

The server system consumes certain amount of time and computation making the script files available over the network. Before demonstrating the proposed idea of using Keyboard Buffer, we discuss an alternate that was put into action. The use of

Commander Agent completely eliminated the extra time and computation of the server. However, the server has to make the files available over network if the administrator needs to install software's or play any audio/ video lectures/files etc. The Commander Agent's prompt is shown in Figure 4.

Fig. 4. The Commander Agent accepting one or more Commands

The administrator can input the system commands that they wish to execute on one or several hundreds of remote systems. The Commander Agent is also capable of carrying more than one system command as it uses an array to store them. On successful execution of all the commands executed either in serial or parallel, the Commander Agent reports to the server with minimal status information.

The basic lines of code used for the development of Commander Agent are shown below.

```
// Declaration
..
{
//Execution
Process commanderAgent =
Runtime.getRuntime().exec("command or array of
commands");
// read the output if necessary
commanderAgent.getOutputStream().close();
BufferedReader reader = new BufferedReader(new
InputStreamReader(commanderAgent.getInputStream()));
...
// Termination
```

Commander Agent saves the valuable amount of server resources. The number of Commander Agents created and dispatched by the server will be equal to the number of remote systems selected by the system administrator. The overall performance of the system is assumed to be improved to a greater extent with the Commander Agent.

4 Use of Keyboard Buffer by the Agents

This section demonstrates the idea of using the Keyboard Buffer by the agents to perform system administration activities at the remote systems. The revolutionary concept of using Keyboard Buffer will exponentially increase the performance of the agents, which has been proposed for the first time.

A keyboard which is the primary input device used in all computers allow the user(s) to enter data or instructions. The keyboard controller detects the key-press and determines the scan code. The scan codes are stored in memory and then passed to processing unit for further processing. The functioning of keyboard can be briefly described in the figure 5.

1. The key is pressed on the keyboard by the user.
2. The keyboard controller determines and sends the scan code for the key to the keyboard buffer.
3. The keyboard controller sends an interrupt request to the system software (Operating System).
4. The system software responds to the interrupt by reading the scan code from keyboard buffer.
5. The system software passes the scan code to the CPU.
6. The CPU processes the data/ instructions and may send the results to output unit for display purpose.

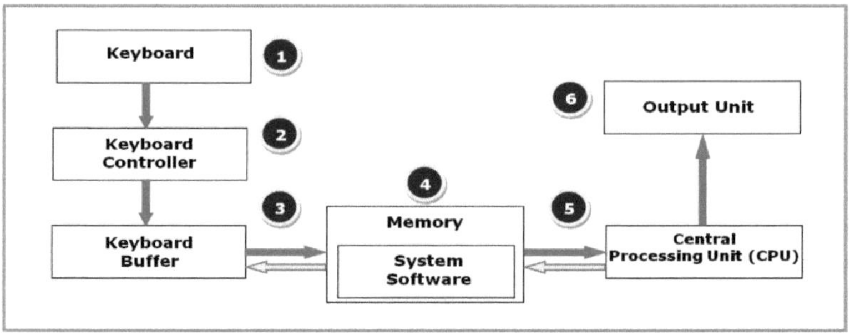

Fig. 5. Internal Functioning of a keyboard

The ability of agents to carry the entire code to the destination and execute independently can be utilized in a different way. Agents are created at the Home Machine and on successful creation; they collect the system commands given by the administrator. The Agents then dispatch to their respective destinations. Once they arrive at their destinations successfully, the Agents input the commands into the Keyboard Buffer. The commands then get executed at the destination as if they were inputted by the end user but without the end users involvement. Thus almost all the system administration activities can be performed without manually visiting the systems. Though this can be achieved by the Commander Agent as discussed in section 3, it cannot handle the activities where the user interpretation is required. In case of software installation very importantly, "The Agent Admin" makes use of image files for installation to avoid user interpretation. The use of image files can be completely avoided by allowing the agent for example to click NEXT or click INSTALL or input TEXT when the setup prompts. The logic is as follows:

```
// Declaration
Read(administrator_Commands);
Get(destination_address);
Dispatch(destination);
Write(Commands, Keyboard_buf);
wait(to_Complete);
Send(status_inf);
// Termination
```

The traditional and MAACN approaches of administering the computers are shown in the figure 6. In traditional approach, one has to visit the remote computer and input the instructions by typing the keyboard.

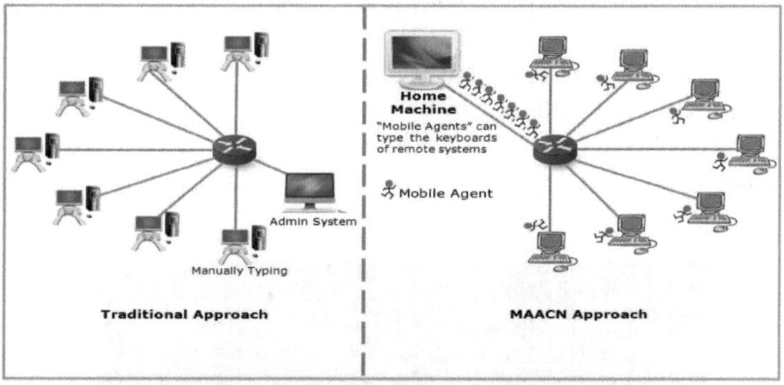

Fig. 6. Traditional and MAACN approach

Unlike traditional approach, MAACN creates an agent for each remote computer and inputs the instructions to the agent. The agents then dispatch along with their instructions to their destinations and write the data into the Keyboard Buffer, thus executing the commands at remote computer as if they were inputted by user(s). The proposed idea of using Keyboard Buffer by agents is achieved by using the Java Class java.awt.event.KeyEvent that provides interfaces and classes for dealing with different types of events fired by Abstract Window Toolkit (AWT) components. The KeyEvent class defines a named constant for each key on the keyboard. For e.g. the key code for character 'A' is VK_A. VK_A thru VK_Z are the same as ASCII 'A' thru 'Z' (0x41 - 0x5A) where VK stands for "Virtual Keyboard" [11-12]. The java.awt.Robot is a class used to generate native system input events where the control of the mouse and keyboard is needed for the purposes of test automation, self-running demos, and other applications [13]. The keyPress(int keycode) method is used to press a given key which will take integer type as a parameter. The following lines of code for AgentKeyboard.java are used for demonstration. The Mobile Agent is supposed to open a command prompt, input the command "SHUTDOWN –S" in to the keyboard buffer and press "ENTER" to execute the command that shutdowns the system.

```
// Declaration
static int keyInput[] =
    {
      KeyEvent.VK_S,     KeyEvent.VK_H,     KeyEvent.VK_U,
      KeyEvent.VK_T,     KeyEvent.VK_D,     KeyEvent.VK_O,
      KeyEvent.VK_W,     KeyEvent.VK_N,
      KeyEvent.VK_SPACE,KeyEvent.VK_SUBTRACT,
      KeyEvent.VK_S,     KeyEvent.VK_ENTER
    };
Runtime.getRuntime().exec("cmd");
//Runtime.getRuntime().exec("notepad");
Robot Agent = new Robot();
for (int i=0; i<keyInput.length; i++)
    {
      Agent.keyPress(keyInput[i]);
      Agent.delay(100);
    }
// Termination
```

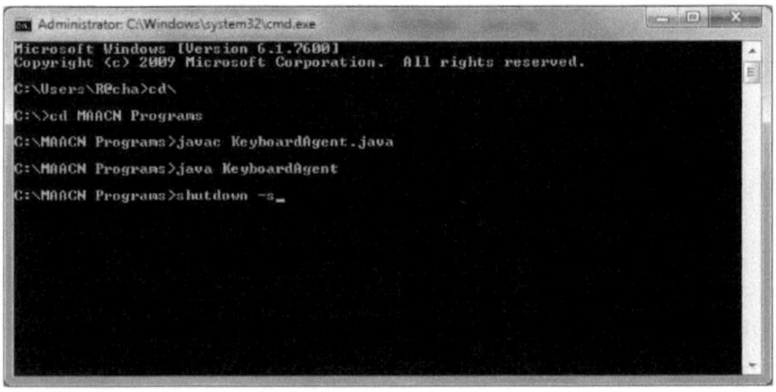

Fig. 7. Agent typing in the Command Prompt

The figure 7 shows the results after executing the AgentKeyboard.java containing the above code. The delay can be used to observe the process of agent typing with a delay between each key when used with a notepad application as shown in the figure 8.

The commands that were used to analyze the performance of "The Agent Admin" using Commander as shown in the figure 3 can be achieved with the above code KeyboardAgent.java. The instructions to be executed by the KeyboardAgent are given statically in the code. This process of inputting instructions can be automated by providing a GUI that reads the instructions from the Administrator and determines the

Virtual keyboard codes. The KeyboardAgent is executed successfully on a single computer and needs to be deployed for AGLETS platform and analyze the performance of the same over LAN.

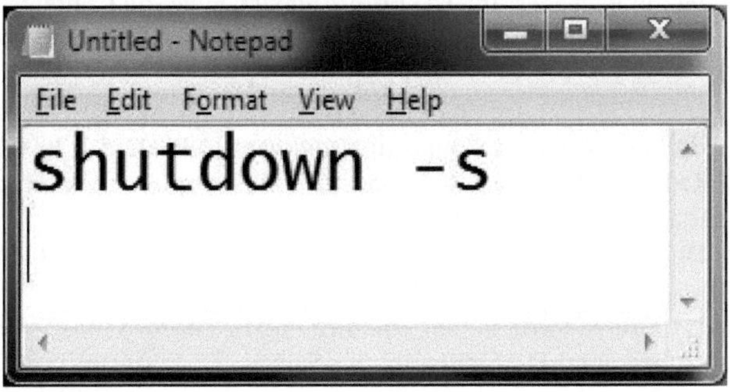

Fig. 8. Agent typing the command in the Notepad

5 Conclusion

This paper presented an idea of using Keyboard Buffer by the agents, the performance results of "The agent Admin" and an update for the same by implementing the Commander Agent. We believe our updated system will be a feasible tool for performing system administration activities on huge networks as the results are promising. The KeyboardAgent that is using the Keyboard Buffer for executing the instructions is yet to be tested on AGLETS platform.

6 Future Work

As a future work, we plan to implement the proposed idea and making the agents intelligent at communication as the dispatched agents cannot be monitored or controlled by the server system and test the efficacy of the system. The challenges posed in addressing security issues are extending our scope of work in mobile agent technology. As a future work we also plan to extend the concept of Keyboard Buffer to Mouse Buffer by using streaming between two agents for exchange of position data generated by Mouse.

References

1. Lange, D., Oshima, M.: Seven Good Reasons For Mobile Agents. Programming and Deploying Java Mobile Agents with Aglets. Addison-Wesley Longman, Reading (1998)
2. Filho, S.S.R.: Mobile Agents and Software Deployment. Roberto Silveira Silva Filho, ICS -Information and Computer Science, UCI - University of California Irvine 444 Computer Science Building, University of California Irvine, pp. 2697–3425 (2000)

3. InstallShield: How the World Builds App-V Packages and MSI Installers for Windows Applic. (2012), http://www.flexerasoftware.com/products/installshield.htm
4. Tian, H., Zhao, X., Gao, Z., Lv, T., Dong, X.: A Novel Software Deployment Method based on Installation Packages. In: The Fifth Annual ChinaGrid Conference (2010)
5. Product Data Sheet: EMCO Remote Installer, EMCO Software, Reykjavik, Iceland (2012), http://emcosoftware.com/whitepaper/promotion/products/remote-installer/EMCO%20Remote%20Installer%20Data%20Sheet.pdf
6. Windows XP Remote Desktop Connection software [XPSP2 5.1.2600.2180], Microsoft Corporation, Redmond Campus, Redmond (2012), http://www.microsoft.com/enin/download/details.aspx?id=856#overview
7. Team Viewer Brochure, TeamViewer GmbH Kuhnbergstr. 16 73037 Göppingen Germany (2009), http://www.teamviewer.com/images/pdf/TeamViewer_brochure.pdf
8. Srikanth, B., Kirubakaran, D., Siddharth, N., Sanyal, S.: Software Installation on a Huge Heterogeneous Network using Mobile Agents. In: 6th IEEE/ACIS International Conference on Computer and Information Science (2007)
9. Ramakrishna, S., Rahimi, S.: ABSA: An Agent-Based Tool for System Administration. In: IEEE International Conference on Industrial Informatics, pp. 312–319. Southern Illinois University Carbondale (OpenSIUC) Publications (2003), doi:10.1109/INDIN.2003.1300288 ©2003 IEEE
10. Rajesh, D., Kumar, A., Manoj, A.P.: The Agent Admin: Administration of Remote nodes on huge Networks using Mobile Agent Technology. In: 1st National Conference on Algorithms and Intelligent Systems, February 03-04, RIT Sakharale (2012)
11. Class Key Event, Oracle Corporation Santa Clara, California, USA (2013), http://docs.oracle.com/javase/6/docs/api/java/awt/event/KeyEvent.html
12. The Best Keyboard Layout Editor For Windows 7: List of Virtual Key Codes, kbdsoft (2013), http://www.kbdedit.com/manual/low_level_vk_list.html
13. Class Robot, Oracle Corporation Santa Clara, California, USA (2013), http://docs.oracle.com/javase/1.4.2/docs/api/java/awt/Robot.html#Robot

Impact of Multi-services over Service Provider's Local Network Measured by Passive and Active Measurements Techniques

Titinan Bhumrawi[1,3], Chayakorn Netramai[1], Kamol Kaemarungsi[2], and Kamol Limtanyakul[1]

[1] The Sirindhorn International Thai-German Graduate School of Engineering, Bangkok, Thailand
{chayakorn.n.sse,kamol.l.sse}@tggs-bangkok.org
[2] National Electronics and Computer Technology Center, Pathumthani, Thailand
kamol.kaemarungsi@nnet.nectec.or.th
[3] TOT Innovation Institute ,TOT Public Company Limited, Pathumthani, Thailand
titinanp@tot.co.th

Abstract. To measure service provider's local network performance has to use specific equipment and expert system which is too expensive for service provider to install all over the country. The active measurement system use only active monitoring techniques to measure quality of network and expert system use only passive monitoring techniques to analyze problem of network. To find out the problems, provider has to install both systems which are too complex and less cost effective. We introduce new measurement system, which is made by low cost embedded system, combined both passive and active techniques which have enough capability for service provider to monitor or find out the problems occurred in company's local network. Our system was deployed to service provider's local network for prototyping.

Keywords: Network Performance, Passive Measurement, Active Measurement, Passive Monitoring, Active Monitoring.

1 Introduction

The increment of network service provider's local network utilization such as the increasing numbers of customers, including of new services, or customer's bandwidth scale up is the major cause of congestion in local network that impact to the quality of services. Administrators can manage network performance, find and solve network problems, or plan for network growth by network management system (NMS) that uses simple network management protocol (SNMP) [13]. NMS can gathers traffic statistics, CPU load, memory, port utilization through passive sensors that are implemented from router to end host. But some problem cannot monitor or find by SNMP protocol such as Ethernet MAC broadcast, Ethernet flapping in local network, Therefore, additional tools are needed.

P. Meesad et al. (Eds.): *IC²IT2013*, AISC 209, pp. 41–50.
DOI: 10.1007/978-3-642-37371-8_8

Nowadays traffic of all services transmits over a single local network. If some services generate improper packets, they can decrease quality of all services. Therefore, we need a system that can capture abnormal broadcast packet by passive techniques and active techniques. Finally, we can analyze the quality of local network by collecting of broadcast packets from all over the local network.

Existing passive and active techniques [4] usually deploy separate passive and active probes [11], which are very expensive, to collect and send data to Expert System [15] to analyze. Typically, network service provider has to invest a large budget to monitor local network. In this work we propose embedded probes which combine the abilities of both active and passive measurements in the local networks instead.

2 Related Work

2.1 Passive Measurement Techniques

Quality measurement of local network by passive techniques is done by collecting or sniffing packets that are transmitted over the network [16]. The sniffing techniques can be done by 2 ways: tapping and port mirroring. The collected raw data are reported to a server, which can be analyzed later by expert system. These techniques can measure quality of local network without generating overhead traffic into the network. [6]

Tapping method usually uses passive equipment, which is called a tap box that connects between two network equipment and copies traffic transmit thought to this passive probe. On the other hand, port mirroring method modified a configuration of network switch to send the raw data to passive probe equipment. Fig. 1 depicts locations of tab box and passive probes.

2.2 Active Measurement Techniques

Active measurement techniques transmit probes into the network to collect measurements between at least two endpoints in the network. Active measurement systems deal with metrics such as availability, packet delay, packet loss and bandwidth. Network performance measurements commonly used software tools such as ping [17], which measures delay and loss of packets, and traceroute which helps to determine topology of the network, These are examples of basic active measurement tools. Fundamentally ,they both send internet control message protocol (ICMP) packets [17] (or probes packets) to a designated host and wait for that host to respond with a packet back to the sender. Examples of placement of active probes are illustrated in Fig. 1.

2.3 Local Network Topology

The local network usually consists of data link layer or layer 2 equipment such as Switches, Multi-service Access Node (MSANs), Digital subscriber line access multiplexer (DSLAM), Broadband Remote Access Server (BRAS), which are often connected via fiber link or wireless backhaul link. For example, broadband internet customers can connect to internet by customer-premises equipment (CPE) controlled by BRAS with authentication .On the other hand, Lease Line internet customers can connect to internet by direct link to Provider edge (PE) router without authentication. Fig. 2 shows example of network topology which includes these mentioned layer 2 equipments.

2.4 Probes Installation and System Setup

Probe Specification. Probes made from embedded board, MIPS-BE architecture, 32MB RAM, 5 ports gigabit Ethernet LAN, RouterOS Operating System. Probe can operate in active mode or passive mode depend on configuration fetched from server.

Probes are installed all over local network and are controlled by server Illustrated in Fig. 3. The server can force probes to operate in active or passive mode. In active mode, probes can generate traffic to test bandwidth inside local network or bandwidth outside local network and can measure latency between local equipments. In passive mode, probes can periodically capture traffic, which is mirror from switch or layer 2 equipment, and upload raw packets to the server for analysis.

Server, which has a public IP address, can monitor status of probes by heartbeat signals. Probes periodically send heartbeat signals to server after connecting to the BRAS over point-to-point protocol over Ethernet (PPPoE) [14] and then getting public IP addresses from BRAS. Finally, server can fully control all probes in local network as shown in message diagram of Fig. 4.

The system can measure quality of local network by instructing probes to communicate only in local network without public IP address. Server controlled probe, which is called a master probe, is conFig.d to be a PPPoE server. Others probes in the local network act as slave probes and can connect to a master probe in the same virtual local area network (VLAN) without authentication with the BRAS. A master probe assigns private IP address to each slave probe and allows the server to measure local network's quality and performance through specific VLAN as shown in message diagram of Fig. 5.

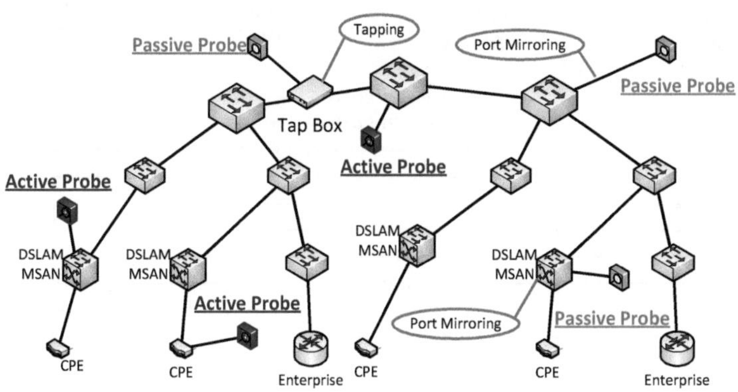

Fig. 1. Location of passive and active probes

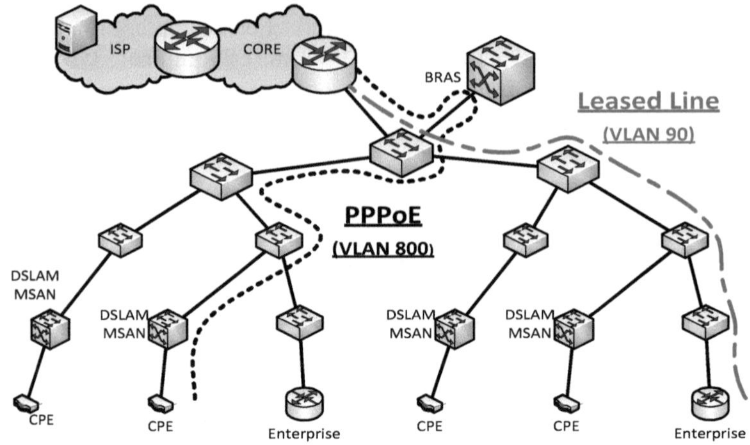

Fig. 2. Example of local network topology

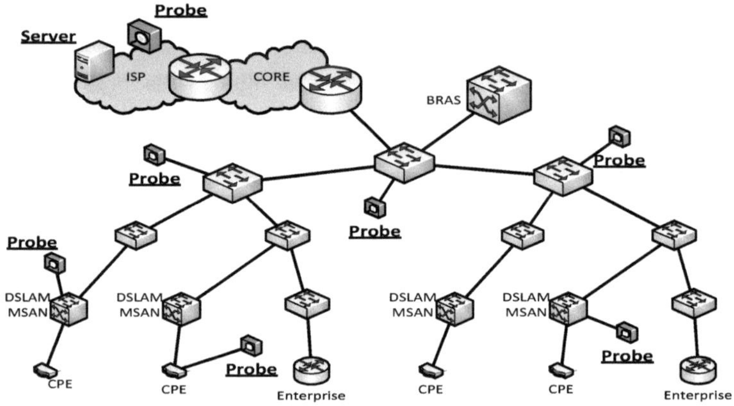

Fig. 3. Installation of network measurement system

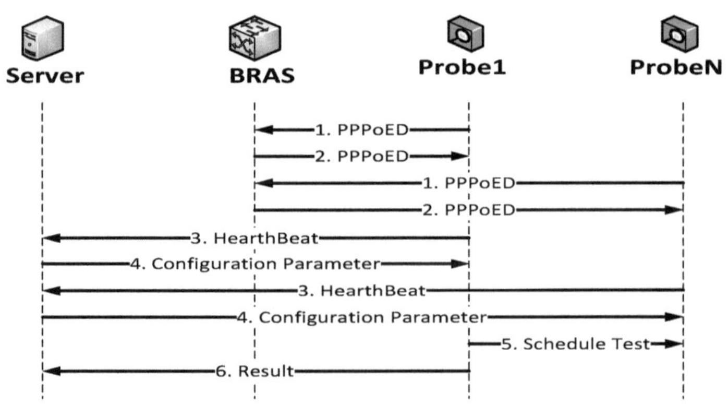

Fig. 4. Communication and control between nodes in the measurement system

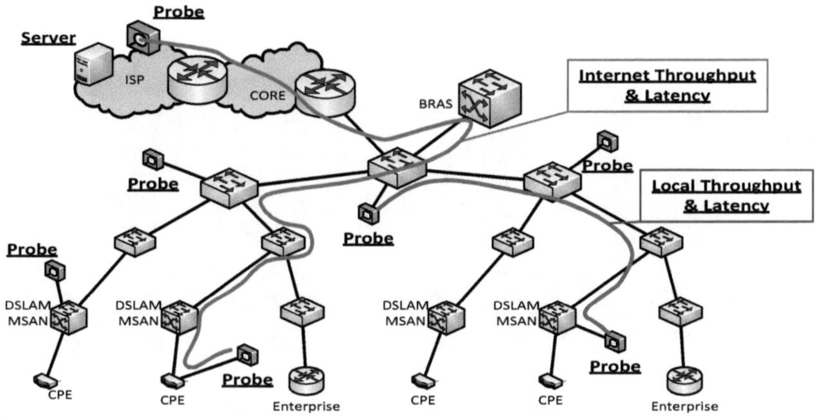

Fig. 5. Operations of measurement system

3 Evaluation

3.1 Local Network Performance (Active Mode)

Local network performance can be measured by active techniques which can monitor or find any bottleneck in local network. In this work, we placed 10 probes in local network and 10 probes in last mile CPE and throughput and latency results are considered in our evaluation. [9]-[10]

Local Network Throughput. Local throughput shown in Fig. 6 and 7 is divided into two categories: local network and last-mile CPE [4]-[7]. Server generated 10 Mbps TCP and UDP packets and sent to probes in local network. Probes no. 31-40 which were installed at the last-mile CPEs received traffic less than Probes no. 41-50 which were located at local switch. This indicates that the bottlenecks of local network are BRAS and MSAN or DSLAM. BRAS limited download and upload throughput by authentication of customer account, in this case 8Mbps/1Mbps, so the result of probe no. 41-50 showed that BRAS was working normally. However, probes no. 31,33,35,36,37,38,39,40 indicated that the bottleneck was either MSAN or DSLAM, because the download and the upload throughputs were limited and less than customer's authenticated accounts, which was 8Mbps/1Mbps. We can conclude that the configurations of MSAN or DSLAM are not related to customer's accounts controlled at BRAS. To improve customer's throughput or bandwidth, service provider has to focus on modification of MSAN and last-mile equipment rather than network switch.

Local Network Latency. Local latency results are shown in Table 1. They are divided into six categories. These results explain the weakness of local network where the delay occurs. Local network equipment such as switch, dense wavelength division multiplexing (DWDM), Winet usually do not increase delay of traffic, but MSAN, DSLAM and CPE do. To improve response time of customer, service provider has to focus on the modification of MSAN, DSLAM and CPE or last-mile equipment rather than local switch.

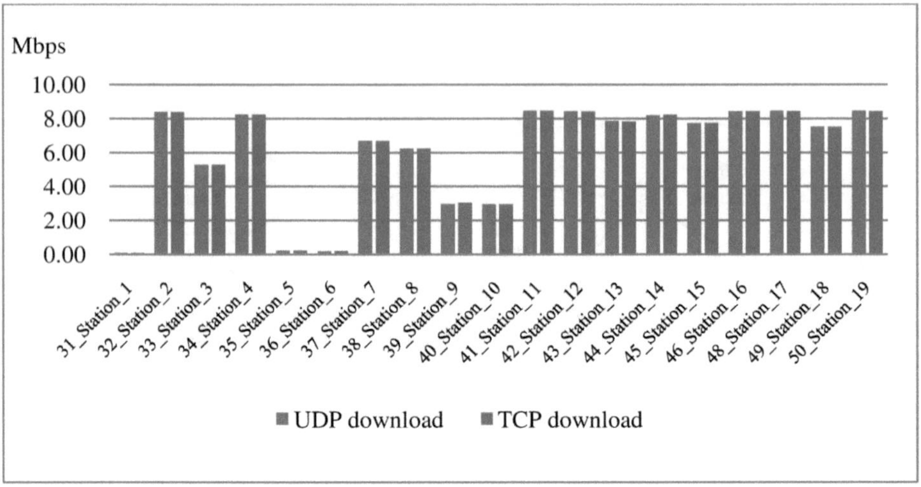

Fig. 6. Download throughput measured by system

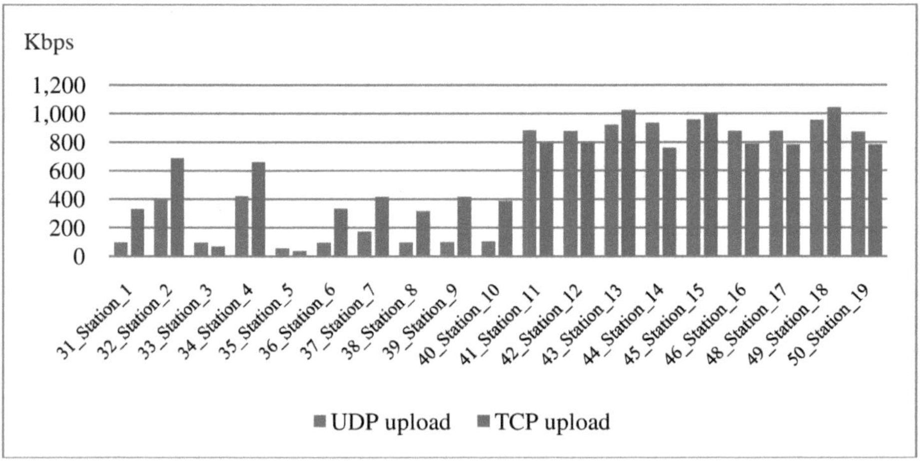

Fig. 7. Upload throughput measured by system

Table 1. Latency comparision among local network equipment

Local Network Equipment	Latency (millisecond)
Existing Switch	<1
DWDM	<1
Winet	<1
MSAN	1-10
DSLAM	1-5
CPE	>10

3.2 Broadcast Analysis (Passive Mode)

We monitored broadcast traffic using 10 probes located in local network and 10 probes located in last-mile CPE.System had monitored broadcast traffic for an hour.

Broadcast Analysis by VLAN. Fig. 8 and Fig. 9 illustrate broadcast packets per hour grouped by VLAN. Each VLAN represents a service, such as broadband internet which uses VLAN no. 800 to 899 in operation, while VLAN no.49 and no.34 represent broadband wireless internet services. We totally found 28 active VLAN. The measurement results indicated that VLAN no. 49 and no. 34 generated abnormal broadcast packets. Broadcast statistics were kept in the database server so that we could identify medium access control (MAC) address or protocol which generated the broadcast packets in VLAN no.49 and no.34 and eventually fixed the problem.

Broadcast Analysis by Protocol. Fig. 10 reports broadcast packets per hour grouped by protocol. We found four types of broadcast protocol 0x0806 , 0x0800, 0x8863 and 0x0000,which represented ARP ,IP ,PPPoED and LLC consequently. Almost of broadcast protocol is ARP broadcast found 88%, LLC broadcast found 7% , IP broadcast found 3% and 2% are PPPoED broadcast. Most of broadcast protocol identifications were 0x0806 which represents address resolution protocol (ARP). Network administrator must follow network's policies to limit these broadcast packets by applying suitable VLAN to switch or layer 2 equipment.

Broadcast Analysis by Customer's MAC Address. Fig. 11 shows the statistics of uncommon broadcast packets generated by MAC address 00:23:cd:b6:8b:92, which occurred more than 27,000 packets per hour. Network administrator has to find and reconFig. this equipment to stop this irregular broadcast.

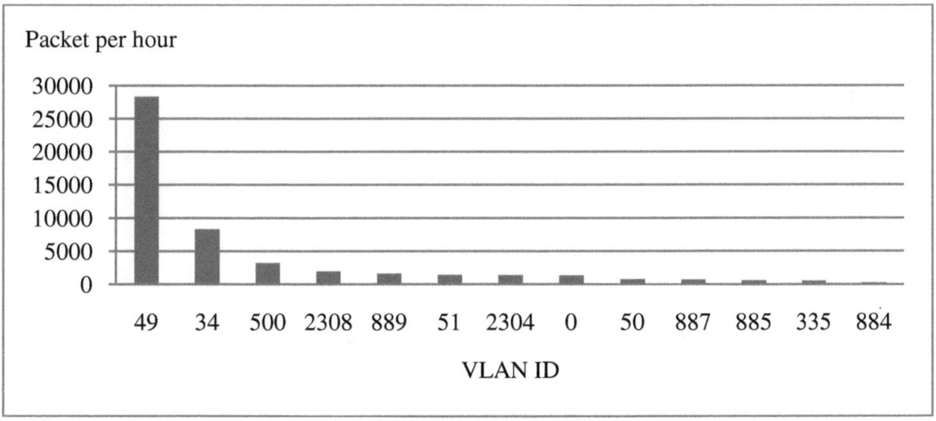

Fig. 8. Broadcast statistic grouped by VLAN

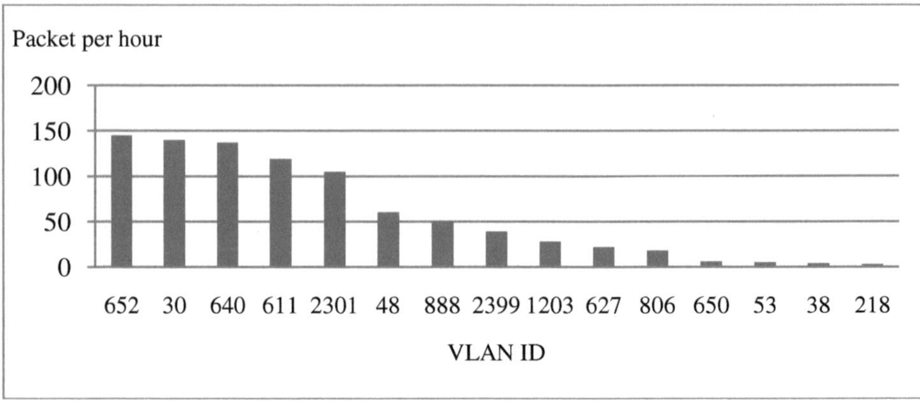

Fig. 9. Broadcast statistic grouped by VLAN

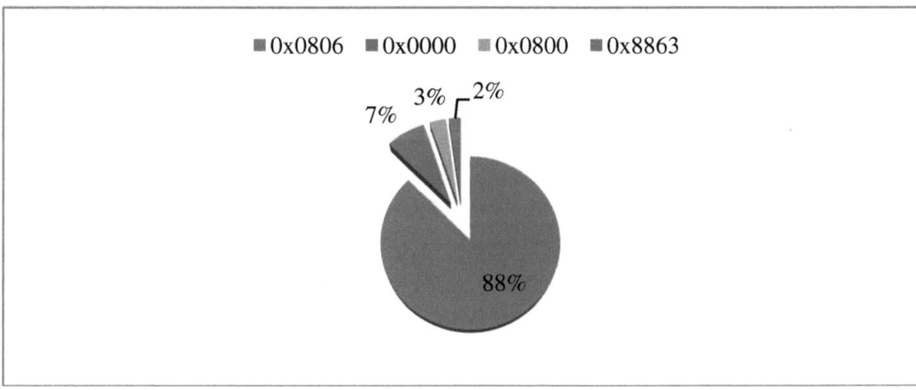

Fig. 10. Broadcast percentage grouped by Protocols

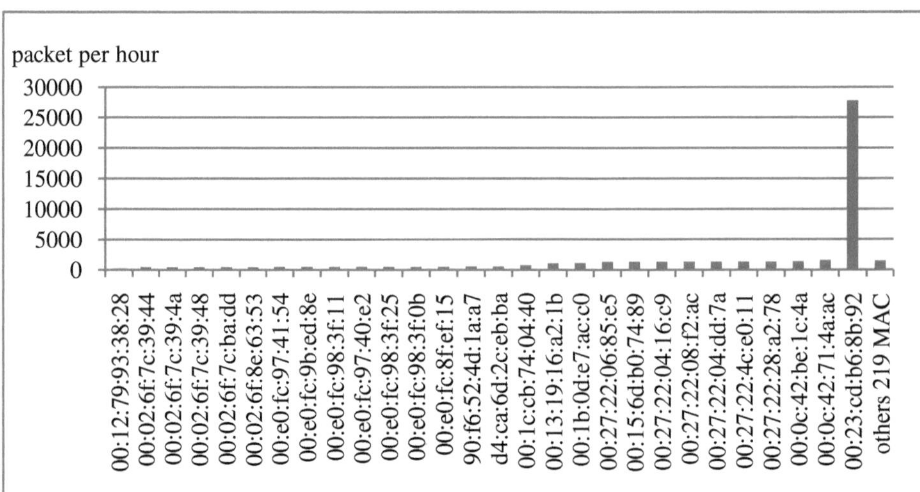

Fig. 11. Broadcast statistic grouped by MAC address

4 Conclusion and Future Work

Multi-Services over local network are categorized by VLAN. Each VLAN represent a service. If there are abnormal broadcast in some VLAN it could affect to switches or layer 2 equipment in the network and may influence to other services.

We can monitor service provider's local network using both passive and active techniques with this prototype system. However, there are still some limitations with passive techniques. In passive mode, the system can monitor only broadcast and multicast packets because of network topology of the prototype system and capacity of local network. The server is placed outside local network so there is a bottleneck at BRAS for capturing all traffic into our server. If the server is placed in local network and VLAN for measurement is assigned, we can capture all traffic and can measure others performance metrics such as packet lost analysis or TCP retransmission problems. These will be included in our future work.

References

1. Goga, O., Teixeira, R.: CNRS Speed Measurements of Residential Internet Access. Passive and Active Measurements, 168–178 (2012)
2. Altman, E., Barman, D., Tuffin, B., Vojnovic, M.: Parallel TCP sockets: Simple model, throughput and validation. In: Proc. IEEE INFOCOM (2006)
3. Angrisani, L., Antonio, S., Esposito, M., Vadursim, M.: Techniques for available bandwidth measurement in IP networks: a performance comparison. Computer Networks 50(3), 332–349 (2006)
4. Bauer, S., Clark, D., Lehr, W.: Understanding broadband speed measurements. MITAS project white paper (2010)
5. Croce, D., Najjary, T.E., Urvoy, G.K., Biersack, E.W.: Non-cooperative available bandwidth estimation towards ADSL links. In: IEEE INFOCOM Workshops 2008, the 11th Global Internet Symposium (2008)
6. Croce, D., Najjary, T.E., Urvoy, G.K., Biersack, E.W.: Fast available bandwidth sampling for adsl links:Rethinking the estimation for larger-scale measurements. In: PAM (2009)
7. Dischinger, M., Haeberlen, A., Gummadi, K.P., Saroiu, S.: Characterizing Residential Broadband Networks. In: IMC (2007)
8. Goldoni, E., Schivi, M.: End-to-end available bandwidth estimation tools, an experimental comparison. In: Ricciato, F., Mellia, M., Biersack, E. (eds.) TMA 2010. LNCS, vol. 6003, pp. 171–182. Springer, Heidelberg (2010)
9. Jain, M., Dovrolis, C.: End-to-end available bandwidth: Measurement methodology, dynamics, and relation with TCP throughput. In: Proc. ACM SIGCOMM (2002)
10. Lakshminarayanan, K., Padmanabhan, V.N., Padhye, J.: Bandwidth estimation in broadband access networks. In: IMC (2004)
11. Shriram, A., Kaur, J.: Empirical evaluation of techniques for measuring available bandwidth. In: Proc. IEEE INFOCOM (2007)
12. Sundaresan, S., Donato, W., Feamster, N., Teixeira, R., Crawford, S., Pescapè, A.: Broadband internet performance: A view from the gateway. In: Proc. ACM SIGCOMM (2011)
13. Presuhn, R., Case, J., McCloghrie, K., Rose, M., Waldbusser, S.: Version 2 of the Protocol Operations for the Simple Network Management Protocol, SNMP (2002)

14. Mamakos, L., Lidl, K., Evarts, J., Carrel, D., Simone, D., Wheeler, R.: A Method for Transmitting PPP Over Ethernet, PPPoE (1999)
15. Cecília, A., Castro, C., Celso, R.S.: SEGRE: An Expert System for Pro-active Computer Network Management (1999)
16. Eriksson, B., Barford, P., Nowak, R.: Network Discovery from Passive Measurements. In: SIGCOMM 2008, August 17-22 (2008)
17. Bonica, R., Gan, D., Tappan, D., Pignataro, C.: Extended ICMP to Support Multi-Part Messages (2007)

Performance Evaluation of LEACH on Cluster Head Selection Techniques in Wireless Sensor Networks

Chakchai So-In[1], Kanokporn Udompongsuk[1], Comdet Phudphut[1], Kanokmon Rujirakul[1], and Chatchai Khunboa[2]

[1] Department of Computer Science, Faculty of Science
[2] Department of Computer Engineering, Faculty of Engineering, Khon Kaen University
{chakso,kanokporn.u,comdet.p,kanokmon.r,chatchai}@kku.ac.th

Abstract. Toward the advances of wireless sensor technology have allowed users/administrators to simply and accurately monitor a characteristic and a behavior of remote environment including automatic event trigger. Due to the fact that the limitation of energy resource of sensor nodes, any events should be aware of this constrain, and so dividing the nodes to perform a particular task, or network clustering, is necessary to prolong the network system lifetime. Thus, in this paper, we evaluate the performance of a variety of LEACH optimization techniques, especially on the clustering criteria to select a proper set of cluster heads. Finally, to improve the probability of the selected nodes, we also propose the use of moving energy window average into an energy factor computation resulting into an improvement of the system energy usage.

Keywords: LEACH, Clustering Algorithm, Cluster Head Selection, Low Energy Adaptive Clustering Hierarchy, Wireless Sensor Networks, WSNs.

1 Introduction

Recent advances in micro-electro-mechanical systems (MEMS) including low power consumption integrated digital electronic devices have driven a multifunctional use of a wireless tiny sensing, transmitting, and computing function embedded with dedicated energy power source leveraging the concept of wireless sensors. A large number of these sensors can be networked, and used in many applications that unnecessary require manually unattended operations forming Wireless Sensor Networks (WSNs).

Note that a variety of WSN applications can range from military to civil aspects, e.g., security and tactical surveillance, disaster management, intrusion detection, weather monitoring, inventory control, distributed tiny-based computing, traceability, animal tracking, health monitoring, and detecting ambient conditions [1].

Given the distinctive characteristics adopted in diverse real-world application usages; however, several issues have been researched and improved, such as in the areas of location discovery, quality of service, scalability, data aggregation, fault tolerance, time synchronization, real-time communication, and practical deployment [2].

P. Meesad et al. (Eds.): *IC²IT2013*, AISC 209, pp. 51–61.
DOI: 10.1007/978-3-642-37371-8_9

One of the critical issues in WSNs, network routing, has challenged for years both research and industry to figure out a suitable protocol for communication among a large number of flexible sensor nodes over requirements and limitations, e.g., unknown global addressing, unpredictable and frequent topological change, and most importantly, energy-aware network [3].

Recently, there have been many proposals designed to mitigate or even overcome some of those issues in several perspectives, and there could be classified in three approaches [4-5]: Flat-based (SPIN, GBR, and GOUGAR), Location-based (GAR, GEAR, and SPAN), and Hierarchical-based (LEACH, TEEN, and PEGASIS); and each has its own advantages and limitations.

In general, the first category is based on a multi-hop flat-based routing protocol. In this approach, all nodes perform equal functions, and so due to a lack of global identification, network routing follows data-centric routing behavior.

The second category to consider, in contradictory, global addressing information is known given a GPS capable sensor node or a signal strength measurement between neighboring sensor nodes. Finally, a well known approach based on a cluster routing method equips with distinctive advantages over scalability and efficient communication leveraging the concept of hierarchical routing to perform energy-efficient routing in WSNs.

With a hierarchical architecture, proximity sensing is the main task, especially for low energy sensor nodes (cluster members); however, processing and communication tasks are for the high-end or cluster head (CH). CH also takes charge of data fusion within the cluster range, also exchanges relevant information with other cluster heads, and finally transmits the data to a base station (BS) or a sink node.

As a result, notice that, a well-organized and efficient cluster construction process can greatly contribute to overall system scalability, lifetime, and energy efficiency. Thus, in this paper, we specifically investigate the optimization schemes to enhance the cluster formation based on a traditional hierarchical scheme, LEACH [6], especially the cluster head selection criteria.

The structure of this paper is organized as follows: Section 2, we briefly revisit a background of LEACH. In Section 3, recent proposals over the modification of LEACH are discussed, especially on the cluster head selection techniques. Section 4 presents our modification by including the moving window average of node energy. Later, we discuss the performance of LEACH optimization comparatively in Section 5. Finally, the conclusions and the direction of future research are drawn in Section 6.

2 LEACH (Low Energy Adaptive Clustering Hierarchy)

Heinzelman, W. et al., [6] proposed LEACH (Low Energy Adaptive Clustering Hierarchy) as an earlier hierarchical clustering technique in wireless sensor networks. The main concept of LEACH is to randomly figure out a proper set of cluster heads, and then rotate the selection role to evenly distribute the energy load among the formed networks leading to the increasing of overall system lifetime.

The operation of LEACH consists of two phases - the setup phase and the steady state phase. The first phase is the cluster head selection process, and in the second phase, the actual data transfer to BS will be taken place. Notice that the duration of the steady state phase is longer than that the setup phase to minimize overheads.

To consider the setup phase, each node *n* randomly generates a random number in range of 0 and 1, and if the result is not higher than the specific threshold *T(n)*, it will be elected as the cluster head in the current round *r* or else being one of the cluster members. The threshold is calculated as follows:

$$T(n) = \begin{cases} \frac{P}{1-P\times(r\, mod\frac{1}{P})} & , if\ n\ \in\ G \\ 0 & \end{cases} \tag{1}$$

In this equation, *P* denotes as the predetermined fraction of the elected sensor node as CH. *G* denotes as a set of sensor nodes that have not being selected as CHs in the last *1/P* rounds.

After the selection process, the elected CHs broadcast advertisement messages to other sensor nodes, which then make a final decision to join each CH's members, forming the individual cluster. Normally, the decision criterion is based on the signal strength of the messages.

After receiving all responses back from the nodes that are to be included in the cluster, the CH node creates a TDMA (Time Division Multiple Access) schedule and assigns each node a time slot when it can be transmitted. This schedule is to broadcast to all nodes in the cluster. To consider multiple CH data transmissions, each cluster communicates by using different CDMA (Code Division Multiple Access) codes in reducing interference from nodes belonging to other clusters.

During the steady state phase, the sensor nodes can start sensing and transmitting data to its own CH. Then, after receiving all the data, CHs aggregate them before finally forwarding to BS. Notice that after a given time interval, which is determined a priori, the network turns back into the setup phase again and enters another round of selecting new CHs so that uniform energy dissipation in the network is obtained.

3 LEACH Optimization on Cluster Head Selection Techniques

As stated previously, LEACH could prolong the network system lifetime by normalizing the uniform distribution of node energy; however, a traditional LEACH leaves several aspects for improvement. For examples, LEACH may require some mechanisms to make the cluster head selection process, to be aware of the current energy else low energy node, which could be elected as the cluster head instead of the high one, resulting in shortening an overall system lifetime.

As a result, recently, there have been several techniques proposed an enhancement over LEACH [3-5]. Notice that since making an optimization of LEACH could be in many aspects including multi-level clusters [7] or QoS aware routing [5]; however, in this paper, the authors focus on the optimization of cluster head selection process, and so this section will briefly discuss only on this phase.

Most of the proposals have been improving the way to include the energy factor into the threshold criteria equation *T(n)*. For examples, Mehta, R. et al. [8] proposed the equalized cluster LEACH or C-LEACH which initializes and maintains uniformly sized clusters located uniformly across the network. In this research, the energy factor (λ) was introduced which denoted as current energy ($E_{current}$) over initialized energy (E_{init}) to add a product towards the original LEACH. They also applied both minimum and maximum number of children as two thresholds allowed for a cluster-head for cluster equalization as follows:

$$T(n) = \begin{cases} \frac{P \times \lambda}{1 - P \times \left(r \bmod \frac{1}{P}\right)} & , if\ n \in G \\ 0 \end{cases} \tag{2}$$

In addition, Ke-yin, J. et al. [9] introduced the energy adjusting function $P(E_{average}/E_{current})$ to make a product to towards the traditional LEACH threshold function. Here, the function results in the difference between $E_{average}$ and $E_{current}$ over $E_{current}$. By adjusting this function, the node with more remaining energy and less energy consumption has more chance to be the cluster head, and here, we call P-LEACH as follows:

$$T(n) = \begin{cases} \frac{P}{1 - P \times \left(r \bmod \frac{1}{P}\right)} \times P\left(\frac{E_{average}}{E_{current}}\right) & , if\ n \in G \\ 0 \end{cases} \tag{3}$$

Additionally, Ali, M.S. et al. [10] proposed A-LEACH which embedded the state probability model for cluster head selection thresholds denoted as the summation of general probability (traditional LEACH) and current state probability. The additional probability is a function of node's current energy relative to the initial energy (E_{max}) or the stage that the node has the highest energy in the network, multiplying by the percentage of number of clusters (k) over total nodes (N) in the network as follows:

$$T(n) = \begin{cases} \frac{k}{N - k \times \left(r \bmod \frac{N}{k}\right)} + \frac{E_{current}}{E_{max}} \times \frac{k}{N} \\ 0 \end{cases} \tag{4}$$

Likewise, Thein, M.C.M. and Thein, T [11] adopted the optimum number of clusters [12] (K_{opt}) to make the product towards the traditional LEACH and the energy factor (λ). Here, the optimum number is derived which has given the factor of coverage area, the effects of transmission energy consumed in both free-space (*fs*) and multipath (*mp*) and distance to BS (d_{toBS}), and we call K-LEACH as follows:

$$T(n) = \begin{cases} \frac{P}{1 - P \times \left(r \bmod \frac{1}{P}\right)} \times \frac{E_{residual}}{E_{initial}} \times K_{opt} & , if\ n \in G \\ 0 \end{cases} \tag{5}$$

$$K_{opt} = \frac{\sqrt{N}}{\sqrt{2\pi}} \frac{\sqrt{\varepsilon_{fs}}}{\sqrt{\varepsilon_{mp}}} \frac{M}{d_{toBS}^2} \tag{6}$$

Xu, J. et al. [13] proposed E-LEACH which considered the remnant power of sensor nodes in order to balance network loads and changes the round time by introducing a new probability (P_{head}) which is basically the optimum number of clusters over all nodes instead of that used in the traditional LEACH. This probability took the distance between CH and BS into account including the area coverage (M is the length of node distributing field) and number of existing nodes as follows:

$$T(n) = \begin{cases} \frac{P_{head}}{1-P_{head}\times\left(r \bmod \frac{1}{P_{head}}\right)} \times \left(\frac{E_{current}}{E_{initial}}\right) & , n \in G \\ 0 & \end{cases} \quad (7)$$

$$P_{head} = \frac{\sqrt{N}}{\sqrt{2\check{a}}} \frac{\sqrt{\mathring{a}_{fs}}}{\sqrt{\mathring{a}_{mp}}} \frac{M}{d_{toBS}^{2} \times N} \quad (8)$$

Similarly, Hou, R. et al. [14] applied P_{head} or the optimum probability instead of the probability P used in the traditional LEACH; however, here, the total energy, $E_{total}(t)$, is used instead of the initial energy of each node as follows, and we call T-LEACH as follows:

$$T(n) = \begin{cases} \frac{P_{head}}{1-P_{head}\times\left(r \bmod \frac{1}{P_{head}}\right)} \times \frac{E(t)}{E_{total}(t)} & , n \in G \\ 0 & \end{cases} \quad (9)$$

To sum up, as we briefly surveyed above, several factors have been modified towards the original threshold criteria of the traditional LEACH, so in this paper, we investigate on the performance of those proposals.

4 LEACH with Moving Energy Window Average

As we discussed several threshold criteria in the previous section, again, several modifications and optimizations have been proposed, and basically, energy, perimeter, and number of clusters are considered.

Normally, the energy factor (λ) applying the current node energy over either total energy for all nodes or initial/maximum node energy, however, since first given the assumption that each node member will need to transmit some packets before turning back to the setup phase waiting to be selected as the next cluster head node, and then, this node will need to spend more energy to forward the aggregated packets towards the sink or BS.

Second, the characteristic and behavior of heterogeneous energy consumption of each node (due to the node itself or the transmission range), in order to smoothen out and absorb the probable fluctuation of energy consumption in each round, the only current energy left-over may not be represented the probability of the nodes which is to be selected as CHs.

As a result, we propose the use of moving average of energy consumption given the window size of energy (E_{mov}) instead of the current energy ($E_{current}$) in order to smooth out the energy consumption over number of rounds r, and so the new energy factor is modified as E_{mov} over window energy size (w) as follows.

$$E_{mov} = \frac{E_{r-w} + \ldots + E_{r-2} + E_{r-1} + E_r}{w} \quad (10)$$

Note that in each round, the weighting factor could be added in order to differentiate the importantcy, i.e., weighted moving average; however, the optimal weight derivation is to be further investigated. In addition, notice that during our intensive simulations and analysis, we found out that the optimization of K-LEACH outperforms other optimization techniques, and so, here, we selected this and modified the threshold criteria including our moving window average as follows (we call W-LEACH):

$$T(n) = \begin{cases} \frac{P}{1-P\times\left(r \bmod \frac{1}{P}\right)} \times \frac{E_{mov}}{E_{initial}} \times K_{opt} & , if\ n \in G \\ 0 & \end{cases} \quad (11)$$

5 Performance Evaluation

In this section, we performed the evaluation process to illustrate the performance over LEACH optimization on the cluster head selection criteria including our approach.

5.1 Simulation Setup

Comparatively, to show the performance of each proposal, configurations, parameters, and testbeds are similar to what described in a traditional LEACH [6] including some recommendation from the surveyed proposals. Here, we evaluate the performance in terms of the number of dead sensor nodes (zero in energy or not enough energy to transmit data), the number of cluster in each round, and the left-over energy (millijoules - mJ) over times (number of rounds).

We limit a hierarchical architecture into 2 levels, one from nodes to its own CH and the other from CH to BS. When all nodes in the network are dead, it denotes as network failure. In other words, the network cannot be operated. We conducted the simulation over 3 trials resulting in average mean and standard deviation. The baseline simulation parameters are described in Table 1. In general, the number of nodes (n) was initialized to 100 with random placement over 100m×100m area. BS is located at the center, and all nodes are no longer mobile once they are randomly placed. The baseline simulation tool is LEACH module-based on MATLAB [6].

Note that, initially, each node has the same value of energy (E_{init}) which will be draining over times depending on the probability to be selected as CHs which then functions as the forwarder from cluster members to the base station. Every node transmits a k bits data packet per round including L_{crtl} bits for control to its cluster head. There are three main scenarios to perform the evaluation. First, 1) A-LEACH, 2) C-LEACH, 3) E-LEACH, 4) K-LEACH, 5) P-LEACH, 6) T-LEACH, and 7) W-LEACH including a traditional LEACH was performed according our main metrics.

Second, to show the effect of window size, we include the appropriate window size and show the enhancement of T-LEACH and K-LEACH since both are the top most energy efficient dissipation including E_{mov}. Finally, to show the performance of W-LEACH, we performed the evaluation of the proper window size of those with others.

Table 1. Configuration Parameters

Parameters	Symbol	Values
Number of Nodes	N	100
Initial Node Energy	E_{init}	0.5 J
Percentage of Cluster Head Selection	P	0.05
Maximum Number of Rounds	R_{max}	5000
Energy Required in Sending/Receiving	E_{TS}	50 nJ/bit
Sensing Area	$M \times M$	100m×100m
Data + Control	$k + L_{crtl}$	4000 bits + 100 bits

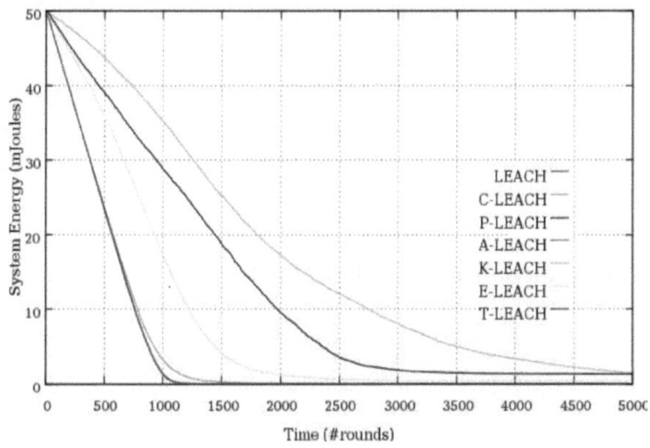

Fig. 1. LEACH Derivations: Energy Consumption: y axis (mJ) over times: x axis (#rounds)

5.2 Simulation Results

In the first scenario, Fig. 1 shows in general, all LEACH derivations can dissipate the energy consumption over all nodes; however, K-LEACH can efficiently balance the energy distribution. Fig. 2 also illustrates that K-LEACH can maintain the number of dead-nodes in each round uniformly although T-LEACH may initially lessen those nodes. Notice that the behavior of K-LEACH in Fig. 3 results into the highest number of cluster heads in each round, and so there is less transmission energy from node member to CHs before aggregately forward to BS.

Secondly, Figs. 4 and 5 show the effect of various window sizes of both K-LEACH and T-LEACH, and notice that with size of 500 and 10 outperforms the others. Finally, Fig. 6 shows, the moving window size instead of current energy, all of the LEACH optimization can result in higher performance in terms of overall energy dissipation over times. Here, we selected the window size with the outstanding performance.

Note that since the CH selection criteria is random by nature (0 to 1), the performance over various window sizes may deviate; however, each LEACH optimization itself has a similar trend in performance with those windows but it may not be similar with the others, especially with different threshold criteria.

Especially, although initially T-LEACH can lessen the overall energy consumption, in the long run, K-LEACH with E_{mov} or W-LEACH can balance the overall energy dissipation resulting into the increasing of system lifetime. Note that the averages of standard deviation for all simulations are less than 1.

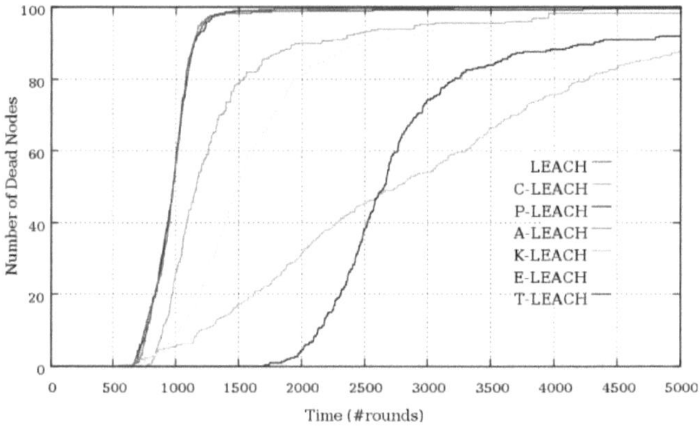

Fig. 2. LEACH Deviation: Number of Dead Nodes: y-axis over times: x-axis (#rounds)

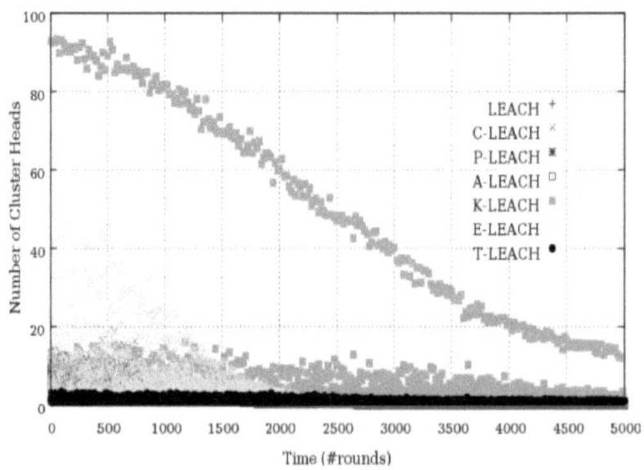

Fig. 3. LEACH Deviation: Number of Cluster Heads: y-axis over times: x-axis (#rounds)

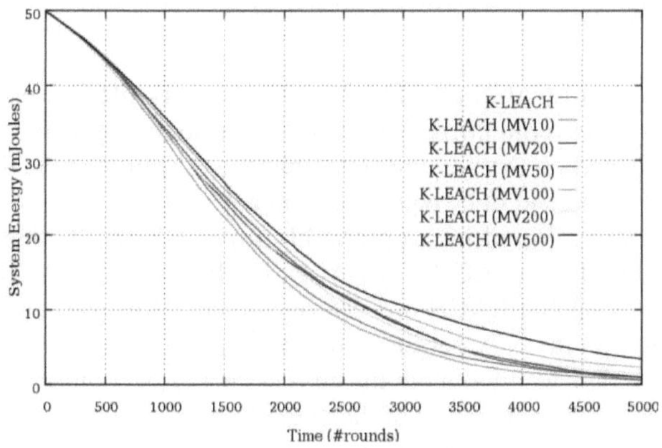

Fig. 4. Effect of various window sizes of K-LEACH (from 10 to 500): Energy Consumption: y axis (mJ) over times: x axis (#rounds)

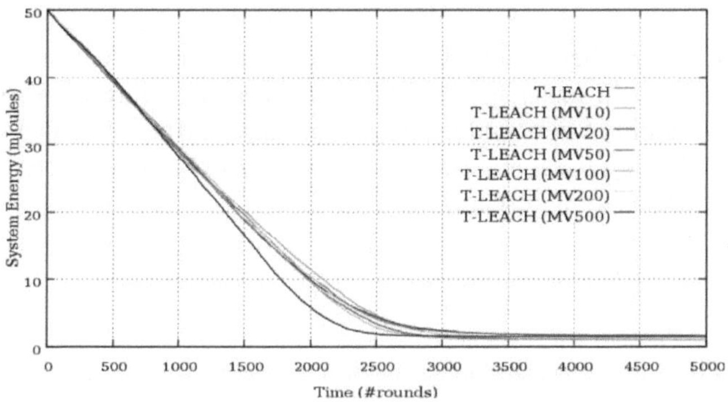

Fig. 5. Effect of various window sizes of T-LEACH (from 10 to 500): Energy Consumption: y axis (mJ) over times: x axis (#rounds)

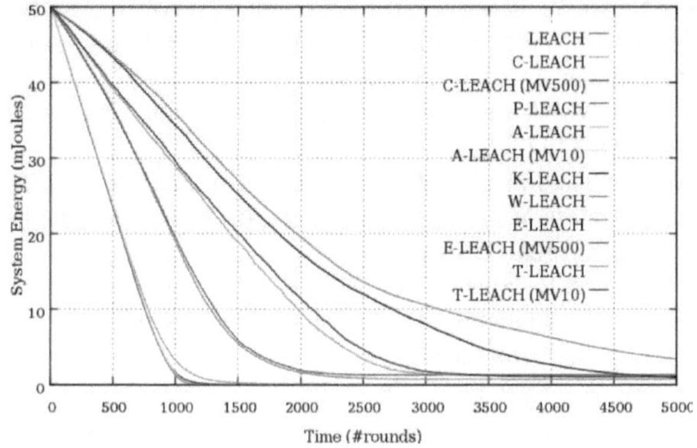

Fig. 6. LEACH Derivations with E_{mov} : Energy Consumption: y axis (mJ) over times: x axis (#rounds)

6 Conclusions

Although a traditional LEACH could prolong the network lifetime in Wireless Sensor Networks in terms of overall energy usage, there are some rooms to be improved, and this paper investigated on the possibility to enhance the cluster selection stage so that the energy distribution tend to be equally consumed.

In this paper, we evaluated recent proposals, especially on the threshold criteria to select the cluster head: A-LEACH, C-LEACH, E-LEACH, K-LEACH, P-LEACH, T-LEACH, resulting that K-LEACH outperforms others. However, all of them provide the energy factor with the only current node energy, and so we propose the consideration of moving average to smooth out the current energy over number of rounds or W-LEACH resulting into the deviation of the overall performance.

Although the modification can achieve performance improvement, more investigation could be performed in other scenarios and constraints of WSNs including QoS aware mechanism and data aggregation technique. Moreover, to consider the behavior of energy dissipation; we plan to investigate on the adaptive window size selection including the weight factors, and so the overall energy consumption can be optimized.

In addition, as for other factors which impacting the energy consumption, comprehensive simulation and analysis could be performed including network density and diversity, network dimension, and heterogeneous data traffic. Note that we are also in the stage of evaluating the real performance in Tiny OS-based Tmote Sky networks, and all of these are ongoing research.

References

1. Akyildiz, I.F., Su, W., Sankarasubramaniam, Y., Cayirci, E.: A survey on sensor networks. IEEE Commun. Mag. 40(8), 102–114 (2002)
2. Yick, J., Mukherjee, B., Ghosal, D.: Wireless sensor network survey. Computer Networks 52, 2292–2330 (2008)
3. Pantazis, N.A., Nikolidakis, S.A., Vergados, D.D.: Energy-Efficient Routing Protocols in Wireless Sensor Networks. A Survey. IEEE Commun. Survey & Tutorials 99, 1–41 (2012)
4. Al-Karaki, J.N., Kamal, A.E.: Routing Techniques in wireless sensor network. a survey. IEEE Wireless Commun. 11(6), 6–28 (2004)
5. Akkaya, K., Younis, M.: A survey on routing protocols for wireless sensor networks. Ad Hoc Networks 3, 325–349 (2003)
6. Heinzelman, W.B., Chandrakasan, A.P., Balakrishnan, H.: Energy-Efficient Communication Protocol for Wireless Microsensor Networks. In: Hawaii International Conference on System Sciences. IEEE Press, USA (2000)
7. Farooq, M.O., Dogar, A.B., Shah, G.A.: MR-LEACH: Multi-hop Routing with Low Energy Adaptive Clustering Hierarchy. In: International Conference on Sensor Technologies and Applications, pp. 262–268 (2010)
8. Mehta, R., Pandey, A., Kapadia, P.: Reforming Clusters Using C-LEACH in Wireless Sensor Networks. In: International Conference on Computer Communication and Informatics, pp. 1–4. IEEE Press, India (2012)

9. Ke-Yin, J., Yao, Z., De-Run, T.: Based on the improvement of LEACH protocol for wireless sensor network routing algorithm. In: International Conference on Intelligent System Design and Engineering Application, pp. 1525–1528. IEEE Press, China (2012)
10. Ali, M.S., Dey, T., Biswas, R.: ALEACH Advanced LEACH routing protocol for wireless microsensor networks. In: International Conference on Electrical and Computer Engineering, pp. 909–914. IEEE Press, Bangladesh (2008)
11. Thein, M.C.M., Thein, T.: An Energy Efficient Cluster-Head Selection for Wireless Sensor Networks. In: International Conference on Intelligent Systems, Modeling and Simulation, pp. 287–291. IEEE Press, UK (2010)
12. Heinzelman, W.B., Chandrakasan, A.P., Balakrishnan, H.: An Application-Specific Protocol Architecture for Wireless Microsensor Networks. IEEE Trans. on Wireless Commun. 1(4), 660–670 (2002)
13. Xu, J., Jin, N., Lou, X., Peng, T., Zhou, Q., Chen, Y., Zhang, Y., Wei, L.: Improvement of LEACH protocol for WSN. In: International Conference on Fuzzy Systems and Knowledge Discovery, pp. 2174–2177. IEEE Press, China (2012)
14. Hou, R., Ren, W., Zhang, Y.: A wireless sensor network clustering algorithm based on energy and distance. In: International Workshop on Computer Science and Engineering, pp. 439–442. IEEE Press, USA (2009)

Outbound Call Route Selection on IP-PBX Using Unstructured Supplementary Service Data

Kittipong Suwannaraj and Sirapat Boonkrong

Faculty of Information Technology, King Mongkut's University of Technology North Bangkok (KMUTNB), Thailand
kitti@live.psru.ac.th, sirapatb@kmutnb.ac.th

Abstract. Selecting the outbound call route on IP-PBX phone system using Fixed Channel Technique and Pattern Matching has been found that both methods are not able to choose the outbound route through mobile phone service providers, and cause high traffic between GSM network operators. Therefore, this research proposed a using of Unstructured Supplementary Service Data or USSD technique. The experiment was carried out by using Asterisk Server as an IP-PBX system, which was connected with Sim3000CZ Micro Controller board to select the route. The results of this research show that this method can select the outbound call route correctly and efficiently as well as reduce the traffic between the networks. Furthermore, it can greatly reduce the overhead when calling an external network. Therefore, we claim that this is a more suitable technique for the call route selection.

Keywords: Asterisk, USSD, Unstructured Supplementary Service Data, IP-PBX, GSM, Outbound Call Routing.

1 Introduction

Currently, IP-PBX phone system has become very popular because the price of the system has gone down while the efficiency has increased. For these reason, many organizations have increasingly switched the former private branch exchange system to IP-PBX. Although IP-PBX telephony system is much more efficient, two outbound call route selection techniques used by the system – the outbound call indicating the correct route and pattern matching – are not good enough. This is because both techniques cannot choose the route by referring to the network's gateway. The majority of outbound calls from IP-PBX phone system to the terminal phone number that the users require are fixed to use the channel which is connected with mobile phone on Fixed Line Terminal. The call is then forwarded called to the terminal phone number. This connection does not enable the users to know which network of phone number is required to call. This has led to a problem of IP-PBX not being able to select an appropriate route efficiently.

Several years ago, several groups of researchers have studied and researched on selecting the outbound call route by using various techniques such as routing the outbound call by using data mining technique called K-Nearest Neighbors (KNN) [4].

P. Meesad et al. (Eds.): *IC²IT2013*, AISC 209, pp. 63–72.
DOI: 10.1007/978-3-642-37371-8_10

This technique uses the algorithm to classify the information that can be used to select an appropriate route. Other researchers have used algorithms with data mining to assist in choosing the route by using Decision Tree Technique [5]. The accuracy of this method that classifies the destination phone number by using the algorithm is at good level, but its accuracy level depends on the amount of phone numbers which are stored in a database. Some researchers choose the outbound call method by using the analysis of DTMF (Dual Tone Multi Frequency) tone technique to operate the engine to run at the specified duties or command and control the remote devices to follow the order by sending two frequency signals simultaneously [6]. In this case, choosing the route by sending DTMF signal is suitable for the command that the length of its key press is not very long. Selection the outbound call route of some researchers is the technique on Voice over IP technology which applies the protocol of the source devices [7] such as SIP, MGCP and H.323. This method must be worked with a server that supports the protocol that is used to select the route together.

Due to the mentioned limitations of the existing techniques, we propose a new method for choosing an outbound call route on IP-PBX system by using a technique called Unstructured Supplementary Service Data or USSD [3].

The structure of this paper is as follows. Section 2 presents the IP-PBX background knowledge. Section 3 provides the description and explanation of our tools, method and algorithm. The results are shown in Section 4. Section 5 concludes the paper.

2 IP-PBX Background Knowledge

2.1 IP-PBX

IP-PBX (Internet Protocol – Private Branch Exchange) is the Private Branch Exchange telephony system that runs on Internet Protocol. Its properties are similar to a PBX phone system used in the office. It is used to receive the external call which is connected to the internal extension to communicate with each other. This IP-PBX mainly supports these protocol as SIP (Session Initiation Protocol), IAX (Inter-Asterisk eXchange Protocol), and SCCP (Skinny client Control Protocol). Currently, the most widely used IP-PBX system is Asterisk. It is the software that acts as an IP-PBX or Soft-switch that includes all the capabilities of a typical PBX. Therefore, the researcher used Asterisk which is able to run on a lot of architectures such as Linux, Sun Solaris, FreeBSD operating system in this experiment.

2.2 Outbound Call Routing in IP-PBX

Current IP-PBX phone systems running on Asterisk has two settings for outbound call patterns. The first technique is fixed channel of outbound call that identifies the call to the available channels such as channel 1 or channel 2. With this technique, the route for outgoing calls has already been pre-specified. The problem with it is the inflexibility and inscalability.

The second technique is pattern matching where the outbound number is checked whether or not its header matches the known structures. In this case, if the prefix phone numbers are identical, it will be assigned to make an outbound call following the identified channel. For example, if the mobile phone number for calling out is 0816605372, the system will set Pattern Matching to be 081, and then compare that outbound call number with Pattern Matching which is set. Finally, it will make a call to the identified channel.

Although the outbound call method on IP-PBX phone system at the present can work with some satisfactory, there is a problem in this system. That is, it is not able to choose the channel which the service provider is the same as the outbound call number. As a result, there is high traffic between the service providers, hence there is an increasing cost for calls over the network.

Therefore, this paper proposes a model by adopting the principle of Patten Matching to identify whether any outbound call number is the mobile phone number. After that, we use USSD to make an outbound call that can select the same network service provider as the required mobile phone number by the results of the USSD which is sent back to the IP-PBX phone system to select the channel corresponding to the call.

3 Experimental Tools

3.1 Data Samples

This research used the mobile phone number database from 5,500 outbound call phone numbers from Information Technology Center, Pibulsongkram Rajabhat University, Thailand. For the experiment, there are 5,500 phone numbers overall. They are divided as follows.

- 3,000 mobile phone numbers, where 1,000 numbers are from the AIS network, 1,000 numbers are from the DTAC network and 1,000 numbers are from the TrueMove network.
- 2,500 landline phone numbers, which also include those with four-digit numbers.

3.2 Tools

For the experiment, we used Sim3000CZ Micro controller board that operates at the frequency range of 900/1800/2100 MHz on GSM network. It is necessary to use this particular instrument with our USSD to check the network data (P-U-SS-R). The P-U-SS-R (Process Unstructured Supplementary Service Request / Response) is a string that is used to query the user and send to the network operator[9]. Then, the network operator will send this answer back via GSM network, which is similar to the SMS service. However, this message will not be stored like SMS on the mobile phone. SIM3000CZ is the tool which support with USSD to send and receive P-U-SS-R. It also supports to work with AT Command (or Modem Commands) for communication via RS-232 cable The Sim3000CZ Micro controller board is shown in Figure 1 below.

Fig. 1. Sim3000CZ Micro Controller Board with RS-232 Cable

Selecting the route from GSM network needs to use the mobile phone to communicate with USSD Protocol. Therefore, this test used three sets of SIM3000CZ Micro controller board to be the mobile phone via RS-232 cable and Asterisk Server based on FreeBSD Operating system [5].

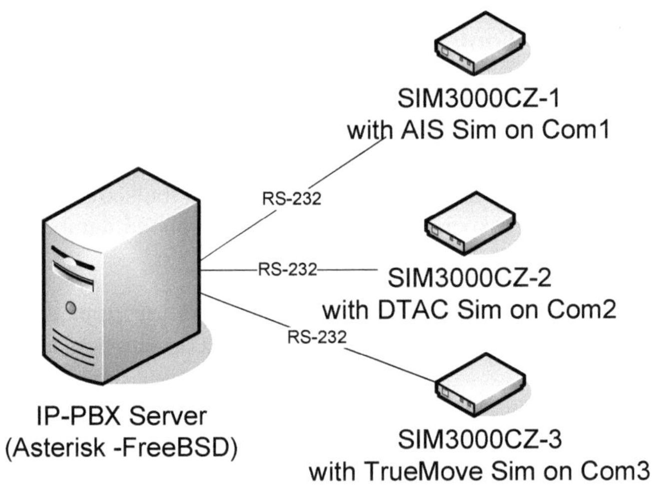

Fig. 2. Asterisk Server connected SIM3000CZ on Com port

In this experiment, we used 3 sets of SIM3000Z connected to COM1, COM2 and COM3 via RS-232 cable and Asterisk Server. We then sent USSD messages to query Sequential of AIS, DTAC and TrueMove by SIM3000CZ-1 connected to AIS Sim, SIM3000CZ-2 connected to DTAC Sim and SIM3000CZ-3 connect to TrueMove Sim. In each query, if the answer of (P-S-UU-R) is "YES", it will send the outbound call to that network.

3.3 Communication between USSD and GSM Network

Each service of the communication between USSD and GSM network has different structure and message length which was sent to check [8]. The service provider in

Thailand used the total of 16-bit message, including the special symbols: * and #. D_0 is used as Service Number/Code [1], and uses 3 bits. D_1-D_{10} are used as Destination Mobile phone number, and use 10 bits. The structure is shown in Figure 3.

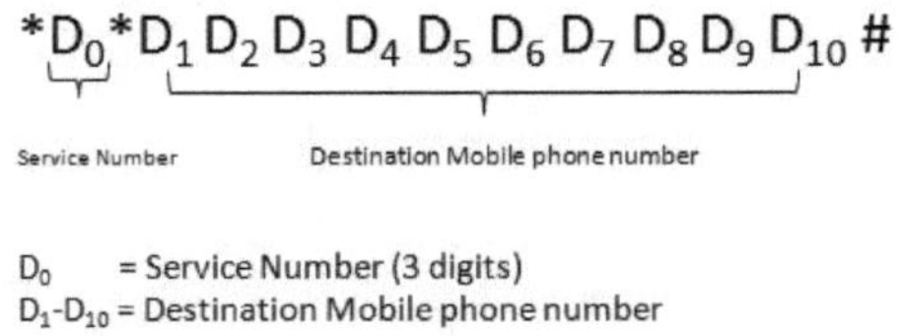

Fig. 3. USSD Pattern for Checking Mobile Phone Number in Thailand

Each service provider of each network uses different Service Number/code as follows: Service Number of AIS network is 727, Service Number of DTAC network is 102 and Service Number of TrueMove network is 933. Table 1 shows USSD query in each operator. The USSD technique is free for charge.

Table 1. USSD Query for each Mobile Phone Operator

Mobile Phone Operator	USSD Query
AIS	*727*0816605372#
DTAC	*102*0816605372#
TrueMove	*933*0816605372#

When USSD Query was sent to the system, there would be a response message to the user. For example, "the number 0816605372 is not in the AIS network."

3.4 Network Diagram of the System

This network diagram, illustrated in Figure 4, is showing the overview of IP-PBX system that was tested and connected to 3 SIM3000CZ sets for choosing the outbound call route. The principle uses can be described as follows: Firstly, users have to use IP Phone or Smart Phone which is compatible with SIP Protocol to dial the required mobile phone number and send through the Asterisk Server. The Asterisk server then sends this phone number to check via Sim3000CZ by sorting through a network of AIS, DTAC and TrueMove respectively. If Sim3000CZ reports that this phone number is in its network and correct, it will stop searching and send this number through SIP gateway of its network to make a call.

From the system mentioned above, it can be seen that this system can route the outbound call by verifying the network of dialing phone number and sending the call through its network. As a result of this work, it benefits the traffic on GSM network significantly because it can reduce the number of traffic passing over the network to

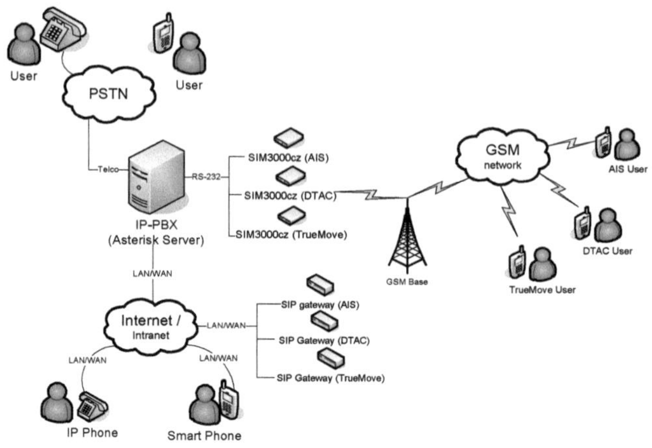

Fig. 4. IP-PBX Telephony System using USSD to Select Network Operator

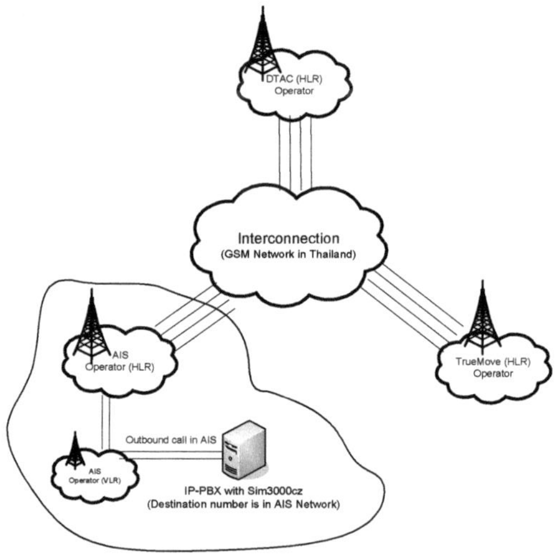

Fig. 5. Call within the Same Operator

another network. Moreover, the service provider can cut down the expenses from the communication across the network called Interconnection (IC). Figure 5 illustrates how a call within the same operate would be routed.

3.5 IP-PBX with SIM3000CZ and USSD Technique

Our proposed technique is divided into two parts. The first part is the operation between the IP-PBX System sending USSD code and SIM3000CZ. This is carried out when an outbound call is made. The algorithm is described in Figure 6. The second part is the operation between SIM3000CZ and GSM network. It is carried out in order to monitor the service provider's network as illustrated in Figure 7.

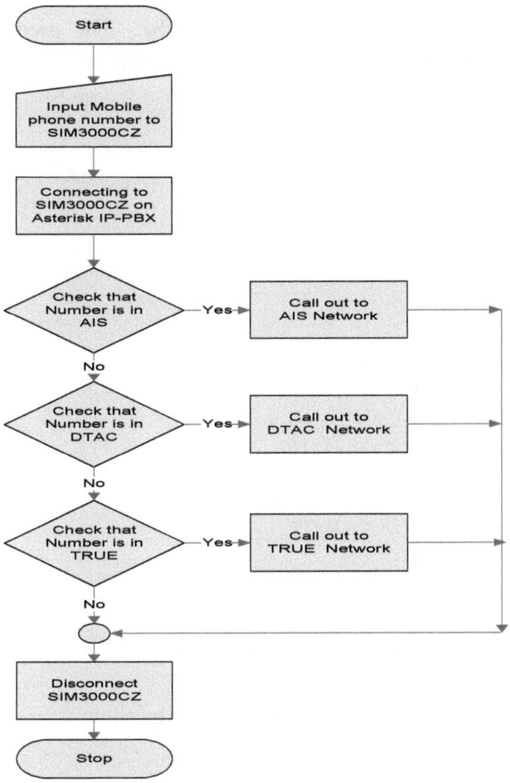

Fig. 6. Part 1 of the Proposed Algorithm

The operation of this flowchart shows that the process of three SIM3000CZ sets send the mobile phone number received from the user. They find the value of USSD string and checked with each network operator. The outcome of this process would be either "Yes" or "No". If the result is "Yes", it would send the outbound call via the devices to that phone number immediately. On the other hand, if the result is "No", it would check with the next SIM3000CZ set for the matching USSD string.

From the sequence diagram in Figure 7, we would like to indicate the step of sending and receiving USSD through P-U-SS-Request message from the applicant (SIM3000CZ) to the SCP(Subscriber Program/Application), and the answers from the system will be sent back P-U-SS-Response string to Subscriber or SIM3000CZ again. This process will start from the process of USSD begins with pressing * SC (Service Code) * DATA via MS (Mobile Software) to check data from the service providers, and then pass that information to the MSC (Mobile Subscriber), VLR (Visitor Location Roaming), HLR (Home Location Roaming) and SCP, respectively. After that the system will check the accuracy of the structure of data and send back to the user again according to (P-U-SS-R) data which was requested. The feedback data is the information which the users require. The feedback answer is a function of the Session which is not stored like Short Message Service (SMS). The speed of USSD is as 7 times fast as regular SMS [9]. The function of USSD can be seen from the Sequence Diagram in above [1].

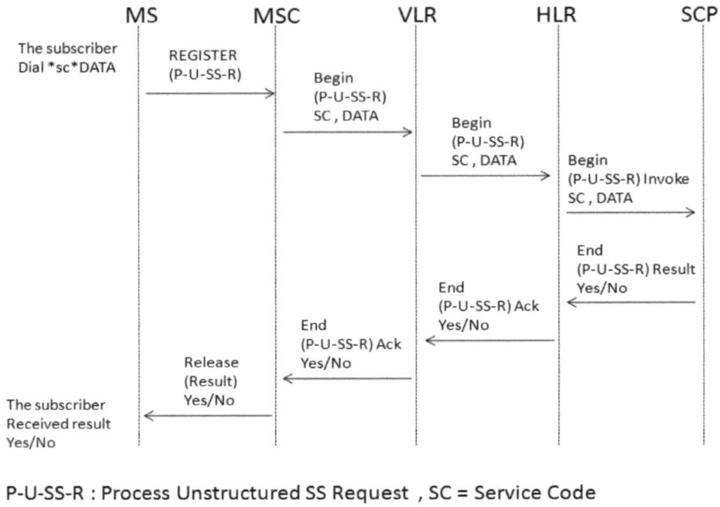

Fig. 7. Part 2 of the Proposed Algorithm

4 Results

This section shows the results of the experiments when using IP-PBX telephony system together with the USSD to route the outbound call, as set up in Figure 4. It is found that the IP-PBX phone system can select the route by checking the network provider for phone number and sending the data to SIP Gateway of network provider correctly. This test used 1,000 telephone numbers of each GSM Operator, as mentioned. The results are shown in Figure 8.

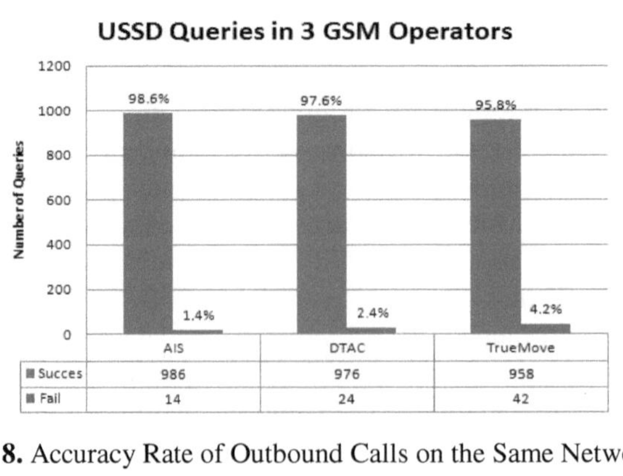

Fig. 8. Accuracy Rate of Outbound Calls on the Same Network

The results in Figure 8 show that the proposed algorithm with the USSD technique has given a high success rate, albeit with small inaccuracies. The errors occurred from the lack of response from the service provider when checking or looking up the information. From the results, it can be seen that we sent 1,000 USSD queries to each network. The results were as follows: The AIS network had 986 query success or 98.6% success rate, and 14 query errors or 1.4% error rate. The DTAC network had 976 query success or 97.6% success rate and 24 query errors or 2.4% error rate. The TrueMove network had 958 query success or 95.8% success rate and 42 query errors or 4.2% error rate.

Table 2. Overhead that Incurred in the Process of Our Proposed Algorithm

Source Network Operators	**Reference Network Operators**		
	AIS (SIM3000CZ-1)	DTAC (SIM3000CZ-2)	TrueMove (SIM3000CZ-3)
AIS	2 sec (2)	-	-
DTAC	2 sec	2 sec (4)	-
TrueMove	2 sec	2 sec	2 sec (6)

From the values of overhead time value in Table 2, it can be seen that every time a USSD query was sent to each SIM3000CZ, it took approximately 2 seconds to query. For example, if a mobile number (Source) which was the number of AIS Operator was transmitted through SIM3000CZ-1 which was AIS Operator, it would lose only two seconds to answer back as "YES". This answer would be able to set the call routing through the AIS Network channel or via SIP gateway which was connected to the AIS on Asterisk Server. However, if this number was the number of DTAC, it would lose 4 seconds to get the answers back to "YES", and if this number was the number of True Move, it would lose 6 seconds.

What has caused longer delay (4 seconds and 6 seconds) is that this algorithm/experiment works, sequentially, by first querying USSD string of the order SIM3000CZ-1(AIS), SIM3000CZ-2(DTAC) and SIM3000CZ-3(TrueMove), respectively. The reason for querying the AIS network first is because it is the biggest network operator with more people than other networks. Therefore, statistically there is a better chance of finding the route there.

5 Conclusion and Future Work

This paper has provided an overview of existing mechanisms used to select an outbound route in the IP-PBX system. We have explained that those existing mechanisms do have their problems, such as lack of flexibility and scalability. Some techniques even lacked the accuracy and efficiency.

We have, therefore, proposed a new method for selecting outbound call route by using unstructured supplementary service data technique or USSD. The method proposed here used USSD as a part of the decision making for which routes are selected.

The results of our experiments have shown that the accuracy in selecting an outbound route with three main mobile operators in Thailand was between 95% and 98.6%. With the accuracy comes a little overhead. Our results have shown that there was some overhead and delay in selecting the outbound route.

We claim that choosing the route by new USSD technique can be used with IP-PBX phone systems. Apart from the results illustrated, it can also be seen that our method would be able to reduce the amount of traffic occurred in internetwork connection significantly. Thus, this would lower the cost for users, too.

From knowledge of this research, we have an idea to bring this selection method into the two-sim or multi-sims mobile phone environment.

Acknowledgement. The researchers are very grateful to Information Technology Center, Pibulsongkram Rajabhat University, Thailand. for the help with the database of mobile phone numbers, all equipments in Sim3000CZ and Asterisk Server used in this experiment.

References

1. Dabas, A., Dabas, C.: Implementation of Real Time Tracking using Unstructured Supplementary Service Data. World Academy of Science, Engineering and Technology 54 (2009)
2. Herwono, I.: Performance Evaluation of GSM Signaling Protocols on USSD. Aachen University of Technology (2009)
3. Sanganagouda, J.: USSD: A communication Technology to Potentially ouster SMS Dependency. In: ARICENT (2011)
4. Suwannaraj, K.: Prediction of Name-based Cellular Service Provider via using K-Nearest Neighbor Data mining techniques on IP-PBX telephony system. In: Naresuan University (NURC8), pp. 355–365 (2012)
5. Suwannaraj, K.: Prediction of Name-based Cellular Service Provider via using Decision tree Data mining techniques on IP-PBX telephony system. In: National Conference Payap University Thailand (2012)
6. Mahler, S.P.: System and Method for Making Simultaneous Outbound Call Using DTMF Tones. USA Patent (2012)
7. Voznak, M., Rezac, F., Pozhon, J.: Applied multi protocol routing in IP telephony. In: Information and Communication Technologies and Services 2010 (EEE), vol. 8, pp. 118–123 (2010)
8. Kassinen, O., Koskela, T., Ylianttila, M.: Using Unstructured Service Supplementary Data Signaling for mobile PEER-TO-PEER innovations. In: The 12th International Symposium on Wireless Personal Multimedia Communication (2009)
9. Gupta, P.: End to End USSD System in CDMA. TATA Tele Service (2010)

Using Routing Table Flag to Improve Performance of AODV Routing Protocol for VANETs Environment

Hassan Keshavarz, Rafidah Md. Noor, and Ehsan Mostajeran

Department of Computer System and Technology,
Faculty of Computer Science and Information Technology,
University of Malaya, Kuala Lumpur, 50603, Malaysia
{keshavarz_hassan,ehsan_mostajeran}@siswa.um.edu.my,
fidah@um.edu.my

Abstract. With the growth of wireless topology during recent decades, routing protocol in Ad-hoc networks has come to the limelight. Routing is defined as a technique of the best route from the source to the destination. VANETs comprise a special subclass of Mobile Ad Hoc Networks (MANETs). One of the hardships with routing protocols in VANET is related to do with HELLO message which brings about a route. Among the most common solutions is one that utilizes Routing Table Flag for checking the route status. By checking the Routing Table Flag (RTF) each node's condition will be specified. Through considering this mode, various acceptable results regarding packet loss, packet delivery ratio, throughput and number of received packets have been achieved. Experiments using NS-2 to measure the packet loss, throughput, number of receive packets and packet delivery ration for three different scenarios are presented.

Keywords: Ad- hoc Networks, AODV, Routing Table Flag, Routing Protocols, HELLO Message.

1 Introduction

With the advent of wireless networks in the recent decades and the demand for use of technology, Ad hoc networks have also come into the limelight. Ad Hoc networks have proven to be a simple, fast, and cheap solution for creating self-organizing networks. This is mainly due to special features present in the Ad Hoc networks such as flexibility, Scalability and self-configuring. Ad-hoc networks lack infrastructures and the reason for this lies in the fact that the connection between the nodes is wireless [1]. No device such as server or Access Point (AP) router exists on the Ad hoc network. The data is sent through active links. When two nodes are placed far apart from each other and the communication range is limited for both nodes, Multi hop topology has to be used [2], [3]. It is under these circumstances that the self-organizing quality has to be facilitated in order to improve the flexibility of the ad hoc network The configuration time required for implementation in the ad hoc network is very short and this is the

P. Meesad et al. (Eds.): *IC²IT2013*, AISC 209, pp. 73–82.
DOI: 10.1007/978-3-642-37371-8_11

main reason why ad hoc networks are used for emergency situations such as Traffic Management, Urban Maintenance, Healthcare, Accident Prevention, viewing online traffic map, emergency handling, urban crime, military conflicts, natural disasters and rescue operation. Wireless network structure is divided into two modes: in the first mode- Infrastructure Mode- for network communication, use Access Point (AP) and Based-Station (BS).

On the second mode –Ad Hoc Mode- as transferred and delivered are via mobile nodes, there is no BS and AP in this mode. In the wireless network, mobile nodes are carriers, and regularly the topology is changeable. These types of networks are called "Mobile Ad Hoc Network (MANET)". In this area the carrier can be Vehicles, cell phones, laptops and so on. If the carriers are vehicles, so this network is "Vehicular Ad-Hoc Network (VANET)". On the other words, VANET is a subset or subclass of MANET. In VANET, nodes are always on the move and because of this the network topology is always changing. These unforeseen changes can break the links between nodes. The nodes' intractable movement and dynamic topology support in VANET has caused the routing protocols to play a vital role in the network performance. Of course other contributing parameters include: Power Supply, Nodes' continual movement, and unforeseen changes.

The remainder of the paper is organized as follows: Section II covers related work about routing protocols in VANET, Section III provides our proposed solution in order to improve routing performance, section IV discusses Simulation and Evaluation, Section VI presents Discussion and Result, and Section VII concludes the paper with outlooks on future work.

2 Related Works

With the growth of wireless topology during recent decades, routing protocol in ad hoc networks has come to the limelight. Routing is defined as a technique of the best route from the source to the destination. Since every node in VANET acts as a router, therefore, routing protocols have to be installed on every vehicle. Routing protocols are divided into different categories in the ad hoc networks. For example Topology-Based routing protocols and Geographic routing protocols [4] or Stateless routing protocols and Statefull routing protocols [5] , On the whole, they can be divided into three main categories: Proactive, Reactive and Hybrid [6]. In the first group- namely Proactive routing protocols, or in other words table-driven- tables are updated through the changes in the network topology and by transmission of update packets by nodes. These update packets cause overhead in the network. We can refer to FSR [7], [8] and DSDV [9], as the most prevalent routing protocols in the table driven group.

The second group (reactive routing protocol) is also known as on demand. This group creates a link with the destination and if a route request exits from the source, then a data packet would be returned. This group has less overhead as compared to the first group. Since the route discovery process is time consuming, the network faces delays. The most famous routing protocols of this group are DFS and AODV.

The third group is known as the hybrids. They are a combination of reactive routing protocols and proactive routing protocols. As an example, we can refer to Zone Routing Protocol (ZPR) [10] which uses proactive routing protocol in intra-zone and reactive protocol in Inter-zone in order to decrease delay and overhead in the network. As mentioned before, VANET is a subclass of MANET and shares many features. As an example, we can refer to using multi hop technology for forwarding data packets. Of course there are many differences between the two; for example, in MANET, nodes can move in any direction but in VANET due to the streets and highways, the vehicles can only move in a predictable manner. Also, due to the fast movement of vehicles, especially in the highway, the VANET topology changes much faster and this makes the routing process to be more complicated than MANET. The protocols used in MANET might have a weaker performance as compared to VANET [11] .Due to this reason, AODV [12] as one of the most used routing protocols in MANET and Ad hoc network needs to be improved in VANET.

Here are so many research in this area but a function which is called Check_Route_Flag () is added to the original AODV routing protocol. This function using in Ad hoc On-Demand Multipath Distance Vector (AOMDV) (Marina et al., 2001) routing protocol and it checks only flags which are in UP status. Another point that the functionality of Ad hoc On-Demand Multipath Distance Vector (AOMDV) affects on packet salvation and path maintenance. In this study we attempted to use Check_Route_Flag function in routing table instead of path. Also, this research discusses about other flag that are IN_REPAIR status and furthermore we should take into consideration to route status because status of repair route are interchangeable and may be set to UP in ti+1 time in ti was in repair mode. To the best of my knowledge, may be this route can recommend shortest and freshness route as a neighbor that has been saved in routing table under Next Hop name.

3 Proposed Solution

With the considerable increase in the number of vehicles on the roads, the road traffic safety becomes more and more important. In addition, VANET is creating a new generation of wireless networks which have a high security and traffic efficiency.

Intelligent Transportation System (ITS) is one of the main VANET applications. Intelligent Transportation System (ITS) includes a set of applications such as Blind crossing, nearby information source and co-operative traffic monitoring. Through accessing information from other vehicles, it can bring about the safest and more efficient roads. As mentioned before, there are numerous differences between VANET and MANET [11], [13]. AODV is one of the path finding protocols in MANET environments. AODV can be affected by the differences and might not perform well in VANET environments. The aim of this study is to achieve the highest performance with the least network overhead. This aim is achieved through adding a function to the original AODV. Since the nodes are always on the move, neighbor discovery becomes important because through identifying the neighbor, we can prevent the dispatch of extra nodes that can cause an overhead and in turn this increases the performance of the network. By this the number of received packets has increased and so the Packet delivery ratio and throughput increases and the number of dropped packets is decreased.

A: Scenario

In this research, three scenarios with 100, 150 and 150 mobile nodes (vehicles) in a flat area are simulated. All the vehicles move in the same direction with speed of vehicles 50 km/h in a highway. The first car is the source node and sends HELLO message to the destination, which is the last car. Each car has 46 meter distance from the following car. The source node sends data packets at 80 kbps that is generated by CBR interval 0.25s which rely on UDP agent.

From the beginning of the program, all nodes are analyzed in a random manner in a 10x10000 meter area and a density of 0.005, 0.010, and 0.015. One of the important parameters in administrating this scenario is that to analyze the accumulated nodes in various sections with the use of Smart-AODV. In this scenario, a highway of about 10000 meters long is considered. In this highway, nodes are capable of moving forward or to the sides during movement. The movement of nodes to the left and right along the length of the highway creates some challenges including mobility for the Smart-AODV protocol. In this study, we have tried to take into account mobility so as to improve Packet Delivery Ratio (PDR), Packet Dropped and number of receive packets. For this scenario, Throughput, Packet Loss, Packet Delivery Ratio, and the Number of Packet Receive are evaluated for Smart-AODV and AODV and compared to each other.

B: Smart-AODV

Due to the fast movement of vehicles and high mobility in Vehicular Ad-hoc Networks (VANETs) environment, reactive routing protocols are more suitable in comparison with proactive routing protocols. As proactive routing protocols send update message regularly, as a result, the network encounter heavy overhead. However, reactive routing protocols suffer from some problems such as route performance in a static way.

This study suggests a dynamic technique by checking routing flag and route status in Ad-hoc On-demand Distance Vector (AODV), which is well-known reactive routing protocol, in order to overcome the mentioned problem in mobile environment like VANET. The modified AODV uses routing flag in routing table. There are a lot of routes in routing table which some of them might be active mode while others are in enactive mode and in repair mode. If the routes is in active mode and routing flag is in repair mode, an originate node disseminates a HELLO message to neighbor. Otherwise the route is in inactive mode and not need send HELLO message. By doing this, the overall overhead of the network mitigates. This technique determines the new path routes with less hop count and gain reasonable performance of the network by increasing the packet delivery ratio and decreasing the packet drop. This study has been attempted to achieve a reasonable performance of AODV protocol with minimum modified in necessary parts of original protocol.

First of all, we defined aodv_rtable::Check_Route_flag with Boolean data type in adov_rtable. Then we created a temporary route and checked the routes through the routing table and if the desired routing flag is set as RTF_UP and RTF_IN_REPAIR, then function returns true value. In essence, this function only selects active route links from available routes in the routing table and finally HELLO message is sent.

In the next step, we will going to aodv.cc and perform two tasks: First, if vehicle or node selects more secure routes while sending HELLO message, consequently packet drop for packets which sent through inactive route will be decreased. Using of this method, we can alleviate the number of waste HELLO messages. Second, when sending HELLO message within HELLO function, it is set by applying a condition for finding destination address which is in HELLO message header. Eventually, we can filter by HELLO message broadcast based on destination address in neighbor list by using two tasks the routing overhead goes down especially in terms of HELLO message. HELLO message broadcasts periodically for neighbored discovery and it causes massive routing packet in scenario. A function which is called Check_Route_Flag () is added to the original AODV routing protocol. This function using in Ad hoc On-Demand Multipath Distance Vector (AOMDV) [14] routing protocol and it checks only flags which are in UP status. Another point that the functionality of Ad hoc On-Demand Multipath Distance Vector (AOMDV) affects on packet salvation and path maintenance. In this study we attempted to use Check_Route_Flag function in routing table instead of path.

Also, this research discusses about other flag that are IN REPAIR status and furthermore we should take into consideration to route status because status of repair route are interchangeable and may be set to UP in ti+1 time in ti was in repair mode. To the best of my knowledge, may be this route can recommend shortest and freshness route as a neighbor that has been saved in routing table under Next Hop name.

4 Simulation and Evaluation

Network Simulator 2 (NS-2.34) was employed to simulate the scenarios for this experiment. The different scenarios were tested with a varying number of nodes. Traffic, type of mobility, MAC and Physical Layer specifications were defined the same in all scenarios. In this paper, the original AODV was observed in NS-2.34 and then compared with Smart-AODV performance metrics. According to results, Smart-AODV generates less routing overhead and provides more reasonable performance than AODV. Table 1 shows the 802.11p specifications communication and table 2 presents network parameters that are used in comparison study between Smart-AODV and the original AODV routing protocols in NS-2 simulation in all scenarios.

Table 1. IEEE 802.11P Specification

MAC Layer		Physical Layer	
CWMin	15	CPThresh	10.0
CWMax	1023	CSThresh	2.5118864e-13 -96 dBm
SlotTime	0.000009	RXThresh	1.0e-12 -90 dBm
SIFS	0.000016	Bandwidth	6.0e6
ShortRetryLimit	*7*	Pt	0.00000025
LongRetryLimit	*4*	Freq	5.9e+9
HeaderDuration	*0.000020*	L	0.00250
SymbolDuration	0.000004	noise_floor	2.51189e-13
BasicModulationScheme	0	HeaderDuration	0.000020
use_802_11a_flag	True	BasicModulation Scheme	0
RTSThreshold	2000	trace_dist	16e
MAC_DBG	1	PHY_DBG	1

Table 2. Simulation Parameters

Network Parameters	Value
Network Simulator	ns-2.34
Routing Protocol	AODV
Simulation Time	200 s
Simulation Area	10 * 1000 m
Number of Nodes	50,100,150
Traffic Source/Destination	Deterministic
DATA TYPE	CBR
Packets Generation Rate	5 packets
CBR interval	0.25 s
Packet Size	100 bytes
MAC Protocol	ieee802.11p WAVE
MAC Rate	1 Mbps
RTS/CTS	None
Transmission Range	46m
Radio Propagation Models	2-ray ground
HelloDYMO Interval	1 s
Length of highway	10000m
Number of lanes	3
Speed of vehicles	50-100km/h
Sensing range	85m

The Random Walk Mobility model was selected for this investigation, and it concerns parameters such minimum and maximum speed of mobile nodes while they are moving around in random directions [10]. This project aimed to provide higher chances for mobile nodes to detect neighbours while the nodes are moving at a default speed in all scenarios.

5 Discussion and Result

In order to provide an unbiased analysis of the obtained data, the present research evaluated the data under the same conditions. It was because of this that the number of sent packages was held constant at 19991 packages for all three scenarios.

A: Throughput

It can be seen Figure 1 that by increasing the number of vehicles, the average of network throughput is better for Smart-AODV and AODV routing protocols. In scenario 1 with 50 nodes, the throughput of Smart-AODV is negligible and it is very near to original ADOV but it has some changes. In scenario two and three which the number of vehicles is 100 and 150 nodes, the throughput Smart-AODV is better versus with AODV original and this difference is remarkable. Figure 4.4 describes the throughput of network for three scenarios. The throughput shows a negligible increase from 7993.80 kbps to 7994.35 kbps for Smart-AODV. This value in scenario 2 for AODV is decreased to 7733.51 due to packet loss against 7994.35 kbps for Smart-AODV.

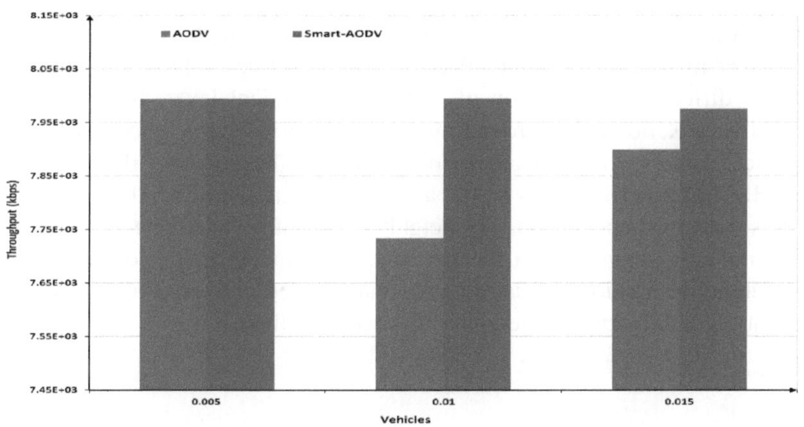

Fig. 1. Smart-AODV vs. AODV Throughput

B: Packet Loss

Physical layer specifications make one of the mandatory parameters that should be configured the same in all scenarios to achieve fair performance analysis. Parameters such as Data Rate with unlike values for instance, can produce a different total number of packet drops. In a fair comparison, the proposed neighbor discovery method can decrease the number of packet drops by inserting new neighbors for the routing packets to use. Neighbors are checked once routing packets such as RREQ and RREP are sent and received. Mobile nodes are able to check a next node's information in a neighbor discovery process by using these routing packets. Figure 2 indicates a lower total number of packet drops in each scenario with the proposed neighbor discove method.

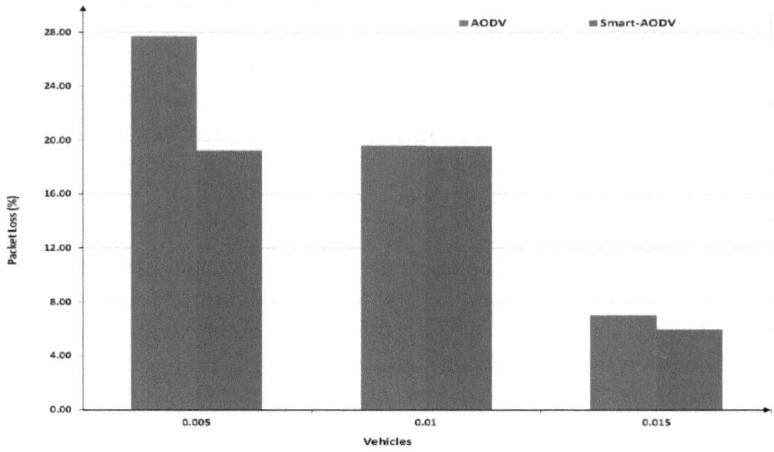

Fig. 2. Packet Dropped for Smart-AODV and AODV

C: Packet Delivery Ratio

Packet Delivery Ratio in this experiment was analyzed under the same MAC and Physical layer mandatory elements. Packet drop, effective on Delivery Ratio, is very sensitive to different specification of MAC and Physical layer. In this experiment Packet Delivery Ratio was improved by using the proposed neighbor discovery method in each scenario. The advantage of calling on the neighbor discovery process between HELLO timer intervals is that a more accurate list of neighbors can be delivered, thus reducing packet drop. The neighbor list is always affected by node mobility, so having new neighbors in the list after movements helps select the shortest, fresher route to destination. Figure 3 shows that Packet Delivery Ratio is enhanced in each scenario via the proposed neighbor discovery method.

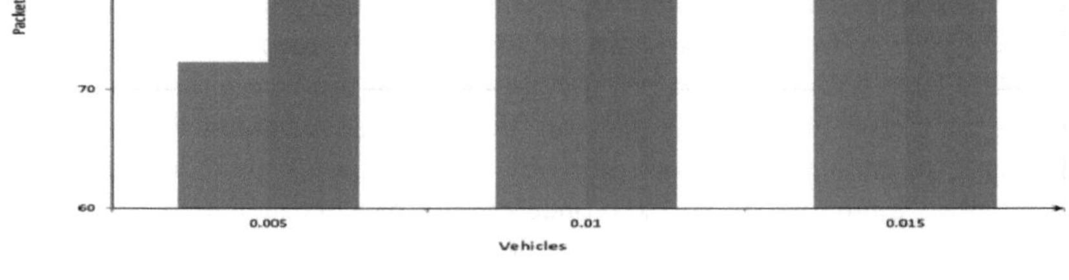

Fig. 3. Packet Delivery Ratio for Smart-AODV and Original AODV

D: Number of Receive Packets

According to Figure 4 the number of received packages in the Smart-AODV protocol is improved especially in the scenario with 50 nodes. For example in this same scenario, from the 19991.00 sent packages according to the original AODV protocol, only 14448.00 packets were received. This number was increased to 16147.00 using Smart-AODV. This shows an improvement in the algorithm original AODV. The reason behind this can be said to be the fact that each node gathers information on the position of the neighbouring nodes through using HELLO message. Since some nodes are In Repair mode, and in the meantime could change to an UP position, this can cause an increase in the number of packages received since the same condition was not taken into account in the original situation.

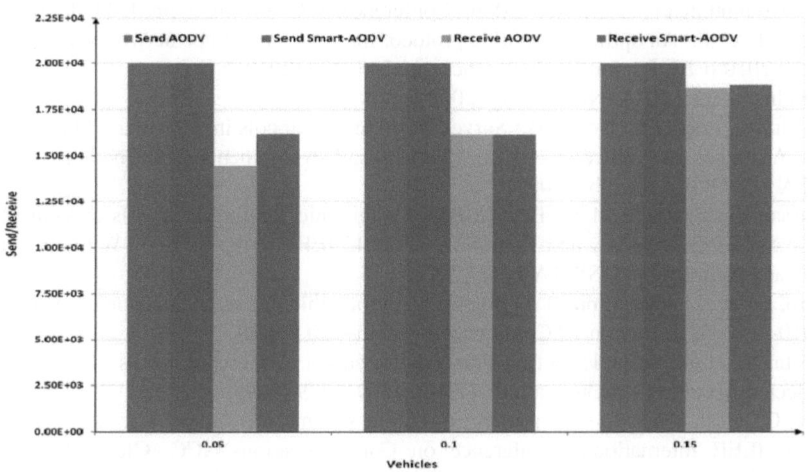

Fig. 4. Number of Send and Receive for Smart-AODV and AODV

6 Conclusion

This paper introduced an innovative, improved neighbor discovery method for AODV called Smart-AODV. It provides reasonable performance by checking Routing Table Flag (RTF). Moreover, the broadcast of HELLO messages is filtered by checking the destination node in the neighbor list to reduce overhead. Simulation results of the proposed neighbor discovery method based on mobility in various scenarios and compared with the original AODV demonstrate that Smart-AODV performs better.

According to these results Average End-to-End is a feature for improving. Smart-AODV provides higher delay time than original AODV that the main reason is neighbor discovery process to be done on route request (RREQ) and route reply (RREP). There is chance to having lower average End-to-End delay like other elements of performance in Smart-AODV. The Smart-AODV routing protocol was simulated in NS-2 and then its performance was analysed and compared to the original AODV in

VANET environment. The performance was evaluated based on four metric: Throughput, Packet drop, Packet Delivery Ratio and Number of Receive packets. The results that were achieved from analysing of the NS-2 trace file show that Smart-AODV enhances the network performance in comparison with the original AODV.

Acknowledgements. I would also like to express my most sincere gratitude and utmost appreciation to Assoc. Prof. Dr. Diljit Singh for providing constructive feedback on my publications.

References

1. Lee, S.J., Gerla, M.: AODV-BR: Backup routing in ad hoc networks. In: 2000 IEEE Wireless Communications and Networking Conference, WCNC, vol. 3, pp. 1311–1316 (2000)
2. Guo, H., et al.: An optimized routing protocol for vehicular ad hoc networks. In: TENCON 2010-2010 IEEE Region 10 Conference, pp. 245–250 (2010)
3. Perkins, C.E.: Ad Hoc Networking (2008)
4. Kevin, C., Lee, U.L., Gerla, M.: Survey of routing protocols in vehicular ad hoc networks. In: Advances in Vehicular Ad-Hoc Networks, Developments and Challenges, UCLA, USA. IGI Global (October 2009)
5. Hassan Keshavarz, R.M.N.: Beacon Base Geographic Routing Protocols in Vehicular Ad Hoc Networks A Survey and Taxonomy. In: 2012 IEEE Symposium on Wireless Technology and Applications, ISWTA 2012 (2012)
6. Guo, H., et al.: An optimized routing protocol for vehicular ad hoc networks. In: TENCON 2010-2010 IEEE Region 10 Conference, pp. 245–250 (2010)
7. Iwata, A., et al.: Scalable routing strategies for ad hoc wireless networks. IEEE Journal on Selected Areas in Communications 17(18), 1369–1379 (1999)
8. Pei, G., et al.: Fisheye state routing: a routing scheme for ad hoc wireless networks. In: 2000 IEEE International Conference on Communications, ICC, Global Convergence Through Communications. Conference Record, vol. 1, pp. 70–74 (2000)
9. Perkins, C.E., Bhagwat, P.: Highly dynamic Destination-Sequenced Distance-Vector routing (DSDV) for mobile computers. In: Proceedings of the Conference on Communications Architectures, Protocols and Applications - SIGCOMM 1994, pp. 234–244 (1994)
10. Haas, Z.J.: A new routing protocol for the reconfigurable wireless networks. In: IEEE 6th International Conference on Universal Personal Communications Record, Conference Record (1997)
11. Ding, B., et al.: An improved AODV routing protocol for VANETs. In: 2011 International Conference on Wireless Communications and Signal Processing (WCSP), pp. 1–5 (2011)
12. Perkins, C.E., Royer, E.M.: Ad-hoc on-demand distance vector routing. In: Proceedings WMCSA 1999. Second IEEE Workshop on Mobile Computing Systems and Applications, vol. 6, pp. 90–100 (1999)
13. Chao, L., Ping'an, L.: An efficient routing approach as an extension of the AODV protocol. In: 2010 2nd International Conference on Future Computer and Communication, pp. V1-95–V91-99 (2010)
14. Marina, M.K., Das, S.R.: On-demand multipath distance vector routing in ad hoc networks. In: Proceeding of the 9th Ninth International Conference on Network Protocols (2001)
15. Naoum-Sawaya, J.: How to Measure the Throughput. Packet Drop Rate, and End-to-End Delay for UDP-based Application Over Wireless Networks (2011)

Wireless Sensor Network Planning for Fingerprint Based Indoor Localization Using ZigBee: Empirical Study

Nawaporn Wisitpongphan

Faculty of Information Technology,
King Mongkut's University of Technology North Bangkok
Bangkok, Thailand
nawapornn@kmutnb.ac.th

Abstract. Technology defined by ZigBee standard is intended for a wide range of ad hoc wireless sensor network (WSN) applications. Among which is the location-aware services which can be applied to both indoor and outdoor environments for locating expensive equipments or tracking any moving objects. While there are many existing localization algorithms, the fingerprint technique which relies on determining target location from an off-line empirical database seems to be the most practical indoor solution using off-the-shelf products. In this work, we present a wireless sensor network planning solution suitable for indoor localization using fingerprint technique. Based on our extensive feasibility studies, we derived several network planning solutions which answer some of the key wireless sensor network design questions such as (1)where to put the router or anchor nodes?, (2) how many routers should we use in designing location-aware WSN?, (3) how often should the end-device node transmit data to the server?, (4) what should be a suitable packet size?, and (5) does mobility have any impact on the network performance?

Keywords: ad hoc, zigbee, 802.15.4, indoor localization, tracking, fingerprint.

1 Introduction

Location-aware service has gained its popularity over recent years due to a fast growing interest in social network application such as Facebook, FourSquare, Instaweather, etc. In telecommunication and networking, however, localization problem has been introduced for more than a decade. While some work has been commercialized for public usage, e.g., Global Positioning System (GPS), many are still under research and development due to several technical challenges especially in an RF wireless sensor network environment. This is because there are many dynamic factors such as operation failure due to loss of battery power or unexpected obstruction in a real RF environment. These uncontrollable events could lead to large localization error and less accurate results.

P. Meesad et al. (Eds.): *IC²IT2013*, AISC 209, pp. 83–92.
DOI: 10.1007/978-3-642-37371-8_12

While much of the research effort has focused on the design and optimization of the localization algorithms, less attention has been on a fundamental challenge of deploying such a network in a real indoor environment. There are, however, a small number of efforts which tried to investigate the practicality of deploying wireless sensor network for an indoor environment for localization purposes [1,2]. Much of the focus of these works is on demonstrating that the proposed localization algorithm work efficiently in an indoor environment. None discusses about the network planning problem which should be the primary factor that has a direct impact on the localization results. In particular, to be able to track or locate a node one should also be concerned about how to place the anchor/router nodes or how much information should be exchanged in the network so that the error is mitigated.

In this work, we focus on implementing an ad hoc wireless sensor networks for indoor localization using ZigBee with IEEE 802.15.4 standard. In order to develop a practical WSN for such purpose, researchers are confined to using off-the-shelf products in the development phase. Consequently, altering the underlying protocols is sometimes not feasible. With this design choice, we are also limited to using available information which is Received Signal Strength (RSS) for the localization process. More specifically, fingerprint localization algorithm [1] seems to be the most practical and feasible option. The area of interest for this study consists of two floors in a building where each floor is approximately 20 x 40 m^2. The remainder of this paper is organized as follows. In section 2, we present the related work in the area. The prototype design of our WSN devices are presented in section 3. The channel model of our environment is presented in section 4, followed by our methodology for planning a network in section 5. The network performance study is presented in section 6 and finally we conclude our findings in section 7.

2 Related Work

RF-based localization algorithms can be classified into either range based or range-free technique. In the range-based scheme, the location of the target node is obtained by estimating the distance to a set of anchor nodes. The classical approach is measuring Received Signal Strength Indicator (RSSI) [3] and converting it to a distance by using theoretical or empirical model. However, converting RSSI to a distance is not trivial since RF loss in most environments (both indoor and outdoor) is unpredictable due to many uncontrollable factors such as obstructions, multipath fading, etc. To minimize errors, some measures Angle of Arrival (AOA) [5,6] and use it calculate the potential target location. By measuring angle at which the signals are received, the distance can be more accurately estimated by solving a simple geometric problem. Another popular approach relies on a time measurement, which is similar to that used in global positioning system (GPS). Example of such approaches are Time of Arrival (TOA) [7], and Time Difference of Arrival (TDOA) [8]. TOA and TDOA approaches are fairly accurate in an environment where there is line-of-sight (LOS). By converting propagation time into a distance and using known signal propagation speed, the location of the target nodes can be easily and precisely realized. However, these methods calls for a very precise measurement and may not be practical for indoor environments where there is typically no LOS.

Alternatively, range-free approach can be further categorized into proximity-based and scene analysis. Proximity-based technique estimates the target node's location from connectivity and network topology information. For example, mobile nodes in a WLAN or WSN are often connected to the closest Access Point (AP) or anchor node. Hence, its location should be in the vicinity of the station it is connected to. A more precise location can then be further calculated by using one of the several techniques which include trilateration, multilateration, triangulation, probabilistic approach, bounding box, or central position [9].

Scene analysis involves an off-line learning phase with extensive Received Signal Strength Indicator (RSSI) data collection for constructing reference database and an online phase which tries to match the observed RSSI with the record stored in the database. During the off-line phase, RSSI values at different locations are recorded with respect to different anchor nodes. These information are used to construct a probabilistic radio map [10] or an RF fingerprint database [1]. During the online phase, the end-device measures RSSI values to different anchor nodes (active mode) or anchor nodes measure the signal received from the end-device and report to the server (passive mode). A set of measured RSSI values are then compared with the existing fingerprint database collected during the off-line phase. Location of the end-device is, therefore, the place at which the RSSI fingerprint matches with the fingerprint measured during the online phase. However, if the radio map is used, the location of the device is the point on the map that maximizes the location probability. While these two scene analysis approaches are time-consuming, it is quite practical for indoor environment where there is little to moderate wireless channel fluctuation.

In this study, we are limited to off-the-shelf products. Therefore, the only information available without any modification made to the hardware/software is the RSSI information. Hence, fingerprint approach is more favorable than the rest of the techniques discussed in this section. As we will show later in Section 4, the channel modeling for indoor environment is quite unfeasible due to various factors such as different furnitures/decorations in each room, mobile factors such as human movement and moving elevators, etc. Hence, any approach which relies on channel modeling is not practical. In this work, we aim to deploy an indoor fingerprint based localization system using ZigBee-based wireless sensor networks.

3 Wireless Sensor Network Prototype Design

The goal of our prototype design is to provide the ZigBee chipset, XbeePro 50mW as is shown in Figure 1(a), with its own power supply and an appropriate covering case for indoor usage. There are three types of devices in our experiments, i.e., Coordinator, Router, and End-Device. In this section, we describe their functions and the corresponding design decision for each device.

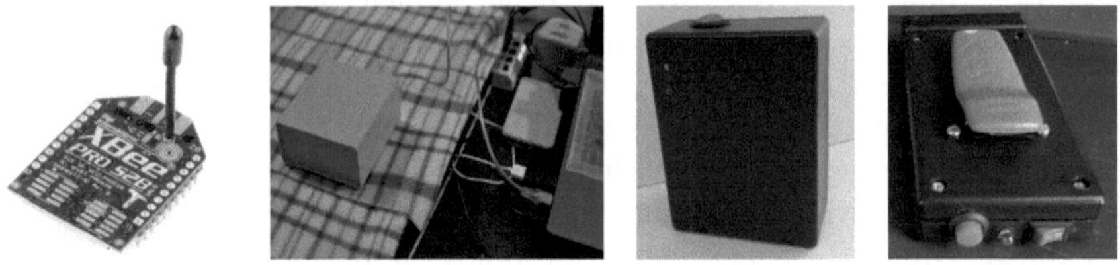

(a) (b) (c) (d)

Fig. 1. ZigBee Prototypes (a) XBee Pro 50mW (b) ZigBee Coordinator (c) Router, (d) End-Device

3.1 Coordinator

The coordinator's job is to probe/gather information to/from all the end-devices and routers. It is also directly connected to the server. Hence, there is no need to design a power supply for a coordinator since we can directly connect it to the server via an Xbee USB Dongle. However, there is still a need for a durable case to protect the device. Therefore, in our design, we simply package the Xbee chipset together with the dongle board inside a plastic case, as is shown in Figure 1(b).

3.2 Router

A router is a device that helps relay messages between coordinator and end-devices. They have to be placed within a building in many different locations to extend the range of the coordinator in order to cover the area of interest. Hence, the routers have to have their own DC 3.3V power supplies. In our design, we use 4 typical AA batteries to supply power and a PIC16F627A microcontroller to manage the packet transfer. Last but not least, users should also be able to speculate the status of the routers from a far distance, therefore, we also put a power on/off switch and some LED lights on the circuit to display the current status of the router, shown in Figure 1(c).

3.3 End-Device

Similar to the design of the routers, end-device also needs its own power supply and covering case. However, the design should be more compact than the router's since it is the only device in the system that can move around. Therefore, additional design we made to the end-device includes a smaller case to make it more compact, a calling button so that users can send a message to the coordinator, and a clip attached to the case so that users can carry the device around the building more easily, as shown in Figure 1(d).

4 Channel Characterization

Once we have the prototypes, the first step in our study was to conduct a channel characterization so that we could find a suitable channel model that could be applied in our localization process. By knowing the physical characteristic of the ZigBee wireless channel, one can roughly use this information to determine the potential locations to install routers. To achieve this, we conducted a channel characterization both with and without Line-Of-Sight (LOS).

4.1 Path Loss Model

The channel model with LOS is conducted on a hallway, with a range of roughly 30 meters, in Nawamintararachinee Building at King Mongkut's University of Technology North Bangkok. The coordinator is placed at one end of the hallway while the end-device is placed at 5, 10, 15, ..., 30 meters away from the coordinator. At each location, the end-device sends 50 messages of size 64 bytes using 5 different transmitting power, i.e., 5 mW, 13 mW, 20 mW, 32, mW and 50 mW, to the coordinator.

Figure 2 shows two examples of the closest theoretical free space model which matches the relationship between distance and the RSSI at different transmitting power obtained from the experiments. Observe that the indoor LOS path loss exponent, according to the Friis formula [3], ranges between 2.37 - 2.95. This range is obviously too large. This could plausibly be because each experiment is conducted at different times of the day. As a result, the amount of interference from WiFi users, which varies during the day, become one of the main uncontrollable factors in our experiment. Therefore, in order to find a suitable path loss exponent for our study, we consider the path loss exponent value from 2.0 to 3.0 and calculate the Root Mean Square Error (RMSE) for the experimental results obtained from each transmitting power. Using this approach, we found that the best path loss exponent for the considered indoor environment is $n = 2.64$ (yield the RMSE of 27.94).

4.2 Link Budget Loss

In order to find the loss due to obstructions within the building, we had conducted an experiment to measure the signal attenuation by placing the coordinator and the end-device 5 meters apart from one another with and without an obstructing object in between. Each experiment was repeated 50 times at 50 mW transmitting power level. Our results show that the wooden wall presents roughly 5 dB loss, while the attenuation through cement walls is approximately 7 dB. The ceiling, on the other hand, results in about 10 dB loss, plausibly because of the the steel rod structure.

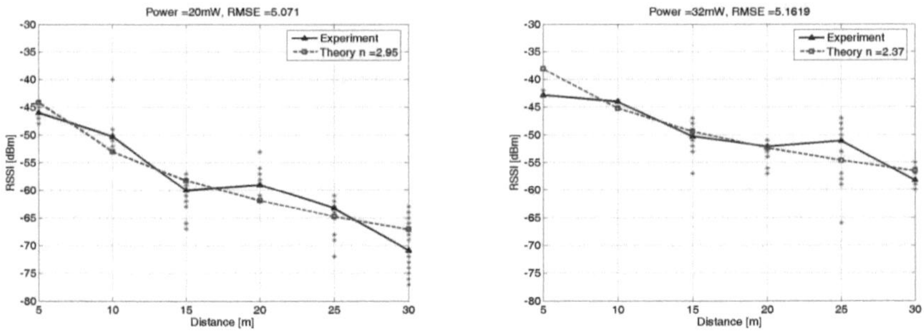

(a) (b)

Fig. 2. Relationship between distance at RSSI at different transmitting power: (a) 20 mW (b) 32 mW

5 ZigBee Network Planning

One of the main challenges in deploying a wireless sensor network is to find the location for placing the router nodes. If the purpose of the WSN is to simply relay a sensor information from end-devices to the coordinator, the location of these routers could be similar to the location of the wireless access point in WiFi network. This is because the goal is to ensure that the end-device can see at least one router at any spots within the area of interest. However, if the goal is to be able to track or locate the end-device, the minimum number of routers that each end-device should be able to contact is at least 3 or more [1]. Therefore, in this section we provide a network planing solutions which solve tracking and localization problems.

5.1 Wireless Sensor Network Site Survey

In our experiments, the area of interest covers the third and the forth floors of the building. Based on our channel model in section 4, we picked six potential locations, normally used for placing the WiFi access points. The considered routers' locations on the third floor are: in the middle (RT3_Middle), at the back end of the hallway (RT3_Back), and on the left wing (RT3_Left). The three locations on the forth floor are: in the middle (RT4_Middle), at the front end of the hallway (RT4_Front), and on the right wing (RT4_Right). We conducted a site survey for each router location where 30 RSSI values are measured and average at each of the 47 end-device locations. Example of the results can be depicted using a color map, as is shown in Figure 3. Excellent signal reception area, with RSSI between -49 dBm and -40 dBm, is depicted by the green regions. The yellow regions represent good reception area with RSSI in between -59 dBm and -50 dBm. The fair and bad reception regions, represented by orange and red colors, are the area where the RSSI value sits in the [-69 dBm, -60 dBm] and [-79 dBm, -70 dBm] ranges, respectively. Finally, the gray regions represent the area where the RSSI value is below -80 dBm.

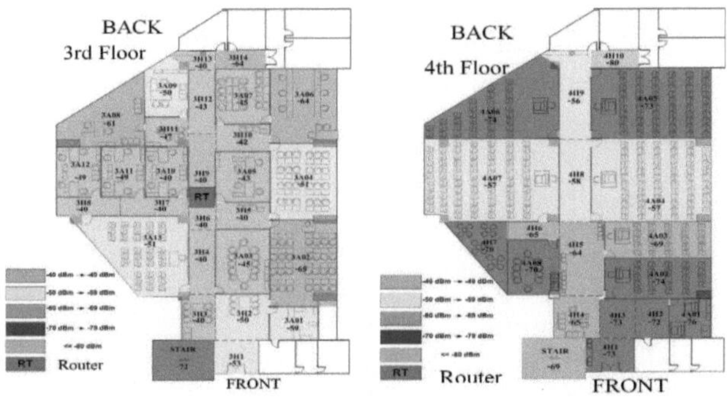

(a) (b)

Fig. 3. Color map of RSSI level on the third and forth floors when the router is placed at RT3_Middle location

5.2 Optimized Router Location Using Maximum Euclidean Distance

According to the fingerprint based localization theory [1] discussed in section 2, a desirable condition for the routers or anchor node placements is for each location to have a unique fingerprint or signal pattern that is very different from all the other patterns from different locations in the area of interest. For example, three routers receives signal from an end-device located at location i and the RSSI values measured from each routers are $RSSI_i^{(1)}$, $RSSI_i^{(2)}$, and $RSSI_i^{(3)}$, respectively. If the end-device moves to a new location j, the signal pattern should not be identical to the pattern observed at location i. The difference between each fingerprint pattern can be converted into a *distance in signal space*, d_{ij} by using the Euclidean Distance formula which is defined as

$$d_{ij} = \sqrt{\sum_{n=1}^{N} \left(RSSI_i^{(n)} - RSSI_j^{(n)}\right)^2} \tag{1}$$

where $RSSI_i^{(n)}$ is the signal strength measured at router n with respect to location i. Similarly, $RSSI_j^{(n)}$ is the signal strength measured at router n with respect to location j. The best set of locations to install routers should result in fingerprint patterns at all locations being very different or being very far apart in a signal space.

Figure 4 shows a cumulative distribution function of Euclidean distances measured between all the location pairs in the best router placement combination scenarios with 3 and 4 routers. We claim that if the router placement is optimal for localization purposes, then the distance between any signal patterns should be greater than a certain threshold. This threshold can be set by an application. The greater the threshold, the higher the accuracy in locating the end-device. For example, if the threshold is set to 10 dB, then the best 3-router combination would be to put one router on the forth floor in the front-end of the

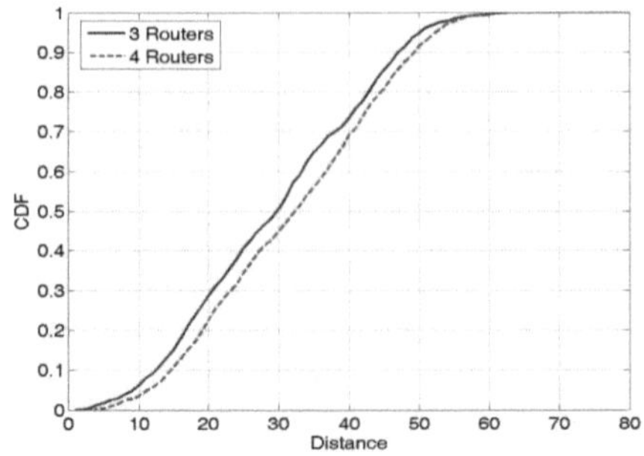

Fig. 4. CDF of the Euclidean distances of all the location pairs

hallway while the other two should be on the third floor in the middle and at the back-end of the hallway. This combination would results in only 67 out of 1081 location pairs (6.2%) being too close to one another in the signal space,i.e., the distance between the pair is less than 10 dB. However, if we increase the number of routers to 4, then this fraction would be reduced to 3.4% or approximately 37 out of 1081 location pairs. According to the simulation results, the optimal 4-router combination is to put two routers on the third floor in the back and on the left-wing. The remaining two will be on the forth floor in the front an on the right wing.

6 Network Performance under Mobility

To evaluate the performance of a network where there are some mobile nodes, we conducted an experiment similar to the setup in section 4. Instead of placing an end-device at the appointed location, we carry the device and walk at a rate of 0.7 m/s and 1.4 m/s. The former rate represents a *slow* human walking rate while the latter is comparable to a *fast* walking rate. Any moving rates faster than 1.4 m/s are equivalent to running speeds and are not considered in our study as it is not very likely to observe a running person in a building.

6.1 Impact of Transmission Rate

In this experiment, we consider a packet of size 100 bytes, maximum payload size for the Xbee chipset, and a transmission ranges from 0.5 to 20 pkts/s. The end-device moves along the hallway covering the distance of 30 meters at the rate of 0.7 m/s and 1.4 m/s. The results are shown in Figure 5(a). According to the results, we observe that transmission rate has a direct impact on the performance in terms of packet delivery ratio (PDR). More specifically, PDR drops significantly as transmission rate exceeds 1 pkt/s. On the other hand, the impact of node mobility will only kick in when the transmission rate is between 1 pkt/s – 5 pkt/s. When the transmission rate exceeds 5 pkt/s, the PDR would

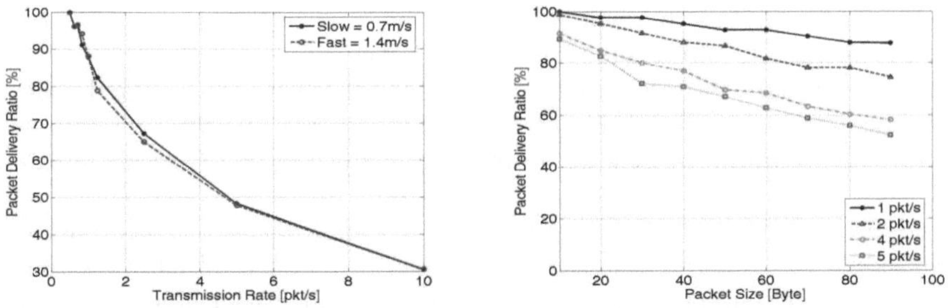

Fig. 5. Impact of (a) transmission rate and (b) packet size

be less than 50% and the performance will be limited to the transmission rate rather than the mobility. On the other hand, when the transmission rate is between 1 pkt/s – 5 pkt/s, the fast moving speed would cause the PDR to drop slightly. However, mobility do not have any impact on the performance when the transmission rate is less than 1 pkt/s. Therefore, one may conclude that the mobility has much less impact on the network performance than the transmission rate. If one were to design any application in a ZigBee network, a practical transmission rate should not be faster than 1 pkt/s.

6.2 Impact of Packet Size

According to the ZigBee standard, the maximum raw data rate is 250 kbps. However, the results in 6.1 shows that the optimal transmission rate is 1 pkt/s when the packet size is 100 bytes. To our surprise, this corresponds to a data rate of only 800 bps. If ones also consider the MAC header (25 bytes) and the ACK frame (5 bytes), the total throughput will be roughly 1 kbps which is less than 1% of the available bandwidth. Therefore, we extend our experiments to consider transmitting different packet sizes at a transmission rate faster than 1 pkts/s in order to find out if the channel can be further utilized when we change this two fundamental paramters. Figure 5(b) shows that in order to achieve at least 90% PDR a suitable packet size for the considered ZigBee chipset should be less than 60 bytes. The optimal transmission rate remains to be 1 pkts/s. We suspect that this extremely low data rate would be due to the overhead caused by the testing software and the mobility. While this may seem to be a a very slow infomation transfer rate, it is sufficient for most wireless sensor network applications which are typically designed to transport small amount sensor data information.

7 Conclusion

In this study, we have conducted extensive feasibility study and derived several solutions that can be used to answer some of the fundamental questions in a location-aware WSN planning. Our studies show that in order to efficiently find a set of router locations that would mitigate the error in the localization process based on the fingerprint technique, network designers have to conduct site survey for every potential router location and compare the distance in signal space of all the location pairs. The best set of locations is the one whose majority of the pairs are separated by at least a certain predetermined threshold value, e.g., 10dB.

In addition, our findings also suggest that any application designed for indoor WSNs should not generate a packet whose size is greater than 60 bytes and the maximum transmission rate should not be greater than 1 packet/second in order to achieve at least 90% packet delivery ratio. If the application can meet this requirement, then the mobility should not have much impact on the network performance. We are currently working on improving our prototype design both in terms of hardware and software to support the indoor localization and tracking application. For practical usage, we expect to at least be able to achieve a room-level accuracy.

References

1. Bahl, P., Padmanabhan, V.N.: RADAR: an in-building RF-based user location and tracking system. In: 19th Annual Joint Conference of IEEE Computer and Communications Societies, pp. 775–784 (2000)
2. Kaseva, V., Hamalainen, T.D., Hanikainen, M.: Range-Free Algorithm for Energy-Efficient Indoor Localization in Wireless Sensor Networks. In: Conference on Design and Architectures for Signal and Image Processing (DASIP), pp. 1–8 (2011)
3. Rappaport, T.S.: Wireless Communication: Principles and Practice. Prentice Hall (1996)
4. Mao, G., Fidan, B., Anderson, B.D.O.: Wireless Sensor Network Localization Techniques. Computer Networks 51, 2529–2553 (2007)
5. He, T., Huang, C., Blum, B.M., Stankovic, J.A., Abdelzaher, T.: Range-free localization schemes for large scale sensor networks. In: 9th International Conference on Mobile Computing and Networking (MobiCom), pp. 81–95 (2003)
6. Peng, R., Sichitiu, M.L.: Angle of Arrival Localization for Wireless Sensor Networks. In: 3rd Annual IEEE Communications Society on Sensor and Ad Hoc Communications and Networks (SECON), pp. 374–382 (2006)
7. Tsui, W.Y., So, H.C., Ching, P.C.: Time-of-arrival based localization under NLOS conditions. IEEE Transaction on Vehicular Technology 55, 17–24 (2006)
8. Savvides, A., Han, C.C., Strivastava, M.B.: Dynamic fine-grained localization in ad-hoc networks of sensors. In: 7th Annual International Conference on Mobile Computing and Networking, pp. 166–179 (2001)
9. Boukerche, A., Olivera, H.A.B., Nakamura, E.F., Loureiro, A.A.F.: Localization Systems for Wireless Sensor Networks. IEEE Wireless Communications 14, 6–12 (2007)
10. Alippi, C., Mottarella, A., Vanini, G.: RF map-based localization algorithm for indoor environments. In: IEEE International Symposium on Circuits and Systems, vol. 1, pp. 652–655 (2005)

CAPWAP Protocol and Context Transfer to Support Seamless Handover

Siti Norhaizum M. Hasnan[1,2], Media A. Ayu[2], Teddy Mantoro[3], M. Hasbullah Mazlan[4], M. Abobakr[1], A. Balfaqih[1], and Shariq Haseeb[1]

[1] Wireless Communication Cluster,
MIMOS Berhad, Kuala Lumpur, Malaysia
{haizum.hasnan,abobakr,balfaqih,shariq.haseeb}@mimos.my
[2] Department of Information Systems,
International Islamic University Malaysia,
Kuala Lumpur, Malaysia
media@iium.edu.my
[3] Advanced Informatics School,
Universiti Teknologi Malaysia, Kuala Lumpur, Malaysia
teddy@ic.utm.my
[4] Universiti Teknologi Malaysia, Johor, Malaysia
hasbullah.mazlan@fkegraduate.utm.my

Abstract. Nowadays, real-time applications become immensely popular and available over wireless devices. An emerging technology IEEE 802.11 Wireless Local Area Network (WLAN) has brought users to have trendy gadgets like smart phones, tablets and laptops. These applications require a fast and efficient handover that guarantee a service of security and Quality of Service (QoS). However, the users will face disruption which can cause high transition delay while running multimedia services. This paper presents an overview of Control and Provisioning Wireless Access Point (CAPWAP) protocol; a solution to overcome security problem during handover within a large network. Furthermore, we conducted a survey in this paper in order to find an algorithm that can support seamless handover in centralized architecture. Overall, the paper ends with further work by suggesting a design of CAPWAP protocol using predictive context transfer in centralized architecture.

Keywords: CAPWAP protocol, delay, handover, wireless LAN.

1 Introduction

In modern times, the mobile revolution has improved every facet of society's lives in developing countries whereby they can use it to access online healthcare, education and government accountability [26]. Wireless devices such as tablet computing, smartphones and the current Bring Your Own Device (BYOD) craze have put a significant emphasis on the corporate wireless LAN network, as reported by Gibbs, "your life is fully mobile" [16]. In fact, an ease of deployment and low cost of

P. Meesad et al. (Eds.): *IC²IT2013*, AISC 209, pp. 93–103.
DOI: 10.1007/978-3-642-37371-8_13

wireless infrastructure network has made computing and communication devices more ubiquitous. Disappointingly, real-time applications tend to have high latency when users are in continuous mobility. Particularly, the use of Wireless Fidelity (Wi-Fi) radio interface while on the move happens in this trend of continuous mobility [5]. This can be a challenge in WLAN because there is no controller when traditional wireless Access Points (APs) turn off their Wi-Fi radios and mobile devices start to search for a new Access Point (AP). Therefore, a centralized architecture network is better in large network due to the simplifying Media Access Control (MAC) layer function in AP by having Access Controller (AC) as a central point in handling heterogeneous APs. This can greatly reduce the possibility of mobile devices disconnected from the network especially in enterprise network [18].

IETF working group has developed a protocol for communication between AC and AP namely CAPWAP protocol for improving the network connectivity. CAPWAP protocol is a standard that come from Lightweight Access Point Protocol (LWAPP) and Secure Light Access Point Protocol (SLAPP) that offer improvement of roaming performance across multiple-vendors [20]. CAPWAP protocol architecture support seamless mobility and can improve roaming efficiency especially in the main delay components of roaming which are scanning and authentication. The CAPWAP protocol has the capability of making decision within the AC due to independent function of AC and its flexibility [13].

The existing handoff performance has been further improved in order to achieve secure and seamless link-layer handoffs which can meet the QoS requirements of real-time multimedia applications. The handoff performance is improved by shortening both the re-authentication latency and the channel scanning delays in the discovery phase [8]. However, discovery phase is delayed by only 90% compared to 802.1x authentication phase which takes longest time more than 90% to authenticate between client and AP [12].

The 802.1x authentication phase is one of handoff phases that works within IEEE 802.1x standard that have a network setup consists of i) a device, called Supplicant, that accesses the network, ii) a switch that the supplicant directly connects, called Authenticator, and iii) an Authentication Server, which has the role of deciding if the supplicant should be allowed to access or not [1]. There are some findings about the relationship between requirements of CAPWAP protocol and procedure of seamless handover for us to come up with a method that can reduce latency in real-time applications. Our contribution in the paper is a proposed method which based on predictive protocol for CAPWAP-Inspired Context Transfer to reduce the handover latency. This method is going to be researched more deeply by understanding the context transfer features.

The next section, Section 2, presents the discussion on related research work and Section 3 highlights on the proposed CAPWAP protocol mechanism. Section 4 describes requirements of CAPWAP protocol in brief and Section 5 presents procedure of seamless handover. Section 6 briefly describes about predictive context transfer and finally Section 7 concludes the paper with further work.

2 Related Work

The related past studies that are being reviewed are categorized into three different areas; WLAN architectures, pre-authentication, and mobility.

2.1 WLAN Architectures

A method of Inter-cell protocol and Intra-cell protocol are proposed to mitigate interference problems in Wireless Local Area Networks (WLANs) especially in deployment of distributed and centralized architecture. The method is focused on dynamic selection channel to overcome scanning phase problem [11]. The authors believe that channel selection is one of the most efficient solutions to alleviate the impact of co-channel interference problems. In fact frequency planning helps to reduce interference. However, it is not efficient enough due to the dynamic changes in the wireless environment and the stochastic characteristics of user traffic and users distributions. The allocation is done with 3 aspects which are; i) policies, ii) protocol that support the exchange of information between APs and the agreement on the next channel assignment and iii) the protocol that support a synchronous switch of the channel by all Stations (STAs) within the cell.

Cluster Chain-based Context Transfer (C^3T) mechanism is used to reduce the Basic Service Set (BSS) transition delay in centralized architecture. IEEE802.11r is a fast BSS transition which is used for higher functionality. It is better than IEEE802.11i (i.e. a standard of security) and IEEE802.11e (i.e. a standard of QoS). The mechanism can have lower signaling cost for context transfer and higher hit rate of station (STA) context. Nevertheless, the authors simply assume that all Basic Service Sets (BSSs) can support the fast BSS transition protocols because IEEE802.11r does not specify how to work in centralized architecture [7]. The disadvantages on this method is big cache is required or it would give impact on hit rate and secondly this method use local MAC which is implemented in centralized architecture.

Another method has been proposed is Fast Handoff Authentication Protocol (FHAP) for mesh networks. The method works when a mesh point (MP) moves among Mesh Key Distributors (MKDs) in IEEE802.11s (i.e. a standard of mesh network) [25]. The method operates within the system where an MP can be a supplicant or an authenticator. When the MP moves into the IEEE802.11s mesh network, the MP will be the supplicant and is authenticated by other Mesh Points (MPs) and Authentication Server (AS) in the network. In the system, AS store all MP's authentication information and certificates while MP stores the AS's certificate. Meanwhile, it is assume that there is a secure channel between AS and MKD. The authors commented that their method has improved authentication latency and lightweight of work load but the security in IEEE802.11s is not specified very well.

2.2 Pre-authentication

Secure roaming between APs can be achieved by a method of pre-authentication that is a fast handoff method to reduce handoff delay during handover process of the mobile from one AP to another AP.

Protocol for Carrying Authentication Network Access (PANA) pre-authentication scheme is designed to carry the authentication information before the actual handover execution from one domain to another domain. The authors suggested having a new approach can substantially lower the time needed for an inter-domain handover by using context transfer compared to PANA pre-authentication. The advantage of this scheme solution is that PANA pre-authentication for nodes from far domains the process may last up to several seconds. In case of nodes are moving fast between domains that have a small overlapping coverage area, it may be too long to assure the fast authentication. However, the available time to complete this handover can be rather short due to the user's movement [17].

Past studies on a Capability Authentication (CapAuth) technique is proposed to reduce the complexity of having context transfer during the handover process. The proposed method is implemented based on the capability of context transfer [10]. This technique is characterized by two key ideas: i) the use of capabilities to achieve user-assisted transfer of context to new access points, and ii) the decoupling of user authentication from the enforcement of single point of access. This technique also enables low latency handovers with minimal communication overhead (especially on the backhaul) and high degree of fault tolerance.

An Enhanced Extensible Authentication Protocol (EAP) based pre-authentication (EEP) scheme is another method proposed for mobile Worldwide Interoperability for Microwave Access (WiMAX) networks. The authors aim to achieve fast and secure inter-Authentication Server Network (ASN) handovers. But, this method vulnerable to Denial of Service (DoS) and replay attacks. The authors aim to overcome the vulnerability of the scheme with much less requirements on the computation and communication resources. The method works by utilizing the following information provided from previous EAP-Transport Layer Security (TLS) mutual authentication and the centralized AS to prevent the vulnerability and reduce the number of cryptographic operations required. Besides, the proposed method can reduce handover delay, where the current handover process specified by the IEEE802.16e have huge bottleneck. [23].

2.3 Mobility

Received Signal Strength Indicator (RSSI) value is a method that used to predict the mobility of user's location system other than Global Positioning System (GPS). Unfortunately, a location system that is based on RSSI values has huge difficulty in terms of deployment due to the fact that the RSSI values and the Mobile Station (MS) of both relationships greatly depend on the propagation environment present between the MS and each AP.

A method of location determination system has been proposed by constructing a radio-map (i.e. radio frequency) by measuring the signal strength of multiple 802.11 beacons at multiple points. Radio-map is used by localization algorithm to extract an estimation of the user's location [24]. The advantage of this method is that it can increase the system accuracy when new APs are detected. It can also record APs' failure to be re-detected at the same position. Besides, the method also can filter unreliable APs in some cases to increase the robustness of the system and decrease the error distance. On the other hand, a trajectory-aware handoff algorithm is proposed based on

RSSI values whereby the algorithm is initiated when the Received Signal Strength (RSS) of Mobile Terminal (MT) is lower than a threshold [22].

Meanwhile, another method of presenting the mobility is on security-based. A Selective Proactive Context Caching (SPCC) technique that work to propagate security context of the mobile client to a selected set of neighboring base stations before re-association occurs. Even though SPCC mechanism successfully reduces the service interruption time and also reduces the packet dropped ratio, overhead of transferring the context information is high when there are more clients visited to the number of APs [21].

Table 1. Summary of Algorithm Schemes for Improving CAPWAP Protocol Performance

Authors	WLAN Architectures			Re-authentication		Mobility	
	Autonomous	Centralized	Distributed	Reactive	Proactive	RSSI	Security
Abusubaih et al. [11]	×	✓	✓	×	×	×	×
Zarimpas et al. [24]	×	×	×	×	×	✓	×
Alfandi et al. [17]	×	×	×	×	✓	×	×
Huang et al. [7]	×	✓	×	×	✓	×	×
Cai et al. [10]	×	×	×	×	✓	×	×
Ei et al. [22]	×	×	×	×	×	✓	×
Ling et al. [21]	×	×	×	×	✓	×	✓
Wei-dong et al. [25]	×	×	✓	×	✓	×	×
Nguyen et al. [23]	×	×	×	×	✓	×	×

Table 1 shows that according to previous research studies, most researchers study on centralized and distributed architectures. Since wireless technologies and the related devices are more advanced, it is necessary to have simpler and easy deployment of network. Besides, the implementation of both architectures is suitable to large network compared to autonomous architecture which is better for a small network. However, centralized architecture will be in our proposed method due to easy handling heterogeneous APs from different vendors and have a better security network within a short time of handover session.

Smooth handover can be achieved by having a CAPWAP protocol which is a new alternative way for communication between AP and AC. Based on past work, re-authentication has been used to reduce handoff latency and there are two approaches namely reactive and proactive. Commonly, researchers focused on proactive approaches such as pre-authentication or proactive key distribution whereby security contexts will be transferred to candidate network entities (e.g., neighboring APs) via inter-AP communications before handoff occurs. Meanwhile, reactive is initiated in response to the completion of a discovery step. Thus, proactive approaches will be used in our proposed method that obviously accommodate prompt and secure handoff. Particularly, the proactive authentication schemes are suitable for centralized network systems.

In this paper, we also proposed a method that focus on security so that AC can act as the central system where the centralized architecture can keep track of all information related to handoff, and even make handoff decision including target indication. Therefore, the central system can transfer client's information to the target network deterministically before handoff.

3 CAPWAP Protocol

CAPWAP protocol is an alternative protocol for communication between AC and AP. The goal of CAPWAP protocol is to allow the interoperability among nodes produced by different vendors of APs. In this objective, the network nodes are divided into two different categories: the Wireless Termination Point (WTP), which roughly corresponds to the AP, and the AC. The requirement of implementation CAPWAP protocol must be in centralized WLAN architectures which can simplify the deployment of large-scale networks, by enabling network-wide monitoring and by improving the management scalability and configurability [2].

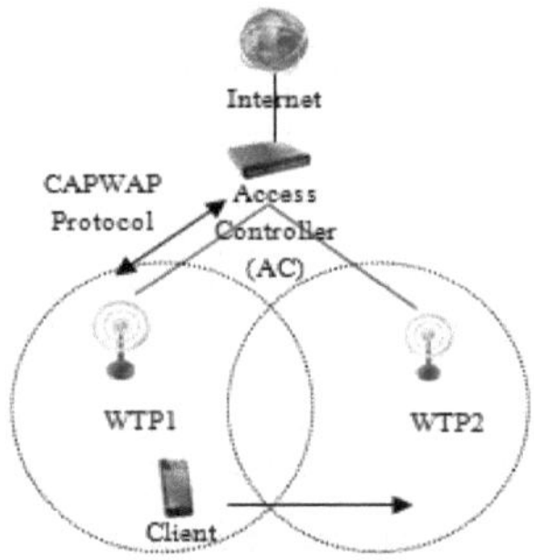

Fig. 1. Hierarchical of CAPWAP architecture

Process of CAPWAP protocol occurs between WTP and AC whereby multiple Wireless Termination Points (WTPs) are connected to the AC via a direct connection either Layer 2 (L2)-switched, or a Layer 3 (L3)-routed network. The important role of AC is it can centrally manage much functionality such as radio frequency (RF) monitors and configuration, WTP configuration and firmware loading, network-wide user database, and mutual authentication between network entities. The advancement of control functions requires a regular exchange of control messages between the WTPs and the AC [6].

Meanwhile, traditional protocols for managing WTPs are either manual static configuration via Hypertext Transfer Protocol (HTTP), proprietary L2 specific or non-existent (when the WTP are not configurable). The communication between the AC and the WTP is provided by exchanging user datagrams protocol (UDP), with an explicit separation between data and control traffic through different UDP ports as illustrated in Fig. 1. CAPWAP control messages, and optionally, CAPWAP data messages, are secured using Datagram Transport Layer Security (DTLS).

In the Local MAC mode, the data frames are locally bridged by the WTP or encapsulated in an IEEE802.2 frame forwarded to the AC. The management frames are processed locally by the WTP and then forwarded to the AP. In the Split MAC mode, both the data and the management frames are encapsulated via the CAPWAP protocol and exchanged between the AC and the WTP [14].

4 Seamless Handover

Continuous wireless connectivity is provided by bridging packets of APs from the wireless domain to an internal network. The common issue for connectivity is caused by mobility. When a device moves and goes out of the range, it will lose the signal from its AP. In that case, the mobile device would try to maintain its wireless connectivity by associating to a new AP.

Handover process consist of four types of phases namely scanning phase, authentication and association phase, 802.1x Authentication phase and 4-way handshake phase which depicted in Fig. 2. Each phase takes different time duration whereby scanning phase is 350ms – 500ms, authentication and association phase is less than 10ms, 802.1x Authentication phase is 750ms – 1200ms and 4-way Handshake phase is 50ms [3].

Open System authentication is the simplest of the available authentication algorithms and it is a null authentication where any station that requests authentication with this phase may become authenticated if 802.11 Authentication type at the recipient client is set to Open System authentication. That is the reason Open System only take less than 10ms.

The communication process of Wi-Fi Protected Access 2 (WPA2)-Enterprise begins with an unauthenticated supplicant (i.e., client device) attempting to connect with an authenticator (i.e., 802.11 access point). The access point responds by enabling a port for passing only EAP packets from the client to an authentication server located on the wired side of the access point. The access point blocks all other traffics, such as HTTP, Dynamic Host Configuration Protocol (DHCP), and Post Office Protocol 3 (POP3) packets, until the access point can verify the client's identity using an authentication server (e.g., RADIUS). Once authenticated, the access point opens the client's port for other types of traffic.

Meanwhile, WPA2-Personal authentication only takes an average time which is 50ms to connect to the AP. WPA2-Personal security methods use a pre-shared key (PSK) or a passphrase for authentication and are designed for small office or home office infrastructure mode networks. The primary roles are to verify the existence of the same Pairwise Master Key (PMK) between the client and the AP and to derive the Pairwise Transient Key (PTK) using the 4-way handshake which consists of four messages from Message-1 to Message-4 as stated in Fig. 2. The 4-way handshake starts when the AP sends Message-1 to the client. Message-1 consist of three parameters: a MAC address of an AP (called AA), a random number chosen by the AP (called ANonce) and a counter to prevent a replay attack (called SN).

Table 2 shows that among three different phases of handoff, WPA2-Enterprise takes the longest time due to the requirement for authentication procedure for IEEE 802.1x between client and AP. Wi-Fi Protected Access (WPA)-Enterprise, also known as 802.1x authentication (IEEE 802.1x), means the encapsulation of the EAP over IEEE802.11 which is an enhancement from standard 802.11i. This authentication offers an effective framework for authenticating and controlling user traffic to a protected network, as well as dynamically varying encryption keys.

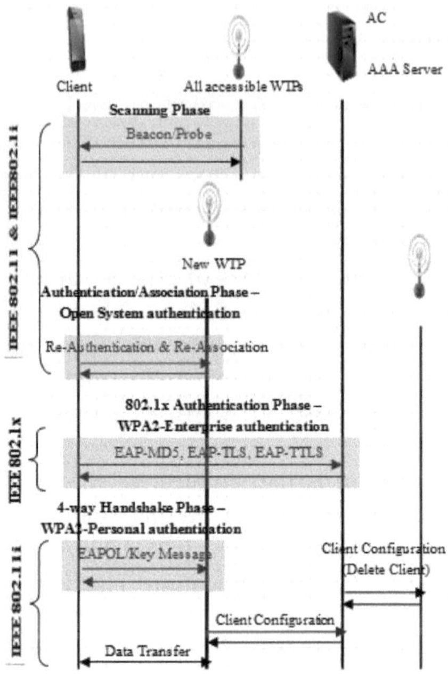

Fig. 2. The timing chart of the 802.11 security procedure seamless handover

Table 2. Summary of Handoff Phases

References	Open System	WPA2-Personal	WPA2-Enterprise	Performance
Kassab et al., [15] , 2005	×	✓	×	30.8ms
Lopez et al., [19] , 2007	×	✓	×	17ms
Ok et al., [9], 2008	✓	×	×	4.024ms
Sarma et al., [4], 2009	✓	×	×	5ms
Balfaqih et al., [13], 2012	×	×	✓	2987.66ms

5 Characteristic of the Proposed Method

Overall, a survey on previous work was conducted in this paper to elaborate deeper on CAPWAP protocol. We are proposing a new method to overcome the security problem that takes long time to authenticate during the handover session. Our proposed method is CAPWAP-Inspired of Context Transfer whereby the criteria of our method is based on accountability, scalability and flexibility.

Our method will be improved on accountability whereby each user who wants to get connected to AP will determine the AP has already authorized by the Authentication, Authorization, Accounting (AAA)/EAP server. The confirmation of the same keying material is done through the AAA/EAP server. Communication for the key of authenticator was handled by the AC using CAPWAP protocol that has been modified. In some systems, explicit authenticator and user mutual authentication is possible. This is desirable due to it greatly improves accountability and users can transfer information safely.

Additionally, our method must also contain scalability where the AC must be able handle each APs that is connected with client. The AC must be compatible with both types of APs namely local MAC and split MAC during the context transfer. The reason is to ensure enough flexibility and sufficient of selection to the next AP.

The flexibility for our method is to enable transferring service context between client and AP without modification on the AP side. All the modification will happen at AC. This is to ensure that the operation of CAPWAP work smoothly. Besides, mobility of users depends on the way of controlling handover decisions. So, our method has to be accurate enough to reduce handoff latency in order to support user's movement during handover process.

6 Conclusion and Future Work

CAPWAP protocol is a new way of communication protocol between AC and AP. The old way is based on standard 802.11f which is Inter-Access Point Protocol (IAPP) that used to work between client and AP. The large deployment of network that caused by increasing numbers of wireless devices give us idea to come out with the new method that can overcome the high latency problem within these days.

Overview of CAPWAP protocol and context transfer led us to CAPWAP-Inspired Context Transfer. The standard CAPWAP protocol will be modified by providing a solution to overcome security problem during handover within a large network. Furthermore, the survey that we conducted in this paper will allow us to do comparison to find an algorithm that can support seamless handover in centralized architecture.

To sum up, our contribution for this paper is we present an overview protocol of CAPWAP and context transfer to be implemented in centralized architecture. The future work is to come up with CAPWAP-Inspired Context Transfer scheme for roaming in secured 802.11 networks.

References

1. Chiornita, A., Gheorghe, L., Rosner, D.: A Practical Analysis of EAP Authentication Methods. In: 9th Roedunet International Conference (RoEduNet), pp. 31–35 (2010)
2. Levanti, A., Giordano, F., Tinnirello, I.: CAPWAP-Compliant Solution for Radio Resource Management in Large-Scale 802.11WLAN. In: GLOBECOM 2007, pp. 3645–3650 (2007)
3. Mishra, A., Shin, M., Arbaugh, W.: An Empirical Analysis of the IEEE 802.11 MAC Layer Handoff Process. SIGCOMM Computer Communication 33, 93–102 (2003)
4. Sarma, A., Gupta, R.K., Nandi, S.: A Zone Based Interleaved Scanning Technique for Fast Handoff in IEEE 802.11 Wireless Networks. In: ISPAN 2009, pp. 232–237 (2009)

5. Li, B., Wang, K.H.: NonStop: Continuous Multimedia Streaming in Wireless Ad Hoc Networks with Node Mobility. IEEE Journal on Selected Areas in Communications 21(10), 1627–1641 (2003)
6. Sarikaya, B., Zheng, X.: CAPWAP Handover Protocol. In: ICC 2006, pp. 1933–1938 (2006)
7. Huang, C.M., Li, J.W.: A Context Transfer Mechanism for IEEE 802.11r in the Centralized Wireless LAN Architecture. In: 22nd International Conference on AINA, Japan, pp. 1–7 (2008)
8. Lee, I., Hunt, R.: A Novel Design and Implementation of DoS Resistant Authentication and Seamless Handoff Scheme for Enterprise WLANs. In: 8th Australian Information Security Management Conference, pp. 1–14 (2010)
9. Ok, J., Morales, P., Morikawa, H.: AuthScan: Enabling Fast Handoff across Already Deployed IEEE 802.11 Wireless Networks. In: IEEE 19th PIMRC 2008, pp. 1–5 (2008)
10. Cai, L., Machiraiu, S., Chen, H.: CapAuth: A Capability-based Handover Scheme. In: IEEE International Conference on Computer Communications (IEEE INFOCOM), pp. 1–5 (2010)
11. Abusubaih, M., Gross, J., Wolisz, A.: An Inter-Access Point Coordination Protocol for Dynamic Channel Selection in IEEE802.11WLANs. In: IEEE Conference on Local Computer Networks (2006)
12. Akhlaq, M., Aslam, B., Khan, M.A., Jafri, M.N.: Comparative Analysis of IEEE802.1x Authentication Methods. In: 11th WSEAS International Conference on Communications, vol. 11, pp. 1–6 (2007)
13. Balfaqih, M., Haseeb, S., Mazlan, M.H., Hasnan, S.N., Mahmoud, O., Hashim, A.: CAPWAP Status and Design Considerations for Seamless Roaming Support. World Academy of Science, Engineering and Technology 68, 2147–2153 (2012)
14. Bernaschi, M., Cacace, F., Davoli, A., Guerri, D., Latini, M., Vollero, L.: OpenCAPWAP: An open source CAPWAP implementation for the management and configuration of Wi-Fi Hotspot. Computer Networks 53, 217–230 (2009)
15. Kassab, M., Belghith, A., Bonnin, J., Sassi, S.: Fast Pre-Authentication based on Proactive Key Distribution for IEEE 802.11 Infrastructure Networks. In: 1st ACM WMuNeP 2005, pp. 46–53 (2005)
16. Gibbs, N.: TIME Mobility Poll: Your Life is Fully Mobile (2012), `http://techland.time.com/2012/08/16/your-life-is-fully-mobile` (accessed on: October 24, 2012)
17. Alfandi, O., Chamuczy´nski, P., Brosenne, H., Werner, C., Hogrefe, D.: Performance Evaluation of PANA Pre-Authentication and PANA Context Transfer. In: 4th ICWMC 2008, pp. 43–48 (2008)
18. Calhoun, P., Montemmurro, M., Stanley, D.: Control And Provisioning of Wireless Access Points (CAPWAP) Protocol Specification. RFC 5414, pp. 1–156 (2009)
19. Lopez, R.M.: Network-layer Assisted Mechanism to Optimize Authentication Delay during Handoff in 802.11 Networks. In: 4th MobiQuitous, pp. 1–8 (2007)
20. Govindan, S., Cheng, H., Yao, Z.H., Zhou, W.H., Yang, L.: Objectives for Control and Provisioning of Wireless Access Points (CAPWAP). RFC 4564, pp. 1–33 (2006)
21. Ling, T.C., Lee, J.F., Hoh, K.P.: Reducing Handoff Delay in WLAN using Selective Proactive Context Caching. Malaysian Journal of Computer Science 23, 49–59 (2010)
22. Ei, T., Wang, F.: A Trajectory-Aware Handoff Algorithm based on GPS Information. Annals of Communications 65, 411–417 (2010)

23. Nguyen, T.N., Ma, M.: Enhanced EAP-Based Pre-Authentication for Fast and Secure Inter-ASN Handovers in Mobile WiMAX Networks. IEEE Journal Transaction on Wireless Communications 6, 2173–2181 (2012)
24. Zarimpas, V., Honary, B., Darnell, M.: Indoor 802.1x Based Location Determination and Real-time Tracking. In: International Symposium on WoWMoN 2006, pp. 1–4 (2006)
25. Wei-Dong, Y., Ji-Zhao, L., Ke, W., Li-Ming, S.: Authentication Protocols to Support Fast Handoff for 802.11s Mesh Networks. In: International Conference on MINES, pp. 644–648 (2010)
26. Kerravala, Z.: Network World: It's Time for Wireless LANs to Evolve, `http://www.networkworld.com/community/blog/its-time-wireless-lans-evolve`

Intelligent Cloud Service Selection Using Agents

Imran Mujaddid Rabbani[1], Aslam Muhammad[1], and Martinez Enriquez A.M.[2]

[1] Department of CS & E, UET, Lahore, Pakistan
[2] Department of CS, CINVESTAV-IPN, D.F. Mexico
imranmrabbani@gmail.com,
maslam@uet.edu.pk,
ammartin@cinvestav.mx

Abstract. One of the most recent developments within computer science is cloud computing which provides services (power, storage, platform, infrastructure etc.). Many clouds provide services are based on cost, efficiency, performance, and quality. Stakeholders have to compromise cost sometimes and performance or quality other times. Provision of the best quality based services to its stakeholders and to impart intelligence, agents can play important roles especially by learning the structure of the clouds. Agents can be trained to observe differences and behave intelligently for service selection. To rank different clouds, we propose a new technique performance factor for the provision of services based on intelligence. The research objective is to enable cloud users in selecting cloud service according to their own requirements. The technique assigns performance factor for each service provided by cloud and ranks it as whole. By doing so, quality of the services can be highly improved. We validate our approach with a case study, which emphasizes the need to rank cloud services of widely spreading and complex domains.

Keywords: Agents, cloud computing, performance factor.

1 Introduction

Computer science is a dynamic field which evolves itself repeatedly, covering all the flavors of computing, power, and storage. One of the most recent developments within Computer science is Cloud Computing which provides services related to computer, allowing users/customers the power of computing, data storage, deployment models, infrastructures, platforms and resources on pay – per – use basis. Its focus is to provide secure, scalable, reliable, sustainable, fault tolerant and sharing of data through web base applications [9].

Cloud computing provides a complete package of services for vendors/consumers, introducing the concept: the payment is function of the usage. The different services provided are: Software as a service (SaaS) like GoogleApp, Salesforce, impels CRM, Infrastructure as a Service (IaaS) like AWS, Flexiscale, Platform as a Service (PaaS) like Google's App Engine, and Microsoft's Azure and Data base as a Service (DaaS) [11, 12, 13, 14] and some others are Amazon's (S3, E2C), Google+, RackSpace,

P. Meesad et al. (Eds.): *IC²IT2013*, AISC 209, pp. 105–114.
DOI: 10.1007/978-3-642-37371-8_14

GoGrid, IBM, RightScale, AT&T etc. Cloud computing can be deployed using four models including public cloud, community cloud, hybrid cloud and private cloud [11, 12, 14]. Private clouds have limited access (within the organization) whereas there is open access in public clouds [20]. To provide these services intelligently we can use agents.

The agent is a type of software that works on the behalf of its owners [16]. Intelligent agent allows people to delegate their work to them, and they do repetitive tasks, learning, compute complex data and make decisions. They act as human beings; can assist in various fields [1, 2]. The agent has features like Autonomy, Pro-activity, Reactivity, Communication and Cooperation, Negotiation and learning [5]. Cloud computing holds a large scope for learning (machine learning). Performance and intelligence will become one of the important requirements for cloud services selection [3]. The joint venture of cloud and agent will be important aspect for providing quality based services to the stake-holders.

Cloud services contain long lists of user defined applications and there is need of comprehensive techniques for cloud service selection [20]. Data storage cloud service is automatically selected using XML schema [21] and other techniques are also summarized in [20]. To rely on single vendor for using its services, stake-holders may suffer with respect to performance as a whole (because some of the services are very best, some are average and some of them are below average). So, to provide best and state-of-the-art services, there should be a mechanism to provide independent services through agents of best quality and suited in all aspects. There are techniques available to select a particular service based on requirements but when there is more than one service against the user requirements then decision making is required. For this, we propose Agent base Cloud Service Provider (ACSP) that provides services for said requirements with ranks. The rank shows the performance of the services in respect of feedback. The proposed technique "Performance Factor for Cloud Services PFCS" maintains the history of the services provided along with their feedbacks and is based on unsupervised learning (like Q-Learning). The PFCS converts the normal services to quality and performance base by giving them a numeric value (performance factor) which identifies its worth for the usage. It improves the overall service and rank of the CSPs as well. The performance factor is based on the feedback calculation performed on different parameters of cloud services like reliability, security, fault tolerance, backup and recovery etc. from user perspectives. The research objective is to enable cloud seekers in selecting service according to their own requirements.

The paper is formulated as follows. The related work is presented in section 2. The proposed algorithm is discussed in section 3. Architecture of the proposed methodology is presented in section 4 and case study in 5. Conclusion and future works are described in section 6.

2 Related Work

In this modern era, clouds are utilized to handle demanding computations and provide huge volumes of data storage in an efficient manner [5] and allow its users to use services on rent basis as much they need and require [8]. Increased use of technology, issues rises like quality, standardization, reliability, cost effectiveness, security, fault tolerance and most importantly intelligence. The agents are used to provide automatic selection of web services on the internet using proxies for service ranking based on feedback that was gathered through third parties [17]. Trust level functions [18] are used for automatic service selection. IMAV [6] is an intelligent multi-agent model for automatic resource allocation based on virtualization especially for mobile devices in cloud computing environment. Web service discovery is done using WDSL, ebXML and UDDI [7]. Intelligent agents increase the performance of service/ information selection and retrieval engine where there is huge volume of data [16].

To provide the best services to the clients, agents play a vital contribution to service se-lection and importantly to impart intelligence in the cloud computing services as there is a huge volume of user calls regarding its services [3]. With the help of machine learning techniques, web services can be allocated automatically along with their necessary resources. The automation is timely needed for providing cloud services (operational, development, storage, scaling, management etc.). [4] The machine learning is the source for managing large resources as Facebook which is known as the largest Hadoop storage cluster in the world. The comparison of different cloud services can be found on [15].

The automation of the cloud services, the learning algorithm can facilitate [3] and is necessary for providing services (operation, developing, manages and dynamic scaling) [4]. It becomes more complex when services are built on open source platform. There are some open source cloud systems available like OpenStack, Open Nebula and Open QRM etc. [5]. In decentralized environment, solving a problem, multi-agent systems are applied which contains many agents providing partial solution by interacting each other [5]. On the other hand, clouds can provide extensive service model for agent base applications for modeling and simulation of complex computer applications and agent can provide intelligence to the clouds for its adaptability, flexibility, auto resource management and service selection etc. It becomes significant that the cloud applications can produce innovative opportunities for their clients and developers for completing their respective tasks. A new field of agent based cloud computing should be set to provide intelligence in their services to their stake-holders [5]. There are techniques used to provide automated cloud resource allocation using micro-agreements [8] and UBIWARE platform [10]. With its three components (Live behavior Engine, S-APL Models and Reusable Atomic behavior (RAB)) and open new options for software design in cloud services for intelligence.

Hence, no such technique is currently available to provide intelligent service selection based on performance factor using the feedback. This lack motivated us to propose an agent base system which could help seekers to look for a cloud which could provide them the service as per their needs. Moreover, autonomously discover rank base services for the clients according to their preferred criteria.

3 Performance Evaluation Factor for Cloud Services

For efficient services to the clients with respect to their requirements, quality attributes are utilized which are indeed for cloud services e.g. security, usability, reliability, backup and recovery etc. The learning environment allows producing best decision making while providing services on demand. The decision making becomes important when there are multiple choices under consideration for choosing best one among them. There are lots of competitors in the market for providing services under the umbrella of cloud. So, there should be a platform where users can check the performance status of CSP's services for its usage. Here we propose a un-supervised learning technique that is similar to Q-Learning [19] named "Performance Factor for Cloud Service (PFCS)". The Q-learning gathers positive reward every time by using learning rate and discount factor to move ahead but in our technique, we collect all types of feedback from past and current users. These feedbacks may be positive / negative. The performance factor is calculated on the basis of feedback provided by its users. The proposed methodology stores and manipulates the information along with the quality attributes available in repository.

PCFS uses parameters of cloud computing like security, reliability, performance, cost effective, saleable, availability etc. It defines a repository called "Parameter Repository (R)" that containing parameters and allows future aspects as well. The processing component of our "Agent based Cloud Service Provider (ACSP)" takes parameters from 'R' one by one and performs its calculation on it. It generates the performance factor of all CSP's services and this collective performance proceeds towards Performance Factor of CSP as a whole. At the time of service delivery ACSP checks the performance factor of each CSP against the defined threshold value and provide service to their stake-holders accordingly.

The performance of the CSP depends upon the performance of all services provided individually. The CSP performance is calculated as.

$$P_{CSP} = \sum_{k=1}^{n} (p_k(s)) \tag{1}$$

Where n = total no. of services provided by CSP.

p_k = performance factor of single service w.r.t all quality attributes in 'R' and is defined as

$$p_R(s) = \sum_{i=1}^{R} (f_i - v_i) \tag{2}$$

Where 'R' = No. of repository attribute belongs to Quality and it is

$$R = \{\text{Security, Reliability, Fault Tolerance, Customizable} \ldots\} \tag{3}$$

f_s = Feedback against the particular quality attribute of the service and is

$$f_s = \frac{fc_1 + fc_2 \ldots fc_N}{N} \quad \& \, 0 < fs < 100 \tag{4}$$

Where N: = total no. of customers and

$$v_i = \text{Company Value} \qquad 0 < v_i < 100$$

It is the value assigned by the CSP as a company threshold value to a service *S*.

3.1 Algorithm

When a user provides his / her preferred requirements, it finds the service providers in the cloud community which matches the demands. Once the CSP has been found, we check each and every service of the CSP one by one and calculate the feedback f_s (related to service). Based on f_s calculates the performance factor p_s according to Eq. 2 and adds it to the CSP's Rank P_{CSP}. After the completion of the process applied on every CSP, we will found the Rank of each CSP. The pseudo code for the above methodology is as follows:

1. Select CSP C
2. Select Service S_i of C
3. Calculate feedback f_s using Eq. 4
4. Calculate performance factor p_s according to Eq.2 with repository parameters
5. Add p_s to CSP rank P_{CSP} value and go to step 2
6. Repeat until all services have been tested.
7. Show CSP and its Services along with Rank and Performance factor.

To illustrate the functionality of the proposed algorithm, let's take an example of three CSP's, say Amazon, Google, and Microsoft. Labeled them with C_1, C_2 ... C_N and each having services like EC2, Google App / doc, CRM, sky Drive etc. say S_1, S_2 ... S_N. There is repository 'R' defined in Eq. (3) and feedback is calculated for every service 'S' against 'R' according to Eq. (4) and hence the performance of the service's' is calculated from the Eq. (2). This step is iterated and performance is calculated against all the attributes defined in the 'R' along with feedback provided by stake-holders / clients and the company value defined by the service provider for each service. The performance Factor of the CSP can be found by adding all the performance factor of different services.

4 Architecture of Agent Cloud Service Provider

The proposed model consists of a public cloud that contains different sub modules as shown in figure 1. These modules work together for achieving a goal of finding performance factor of CSP. Demands of stake-holders are received and necessary actions are performed on the quality attributes available in our repository. After processing, results are shown to the requester along with service's rank and CSP's performance factor. Demander reviews the information shown and made its selection accordingly. The technique is implemented in an agent base cloud as shown in Figure 1. The cloud interacts with different clouds against the stake-holder's demands, gathers services from different clouds, processes them as its internal feature and provides the list of services along with performance factor. The cloud allows the stake-holder to make best decision regarding its business needs. It shows which service is best for him / her and suits requirements. It also displays the trust levels of x-users about each service of a certain cloud. The architecture has the following important modules as shown in Figure 2.

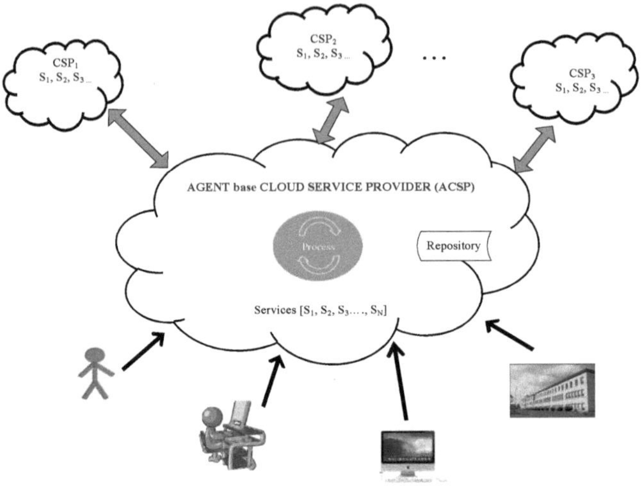

Fig. 1. Agent Base Cloud Service Provider (ACSP)

Ranker: Processing unit of the system that calculates the rank / performance factor of all the services provided by CSP and also performs the rank of the CSP. The ranker utilizes feedback of each service that calculated on quality attributes. It provides numeric value as performance factor to visualize the rank of the service as well as CSP. The Rank and performance factor of all its services is provided to the user interface where it presents to the user as mentioned. It is the inference part of the agent system.

Feedback Unit: It collects the feedback from the x-users of the services found by the Service Selector unit and updates the repository respectively. It contains the average values of the feedback of particular service based on quality attributes. In this methodology, feedback value is used as numeric but in some cases, feedback is available in the form of comments then it is handled using NLP techniques proposed in [22] and it helps to find the negative comments about the service.

Repository: The knowledge database of all the information regarding performance ranks and quality attributes. It also maintains the feedback knowledge in it. All the processing is performed on the data available in the repository. It's like a database repository. It is hub of the agent system and communicates with all parts of the system as shown in Figure 2.

Quality Facilitator: It allows user / service provider to add quality attributes for enhanced quality base service selection. As much the quality parameters in the repository, most quality of service is achieved. After selecting the appropriate attributes from Repository, it sends the array of attributes to next unit as shown in Figure 2.

Service Selector: It is responsible for finding the services matched with the user criterion from different services provider. It receives requirements from User Interface module and finds the services and CSP in the cloud community. Service selection can be done with the help of techniques mentioned in [20, 7]. These found services provided to Quality Facilitator unit to further processing as mentioned in Figure 2.

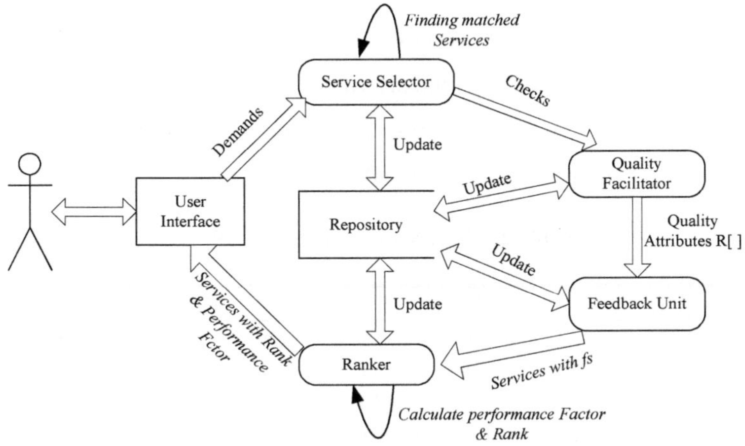

Fig. 2. Design Structure of ACSP

User Interface: The direct interaction of the stake-holders is performed through this unit where user provides its preferences and requirements and feedback as well (if necessary). It is the sensing unit of the agent system. The user's requirements look like as followed.

Table 1. User Requirements through Interface

Parameters	Requirements	Values
Service / Job Name	Data Storage	---
Quality Attribute	Security	YES / NO
	Cost effective	YES / NO
	…	…

5 Case Study

We consider three CSP's having three services each (say S_1, S_2 and S_3) and we operate our formula on one quality attribute i-e (Security) from the repository 'R'. There are different columns in the tables (see Table 1, 2 & 3) where C_1, C_2, C_3, C_4 and C_5 representing the feedbacks from different users. The column *'f'* presents the accumulative average feedback value according to Eq. (4) and *'v'* represents the company value. *'p'* is the performance factor value of the service as per Eq.(2). The last row column *'P'* is the Performance Factor / Rank of the CSP_i as whole. The positive values of 'P' show that these service providers have the higher rank and their services are rated by their users.

Following tables represents CSP's services and their feedback of their services respectively. Consider the Table 3, where in first row, 80, 90, 75, 80 & 70 are the feedback values from five different clients/ users. So, from Eq. (4), the f is calculated as

$$f = (80+90+75+80+70) / 5$$

$$= 79 \text{ (as shown in column 'f' and in first row of Table 3)}$$

$$v = 80 \text{ (company value)}$$

$$p = f - v = 79 - 80 = -1 \text{ (According to Eq. (2))}$$

Performance Factor P = ((-1) + (-6) +0) = -7(adding all column's 'p')

Table 2. Cloud Service Provider CSP_1

Services	C_1	C_2	C_3	C_4	C_5	*f*	*v*	*p*
S_1	65	70	70	75	65	69	65	4
S_2	60	80	75	65	70	70	70	0
S_3	65	80	70	70	70	70	70	2
Services				**Performance Factor P**				**6**

Table 3. Cloud Service Provider CSP_2

Services	C_1	C_2	C_3	C_4	C_5	*f*	*v*	*p*
S_1	80	90	75	80	70	79	80	-1
S_2	90	70	65	75	70	74	80	-6
S_3	65	80	70	70	65	70	70	0
Services				**Performance Factor P**				**-7**

Table 4. Cloud Service Provider CSP_3

Services	C_1	C_2	C_3	C_4	C_5	*f*	*v*	*p*
S_1	70	80	85	65	70	74	80	4
S_2	75	75	70	75	70	72	80	2
S_3	80	85	90	75	65	82	70	7
Services				**Performance Factor P**				**13**

Table 5. CSP's Performance Factor along with thier services

Services	CSP_1	CSP_2	CSP_3
S_1	-1	4	4
S_2	-6	0	2
S_3	0	2	7
Rank	-7	6	13

We operate proposed algorithm on remaining CSP's as well and construct tables 3 & 4. Table 5 shows the summarized form of performance factor and ranks of all services of CSP's and as a whole. The first column presents services of the CSP's and remaining columns shows the calculated performance factors of their respective ser-vices. The last row of the tables represents the Ranks of the CSPs.

The above case study analysis shows that the proposed methodology works very well in the environment where there is need to rank the services based on requirements. From Table 5, one can easily infer the service's status of different CSPs according to its requirements.

Note: The above tables show the simulation for only one quality attribute, in real scenario, it is performed on all attributes available in the repository 'R'.

We evaluate our technique through simulation on the small data. The results show that agents can provide efficient cloud service selection that is based on consumer's feedback.

6 Conclusion

In widely spreading and complex domains of cloud computing, it is necessary to assist cloud users in selection of services offered by this paradigm. Selection may be affected by multiple parameters like cost, efficiency, resource nature, satisfaction, among others. So, we developed a technique based on performance factor to recommend cloud service to its seekers which not only considers the user requirements but also quality of services and feedback of several other clients. We assign performance factor to each service of the cloud using the feedback of users on quality attributes. This leads to rank of the CSP as a whole. Therefore, the service which fulfills the maximum requirements of the user is recommended consequently the surfing time is decreased and service seeker can avail the best of all available cloud services. The propose model enhances the service selection level in efficient manner using the customer's feedback. It also enhances the degree of CSP's stakeholders

For future work, we intend to explore this field more and present the structural and archi-tectural details of the agent base systems that provide best services among the clouds.

References

1. Hanh, T., Thaovy, T.: Intelligent Agent (2012), `http://groups.engin.umd.umich.edu/CIS/course.des/cis479/projects/agent/Intelligent_agent.html`
2. Don, G.: Intelligent Agents: The Right Information at the Right Time, IBM Corporation. Research Triangle Park, NC, USA (1997), `http://www.networking.ibm.com/iag/iaghome.html`
3. Gurmeet, S.: Scope of machine Learning in Cloud Computing (2010)
4. Armbrust, M., Fox, A., Grith, R., Joseph, A.D., Katz, R., Konwinski, A., Lee, H., Patterson, D., Rabkin, A., Stoica, I., Zaharia, M.: Above the Clouds: A Berkeley View of Cloud Computing. Technical Report No. UCB/EECS-2009-28, UC Berkeley Reliable Adaptive Distributed Systems Laboratory (2009)
5. Domenico, T.: Cloud Computing and Software Agents: Towards Cloud Intelligent Services, ICAR-CNR & University of Calabria Rende, Italy

6. Myougnjin, K., Hanku, L., Hyogun, Y., Jee-In, K., HyungSeok, K.: IMAV: An Intelligent Multi-Agent Model Based on Cloud Computing for Resource Virtualization. In: 2011 International Conference on Information and Electronics Engineering, IPCSIT, vol. 6, pp. 199–203. IACSIT Press, Singapore (2011)
7. Shailesh, K.: Chandramohan: Personalized Web Service Selection. International Journal of Web & Semantic Technology (IJWesT) 2(2), 78–93 (2011)
8. Kassidy, C., Martijn, W., Frances, M.T.: An Intelligent Cloud Resource Allocation Service. Agent-based automated Cloud resource allocation using micro-agreements. Faculty of Technology, Policy and Management, Delft University of Technology, Jaffalaan 5, Delft, The Netherlands
9. Singh, A., Malhotra, M.: Agent Based Framework for Scalability in Cloud Computing. International Journal of Computer Science & Engineering Technology (IJCSET) 3(4), 41–45 (2012) ISSN: 2229-3345
10. Sergiy, N., Vagan, T., Michal N.: Mastering Intelligent Clouds. Engineering Intelligent Data Processing Services in the Cloud. Industrial Ontologies Group, University of Jyväskylä, Mattilanniemi, Jyväskylä, Finland
11. Höfer, C.N., Karagiannis, G.: Cloud computing services: taxonomy and comparison. Internet Serv. Appl. 2, 81–94 (2011)
12. White Paper : Introduction to Cloud Computing, by Dialogic Corporation (2012), `http://www.dialogic.com`
13. Network World 2012, Top Cloud Computing Companies List To Watch and Invest in 2012 (May 22, 2012), `http://nanospeck.hubpages.com/hub/Best-Cloud-Service-Providers`
14. Expert Group Report, The Future of Cloud Computing Opportunities For European Cloud Computing Beyond 2010, Public Version 1.0, By European Commission (2009)
15. Cloud Services Comparison (September 26, 2012), `http://www.cloud-computing.findthebest.com`
16. James J.: Using an Intelligent agents to enhancing search engine performance. First Monday 2(3) (1997)
17. Michael, E.M., Munineder, P.S.: Agent-based Architecture for Autonomic Web Service Selection. In: 1st International Workshop on Web Services and Agent Based Engineering. IBM Corporation and NCSU (2003)
18. Michael, E.M., Munineder, P.S.: Agent Based Trust Model Involving Multiple Qualities. In: 4th Int. Joint Conf. on Autonomous Agents and Multi-agent Systems. IBM Corporation and NCSU (2005)
19. Watkins, C.J.C.H.: Learning from delayed rewards. PhD Thesis, University of Cambridge, England (1989)
20. Rehman, Z.U., Hussain, F.K., Hussain, O.K.: Towards Multi-Criteria Cloud Service Selection. In: Fifth International Conference on Innovative Mobile and Internet Services in Ubiquitous Computing, pp. 44–48. IEEE (2011)
21. Arkaitz, R., Marty, H.: An Automated Approach to Cloud Storage Service Selection. In: The Proceeding of Science Cloud 2011. ACM (2011), 978-1-4503-0699-7/11/06
22. Syed, A.Z., Aslam, M., Martinez-Enriquez, A.M.: Sentiment Analysis of Urdu Language: Handling Phrase-Level Negation. In: Batyrshin, I., Sidorov, G. (eds.) MICAI 2011, Part I. LNCS, vol. 7094, pp. 382–393. Springer, Heidelberg (2011)

A New Method of Privacy Preserving Computation over 2-Part Fully Distributed Data

The Dung Luong[1] and Dang Hung Tran[2]

[1] Academy of Cryotographic Techniques, Hanoi, Vietnam
thedungluong@gmail.com
[2] Hanoi National University of Education, Hanoi, Vietnam
hungtd@hnue.edu.vn

Abstract. In this paper, we propose a new protocol of privacy preserving frequency computation in 2-part fully distributed data (2PFD). This protocol are practical than of previous protocol. More specifically, we achieve a protocol that can be done in situations with various number of users and larger than a given threshold.

Keywords: Privacy Preserving, Data Mining, Secure Protocol, 2-Part Fully Distributed Data.

1 Introduction

In this paper, we study problem of privacy preserving frequency computation in 2-part fully distributed setting (2PFD) that exists in various practical applications of privacy preserving data mining (PPDM) as showed in [9]. In this problem, the data set is distributed across a large number of users, and each record is owned by two different users, one user only knows the values for a subset of attributes, while the other knows the values for the remaining attributes. A miner aims to compute the frequency of a tuple of values in the data set. Assume that each user's data includes some sensitive attribute values. To protect users' privacy and also enable learning frequency, our purpose is to design a protocol that enables the miner to learn frequency from all users'data without learning any individual's sensitive values.

Some randomization-based solutions proposed in [10], [3], [8], [2], [1], [13], [6] can be applied to solve our problem. The basic idea of these solutions is that every user perturbs its data, before sending it to the miner. The miner then can reconstruct the original data to obtain the mining results with some bounded error. These solutions allow each user to operate independently, and the perturbed value of a data element does not depend on those of the other data elements, but only on its initial value. Therefore, they can be used in various distributed data scenarios. Although these solutions are highly efficient, their use generally involves a tradeoff between privacy and accuracy, i.e., if we require more privacy, the miner loses more accuracy in the data mining results, and vice-versa. A secure frequency computation protocol in 2PFD proposed in [9], this protocol bases on cryptography method that are able to maintain

P. Meesad et al. (Eds.): *IC²IT2013*, AISC 209, pp. 115–123.
DOI: 10.1007/978-3-642-37371-8_15

strong privacy without loss of accuracy. However, a problem of the frequency computation protocol is that a single client may be able to disrupt the system. Indeed, if in the system with only a user does not send her input, the system crash and the result cannot be found. Although in the semi-honest model we assume that all clients will follow the protocol specification, this may be considered as a strong assumption, especially in large-scale protocols, many technical problems may prevent the participation of clients, such as the network failures. Note that a client does not know a priori who will participate in the protocol, so the obvious fix of constructing the general key values as a function of the number of active participants may not work. Our main objective is to propose a protocol that can be done in situations in which the number of users is various and larger than a given threshold. The achievements will give us more applicability to PPDM problems where Secure frequency computation is a key component such as naive Bayes learning, decision tree learning, association rules mining, etc. For example, considering a real scenario when a miner uses a web-application to investigate a large number of users for his research, a user only needs to use his browser to communicate with the server one time, while he does not have to communicate with the others. We can see that this scenario is quite popular in practice, and thus the proposed method is very significant to many other similar data mining applications.

In relating to our work, the secure multi-party computation problem (SMC) was first proposed by Yao [17] where he gave the method to solve Yao's Millionaire problem that allows comparing the worth of two millionaires without revealing any privacy information of each people. According to theoretical studies of Goldreich [11], the general SMC problem can be solved by the circuit evaluation method. However, using this solution is not practical in terms of the efficiency. Therefore, finding efficient problems specific solutions was seen as an important research direction. In the recent years, many specific solutions were introduced for the different research areas such as information retrieval, computational geometry, statistical analysis, etc. [7], [5], [16].

2 Preliminaries

2.1 Problem Statementjloo

We can re-define the problem of secure frequency computation in 2PFD setting as follows.

Definition 1. Assume that there are n pairs of users (U_i, V_i), each U_i has a binary number u_i and each V_i has a binary number v_i. Let Δ be a set of k pairs of users that is a defined threshold and $\Omega \subseteq \{1,2,...,n\}$ where $|\Omega| \geq k$. The secure frequency computation problem is to allow a miner to compute $f = \sum_{i \in \Omega} u_i v_i$ without disclosing any information about u_i and v_i. In other words, we need a secure protocol for constructing the following function:

$$(u_{i1}, v_{i1}, \ldots, u_{i|\Omega|}, v_{i|\Omega|}) \mapsto \sum u_i v_i$$

Where $u_{ij}, v_{ij} \in \Omega$, the definition notation implies that each pair U_i and V_i provide inputs u_i and v_i to the protocol, and the miner receive output $\sum u_i v_i$ without any other information.

2.2 Definition of Privacy

The privacy of the proposed protocol is based on the semi-honest security model [11]. Here we consider the possibility that some corrupted users share their data with the miner to derive the private data of the honest users. One requirement is that no other private information about the honest users be revealed. Thus the following definition is is similar to the definition in [9]. The difference is the assumption with only the set Ω of users participating to the protocol, hence the definition only requires protocol to be against the collusion of the miner with up to $2k-2$ users .

Definition 2. Assume that each user U_i has a private set of keys $D_i^{(u)}$ and a public set of keys $E_i^{(u)}$, and each user V_i has a private set of keys $D_i^{(v)}$ and a public set of keys $E_i^{(v)}$. A protocol for the above defined frequency mining problem protects each user's privacy against the miner along with t_1 corrupted users U_i and t_2 corrupted users V_i in the semi-honest model if, for all I_1, $I_2 \subseteq \Omega$ such that $|I_1| = t_1 (< k)$ and $|I_2| = t_2 (< k)$, there exists a probabilistic polynomial-time algorithm M such that

$$\{M(f, [u_i, D_i^{(u)}]_{i\in I_1}, [E_j^{(u)}]_{j\notin I_1}, [v_k, D_k^{(v)}]_{k\in I_2}, [E_l^{(v)}]_{l\notin I_2})\}$$
$$\stackrel{c}{\equiv} \{View_{miner,\{U_i\}_{i\in I_1},\{V_k\}_{k\in I_2}} [u_i, D_i^{(u)}, v_i, D_i^{(v)}]_{i\in\Delta}\}$$

where $\stackrel{c}{\equiv}$ denotes computational indistinguishability.

The computational indistinguishability is an important concept when discussing the security properties of distributed protocols [11]. Let $X = \{X_n\}_{n\in N}$ is an ensemble indexed by a security parameter n (which usually refers to the length of the input), where the $X_{i'}s$ are random variables.

Definition 3. Two ensembles, $X = \{X_n\}_{n\in N}$ and $Y = \{Y_n\}_{n\in N}$, are computational indistinguishable in polynomial time if for every probabilistic polynomial time algorithm A,

$$|Pr(A(X_n) = 1) - Pr(A(Y_n) = 1)|$$

is a negligible function in n. In such case, we write $X \stackrel{c}{\equiv} Y$.

2.3 ELGamal Encryption Scheme

In this paper, we use a variant of ELGamal encryption scheme together with the Shamir Secret sharing to build the secure frequency mining protocol. Before describing our protocol in the next section, we briefly review variant ElGamal encryption scheme [12] as follows.

Let G be a cyclic group of order q in which the discrete logarithms are hard. Let g be a generator of G, and x be uniformly chosen from $\{0,1,...,q-1\}$. In ElGamal encryption schema, x is a private key and the public key is $h = g^x$. Each user securely keeps their own private keys, otherwise public keys are publicly known. To encrypt a message M using the public key h, one randomly chooses k from $\{0,...,q-1\}$ and then computes the ciphertext $C = (C_1 = Mh^k, C_2 = g^k)$. The decryption of the ciphertext C with the private key x can be executed by computing $M = C_1(C_2^x)^{-1}$.

ElGamal encryption is semantically secure under the Decisional Diffie-Hellman (DDH) Assumption [4]. In ElGamal encryption scheme, one cleartext has many possible encryptions, since the random number k can take many different values. ElGamal encryption has a randomization property in which it allows computing a different encryption of M from a given encryption of M.

2.4 Shamir Secret Sharing

Secret sharing refers to any method by which a secret can be shared by multiple parties in such a way that no party knows the secret, but it is easy to construct the secret by combining some parties' shares.

Shamir secret sharing is a threshold scheme [15]. In Shamir secret sharing, there are n parties and a polynomial P of degree $k-1$ such that $P(0) = S$ where S is a secret. Each of the n parties holds a point in the polynomial P. Because k points (x_i, y_i) $(i = 1,..,k)$ uniquely define a polynomial P of degree $k-1$, a subset of at least k parties can reconstruct the secret S based on polynomial interpolation. But, fewer than k parties cannot construct the secret S. This scheme is also called (n,k) Shamir secret sharing.

3 Protocol

Assume that only a set Ω user pairs can take part into the protocol. In this section we expand the idea of threshold decryption system [14] for tackling the above problem. For a (n,k) threshold scheme, the basic idea is that a private key is shared among n users, so that only a set of k users can decrypt a ciphertext without explicitly reconstructing the private key. This can be obtained by using a (n,k)-Shamir secret sharing.

In the 2PFD setting, at the beginning of the protocol, we assume that two key seeds x_0 and $p_0 \in [1, q-1]$ are shared among n users U_i and n users V_i by a (n,k)-Shamir secret sharing. Shares owned by U_i and V_i are $x_i = f(i)$ and $p_i = h(i)$ respectively, where $f(x)$ and $h(x)$ are the random polynomials of degree (k-1)$\in Z_q$ such that $f(0) = x_0$ and $h(0) = p_0$. Thus, each user U_i has the key pair (x_i, $X_i = g^{x_i}$) and V_i has (p_i, $P_i = g^{p_i}$). In our protocol, $H = g^{x_0+p_0}$ is announced as the general public key. We can use the Feldman Verifiable Secret Sharing (VSS) among n parties with threshold k to obtain this sharing. The advantage of the Feldman VSS is that the commitments allow us to do various computations related to the secret shares without compromising privacy. The protocol is presented as follows.

- **Phase 1.** Each user U_i does as follows:
  - Randomly choose k_i from $\{1,...,q-1\}$.
  - Compute $C^{(i)} = (C_1^{(i)}, C_2^{(i)}) = (g^{u_i} X_i^{k_i}, g^{k_i})$
  - Send $C^{(i)}$ to the miner.
- **Phase 2.** Each user V_i does the follows:
  - Get $C^{(i)}$ from the miner,
  - Randomly choose r_i and q_i from $\{1,...,q-1\}$,
  - if $v_i = 0$ then compute $R^{(i)} = (R_1^{(i)}, R_2^{(i)}, R_3^{(i)}) = (X_i^{r_i} H^{q_i}, g^{r_i}, g^{q_i})$
  - if $v_i = 1$ then compute

  $R^{(i)} = (R_1^{(i)}, R_2^{(i)}, R_3^{(i)}) = (g^{u_i} X_i^{r_i+k_i} H^{q_i}, g^{r_i+k_i}, g^{q_i})$
  - Send $R^{(i)}$ to the miner.
- **Phase 3.** Each user U_i does as follows:
  - Get $R^{(i)}$ from Miner.
  - Randomly choose y_i from $\{1,...,q-1\}$,
  - Compute $K^{(i)} = (K_1^{(i)}, K_2^{(i)}) = (R_1^{(i)} (R_2^{(i)})^{-x_i} H^{y_i}, R_3^{(i)} g^{y_i})$.- Send $K^{(i)}$ to Miner.
- **Phase 4.** Miner computes $K = \prod_{i \in S} K_2^{(i)}$

• **Phase 5.** The users does as follows:

- Each U_i computes $a_i = K^{x_i}$ and sends a_i to Miner
- Each V_i computes $b_i = K^{p_i}$ and sends b_i to Miner

• **Phase 6.** Miner does as follows:

- Compute $K' = \prod_{t\in T}(a_t b_t)^{\prod_{j\in T, j\neq t}\frac{-j}{t-j}}$
- Compute $d = \dfrac{\prod_{i=1}^{n} K_1^{(i)}}{K'}$.
- Find f from $\{0,1,\ldots,n\}$ that satisfies $g^f = d$
- Output f.

3.1 Proof of Correctness

Theorem 1. The above presented protocol correctly computes the frequency value $f = \sum_{i\in\Omega} u_i v_i$ as Definition 1.

Proof. We show that the miner can compute the desired value f by using the above protocol. Indeed, the first phase of the improved frequency mining protocol implements as the previous protocol (in [9]). Difference from Phase 2 and Phase 3 of the previous protocol, the private keys y_i and q_i of the improved protocol are temp keys that are chosen at the encrypting time. Note that general key Y at Phase 2 and Phase 3 in the previous protocol are replaced by g in this protocol. Before decryption process at Phase 5, the miner computes the product of all the second components at Phase 4 that it received at Phase 3, this product value is $g^{\sum_{i\in\Omega} y_i + q_i}$.

Thus, using Lagrange interpolation formula, with only the set Δ of k user pars involves in protocol, the miner can compute:

$$K = \prod_{t\in\Delta}\left(\prod_{i\in S} K_2^{(i)}\right)^{(x_t+p_t)\prod_{j\in\Delta, j\neq t}\frac{-j}{t-j}}$$

$$= g^{(x_0+p_0)\sum_{i\in\Omega} y_i + q_i}$$

The product of all the first components that miner received at Phase 4 is,

$$K' = \prod_{i\in\Omega} K_1^{(i)} = g^{(x_0+p_0)\sum_{\in\Omega} y_i+q_i}$$

Thus, it is clear that $d = \frac{K'}{K} = g^{\sum_{i\in\Omega} u_i v_i}$. We can obtain f from the equation $d = g^f = g^{\sum_{i\in\Omega} u_i v_i}$.

Note that, in practice, the value of f is not too large, so that the discrete logarithms can be successfully taken (for example $f = 10^5$).

3.2 Proof of Privacy

Theorem 2. The protocol in Subsection III preserves the privacy of the honest users against the miner and up to $2k-2$ corrupted users.

Proof. We show that under the DDH assumption, our protocol preserves each user's privacy in the semi-honest model, and in the case of collusion of some corrupted users with the miner, the protocol still preserves the privacy of each honest user.

In our model, the communication only occurs between each user and the miner, thus the miner receives the messages of all users. Assume that each user can get the messages of the remaining users via the miner, then the information known by the miner and each user are the same during the execution of the protocol. Therefore, it is sufficient to only consider the view of the miner, as follow:

In Phase 1, the miner receives the messages $C_1^{(i)}$, $C_2^{(i)}$ of each U_i. Here $(C_1^{(i)}, C_2^{(i)})$ is an ElGamal encryption of the value g^{u_i} under the private/the public key pair (x_i, X_i), and the value k_i is randomly chosen from $\{1,2,...,q-1\}$.

In Phase 2, the messages $R_1^{(i)}, R_2^{(i)}$ and $R_3^{(i)}$ sent by each V_i are equivalent to An ElGamal encryption $(\alpha H^{q_i}, g^{q_i})$ and the random number g^{β}. Here $\alpha = X_i^{r_i}$ or $g^{u_i} X_i^{r_i+k_i}$; $\beta = r_i$ or $r_i + k_i$. Note that q_i, r_i and k_i are randomly chosen from $\{1,2,...,q-1\}$.

In Phase 3, the messages $K_1^{(i)}$ and $K_2^{(i)}$ sent by each U_i can be represented as An ElGamal encryption.

As well known, the ElGamal encryption is semantically secure under the DDH assumption. So, the view of the miner can be efficiently simulated by a simulator for ElGamal encryptions.

Note that no single participant learns $x_0 + p_0$, but it is only computationally hidden from Discrete Logarithm problem. Moreover, any users can create a simulator for the joint view of any $2k-2$ other users with the miner, this follows immediately from the proof of Theorem 2 in. Thus, this protocol preserves privacy of each user gainst up to $2k-2$ corrupted users.

4 Complexity Estimation

In the improved protocol, the computational cost of each user U_i and V_i from phase 1 to phase 3 is the same as the previous protocol. However, the improved protocol requires that at least k pair of users have to involve in computing the parameter K at Phase 4. Thus the computational complexity of these users increases a modular exponentiation. The computational complexity for miner is nearly equal to the previous protocol. Finally, in the proposed protocol, the total computational cost of each user U_i is 8 modular exponentiations. The computational cost of each user V_i is at most 5 modular exponentiations. The miner uses is $4n$ modular multiplications and at most n comparisons in the third phase.

5 Conclusion

We proposed an improved protocol for secure frequency computation in 2-part fully distributed setting that allows the miner to be able to computing frequencies in situations with various number of users and larger than a given threshold. In the future we will develop this protocol to use in real data mining applications.

References

1. Agrawal, D., Aggarwal, C.C.: On the design and quantification of privacy preserving data mining algorithms. In: Proceedings of The Twentieth ACM SIGMOD-SIGACT-SIGART Symposium on Principles of Database Systems, pp. 247–255 (2001)
2. Agrawal, R., Srikant, R., Thomas, D.: Privacy preserving OLAP. In: SIGMOD 2005: Proceedings of the 2005 ACM SIGMOD International Conference on Management of Data, pp. 251–262 (2005)
3. Agrawal, S., Haritsa, J.R.: A Framework for High-Accuracy Privacy-Preserving Mining. In: Proceedings of the 21st International Conference on Data Engineering, ICDE 2005, pp. 193–204. IEEE Computer Society (2005)
4. Boneh, D.: The Decision Diffie-Hellman Problem. In: Buhler, J.P. (ed.) ANTS 1998. LNCS, vol. 1423, pp. 48–63. Springer, Heidelberg (1998)
5. Clifton, C., Kantarcioglu, M., Vaidya, J., Lin, X., Zhu, M.Y.: Tools for privacy preserving distributed data mining. SIGKDD Explor. Newsl. 4(2), 28–34 (2002)
6. Dowd, J., Xu, S., Zhang, W.: Privacy-preserving decision tree mining based on random substitutions. In: Müller, G. (ed.) ETRICS 2006. LNCS, vol. 3995, pp. 145–159. Springer, Heidelberg (2006)

7. Du, W., Chen, S., Han, Y.S.: Privacy-Preserving Multivariate Statistical Analysis: Linear Regression and Classification. In: Proceedings of the 4th SIAM International Conference on Data Mining, pp. 222–233 (2004)
8. Du, W., Zhan, Z.: Using randomized response techniques for privacy-preserving data mining. In: Proceedings of the ninth ACM SIGKDD International Conference on Knowledge Discovery and Data Mining, KDD 2003, pp. 505–510 (2003)
9. Dung, L.T., Bao, H.T.: Privacy Preserving Frequency Mining in 2-Part Fully Distributed Setting. IEICE Transactions on Information and Systems E93-D(10), 2702–2708 (2010)
10. Evfimievski, A., Srikant, R., Agrawal, R., Gehrke, J.: Privacy preserving mining of association rules. In: Proceedings of the Eighth ACM SIGKDD International Conference on Knowledge Discovery and Data Mining, pp. 217–228 (2002)
11. Goldreich, O.: The Foundations of Cryptography. In: General Cryptographic Protocols, ch. 7, vol. 2. Cambridge University Press (2004)
12. Hirt, M., Sako, K.: Efficient Receipt-Free Voting Based on Homomorphic Encryption. In: Preneel, B. (ed.) EUROCRYPT 2000. LNCS, vol. 1807, pp. 539–556. Springer, Heidelberg (2000)
13. Kargupta, H., Datta, S., Wang, Q., Sivakumar, K.: On the Privacy Preserving Properties of Random Data Perturbation Techniques. In: Proceedings of the Third IEEE International Conference on Data Mining, ICDM 2003, p. 99. IEEE Computer Society (2003)
14. Noack, A., Spitz, S.: Dynamic Threshold Cryptosystem without Group Manager. Network Protocols and Algorithms 1(1), 108–121 (2009)
15. Shamir, A.: How to share a secret. Commun. ACM 22(11), 612–613 (1979)
16. Vaidya, J., Kantarciouglu, M., Clifton, C.: Privacy-preserving Naive Bayes classification. The VLDB Journal 17(4), 879–898 (2008)
17. Yao, A.C.-C.: How to generate and exchange secrets. In: Proceedings of the 27th Annual Symposium on Foundations of Computer Science, SFCS 1986, pp. 162–167. IEEE Computer Society (1986)

Enhancing the Efficiency of Dimensionality Reduction Using a Combined Linear SVM Weight with ReliefF Feature Selection Method

Wipawan Buathong and Phayung Meesad

Faculty of Information Technology,
King Mongkut's University of Technology North Bangkok, Thailand
wipawan.buathong@gmail.com,
pym@kmutnb.ac.th

Abstract. The purpose of this research is to propose a feature selection technique for improving the efficiency of dimensionality reduction. The proposed technique is based on a combined Linear SVM Weight with ReliefF. SVM is used as a classifier. The Leukemia and DLBCL dataset from UCI Machine learning Repository were used for our experiments. We discovered that the combined Linear SVM Weight with ReliefF feature selection technique could provide 100 percent accurate result for the model. There was a significant reduction from 5,147 to 20 dimensional data, which is much more efficient than using Linear SVM Weight or ReliefF alone.

Keywords: Dimensionality Reduction, Feature Selection, ReliefF, Linear SVM Weight, ReliefF-Linear SVM Weight.

1 Introduction

Dimensionality Reduction is about converting data of very high dimensionality into data of much lower dimensionality. It is the process for reducing the number of random variables under considerations, which is considered as an effective downsizing data approach. There are two approaches that have been generally used for dimensionality reduction: 1) feature selection [1-6] (a process for selecting an optimal subset of features or attributes in accordance with an objective function) and 2) feature extraction (a process of creating new features from functions of the original features and use them as representatives for the original features, e.g., sampling and clustering) [7]. The least possibility of important data loss and the accuracy of data classification are two key objectives of dimensionality reduction. Since different data dimensions are not equally important, it is necessary to select an efficient feature selection technique to ensure that the key objectives of dimensionality reduction will not be much affected. Data dimensions downsized from the efficient feature selection technique can be generalized for the majority of data.

P. Meesad et al. (Eds.): *IC²IT2013*, AISC 209, pp. 125–134.
DOI: 10.1007/978-3-642-37371-8_16

ReliefF and Linear SVM are prominent feature selection algorithms. The ReliefF technique is well recognized in its noise-resistance and capability in dealing with different data classes, whereas the Linear SVM technique is acknowledged among research scholars as the technique, which is applicable to all classifiers, including SVM. While a single feature selection technique is widely used for dimensionality reduction, It is assumed that the least possibility of important data loss from dimensionality reduction and may not be provided by only one single feature selection technique. The purpose of this research is to propose an alternative feature selection technique based on a combination of the Linear SVM Weight and ReliefF feature selections with the SVM classifier for improving the efficiency of dimensionality reduction. The efficiency of the SVM Weight, the ReliefF, and the combined SVM Weight with ReliefF feature selections for dimensionality reduction will be measured and compared. The SVM classifier will be used as a feature selection criterion for improving the efficiency of dimensionality reduction.

2 Literature Review

According to Wongkot et al. [2], the efficiency of the Information Gain and Chi Squared feature selection techniques was measured. The two feature selection techniques were used for document classification. The Naïve Bayesian, Support Vector Machine, and Decision Tree classifiers were used together with the feature selection techniques. In the comparison, it was found that the Information Gain was more effective than the Chi Squared. Among the classifiers, Support Vector Machine was discovered by the authors as the most efficient document classifier. In Abdolhossein et al.[8] the ReliefF algorithm was used as a feature selection for Content-Based Image retrieval. It was found that the data retrieval performance using the ReliefF algorithm was faster and more accurate than other algorithms. Xin Jin et al. [9] conducted the experiment on Web Mining by using the Hidden Naïve Bayes classifier and the ReliefF feature selection. It was discovered that the classification accuracy rate is higher. In Yi Zhang et al. [5] the ReliefF and mRMR were used together for gene selection. The result showed that the classification accuracy rate is much higher than a single feature selection and reduced to 30 dimensional data. Based on previous works, they are based on a single feature selection technique for dimensionality reduction, which may not be efficient enough in excluding redundant and irrelevant features. Some important dimensionality may not be selected when a subset is chosen.

A number of classifiers have been proposed from time to time. However, Support Vector Machine (SVM) has been considered by a number of research scholars as the most efficient classifier [1-2], [10]. In Buathong [11], the efficiency of feature selection ranking techniques for dimensionality reduction was measured. The Information Gain, Gain Ratio and Linear SVM Weight ranking techniques were measured using four classifiers, including k-NN, Naïve Bayes, SVM and Classification Tree. It was found that the Linear SVM Weight feature selection technique with the SVM classifier was the most efficient method in providing the best result for dimensionality reduction.

Support Vector Machine (SVM) [12] is a data classifier algorithm that has been applied in various disciplines. It takes a set of input data and applies a simple linear method to the data but in a high dimensional feature space, which is non-linearly related to the input space. The SVM algorithm is consisted of support vectors (training samples), which are the data points that lie closest to the decision surface. The decision function is specified by a subset of support vectors. Two dimensions of data are separated by a linear (hyperplane). The SVM also uses a kernel function that corresponds to a dot product of two feature vectors in some expanded feature space that aims to minimize errors and maximize the margin around the separating hyperplane. The original input space can be mapped to some higher-dimensional feature space where the training set is separable. It is different from other techniques such as Artificial Neural Network (ANN) that aims to minimize the possibility of predictive errors only.

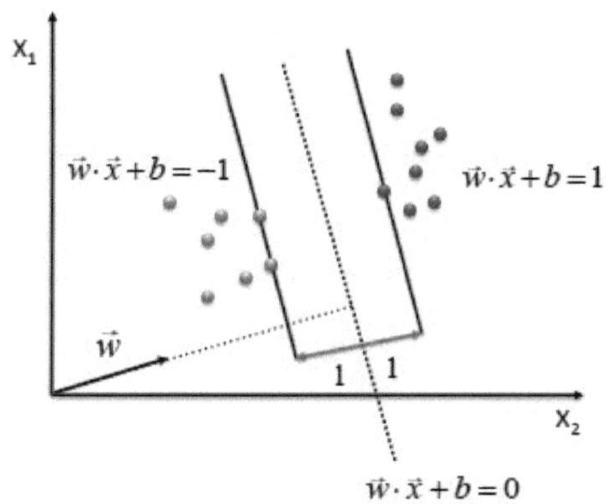

Fig. 1. Two groups of data divided by the linear SVM

There are some occasions that two groups of data cannot be divided using the linear SVM because data are clustered in different positions. There is a need for appropriate tools and techniques for ranking data in higher dimension space. In this case, multidimensional linear classifiers are considered as more efficient than general methods in data classification. The Polynomial kernel classifier is generally used to calculate a linear classifier higher than two degree. The Radial Basis Function Kernel classifier (RBFKC) uses C as a variable for balancing point for measuring the best classification range and the least potential error rate. Our research will be based on the RBFC. The RBFKC equation is listed below.

$$K(\vec{x}_i, \vec{x}_j) = \exp\,(-|\vec{x}_i - \vec{x}_j|\,\sigma^2) \tag{1}$$

Linear SVM Weight is a feature ranking that can be applied to all types of Classifier [6].

Algorithm 1. Feature Ranking Based on Linear SVM Weights
Input: Training sets, (X_i, Y_i), i =1,.....m.
Output: Sorted feature ranking list.

1. Use grid search to find the best parameter C.
2. Train a L2-loss linear SVM model using the best C.
3. Sort the features according to the absolute values of weights in the model.

The ReliefF algorithm was invented by Kira and Rendell in 1992. The algorithm is one of the most successful algorithms for evaluating the quality of features due to its simplicity and effectiveness [10]. In addition, the algorithm is well recognized among researchers in its robustness against noise. The algorithm is mainly used for calculating weight (W) of data feature in reference to random data (R). Given that A, B, and C are three data classes and R is allocated to A, a variable H will be assigned to for the nearest data to A. M_b and Mc will be assigned for the nearest data to B and C respectively. The ReliefF feature selection technique can be applied to the data that contains more than two classes [9].

Algorithm Relief
Input: for each training instance a vector of attribute values and the class value
Output: the vector W of estimations of the qualities of attributes

1. set all weights W[A] :=0.0;
2. $diff(A, dataX, dataY) = \frac{X(A)-Y(A)}{\max(A)-\min(A)}$
3. **for** i:=1 **to** m **do begin**
4. randomly select an instance
5. find k nearest hits H and nearest miss M for each class not class of R
6. **for** A :=1 **to** #all attribute **do**
7. $w[A] := w[A] - \frac{diff(A,R,H)}{m} + \sum_{C \neq class(R)}(P(C) * \frac{diff(A,R,Mc)}{m})$;
8. **end;**

Variables in the ReliefF algorithm can be explained as:
Where:-

m	:	the number of randomly sampled instance
R	:	a randomly sampled instance
H	:	the nearest hit
M_c	:	the nearest miss with C class such that $C(C \neq class(R))$
$w[A]$	:	the weight of feature A
$X(A)$ and $Y(A)$	:	Data X and Y with A feature.
$P(C)$	:	the possibility of getting data with C class

3 Methodology

The methodology used in our research consisted of four steps below.

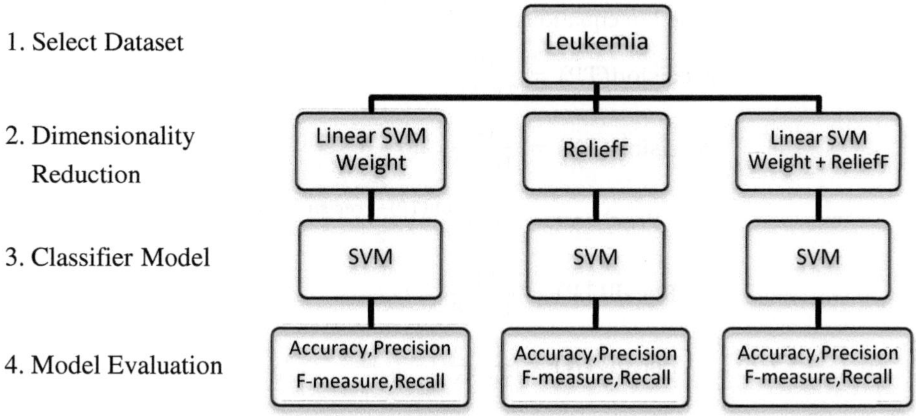

Fig. 2. Research Methods

(1) Select Benchmarking Dataset. The Leukemia and DLBCL dataset from UCI database were used with two data classes, including ALL and AML and DLBCL and FL respectively. There were no missing values in the selected dataset. Table 1 represents details of the dataset.

Table 1. Details of the dataset used in the research

Dataset	Attribute	Instance	Class
Leukemia	5,147	72	- ALL - AML
DLBCL	7,070	77	- DLBCL - FL

(2) Reduce data dimensionality using selected feature selection techniques. The ReliefF, Linear SVM Weight, and the combined Linear SVM Weight and ReliefF techniques will be used for dimensionality reduction.

(3) Apply the SVM classifier. The SVM classifier will be used together with each feature selection technique for dimensionality reduction. The data that has been downsized will be used to construct a classifier model.

(4) Evaluate the efficiency of feature selection techniques. The dataset is classified into two parts: 1) training set (60%) and 2) test set (40%). These two parts of the dataset will be used as a model for measuring the accuracy of data classification. The ten folds cross validation will be applied to each set of data. The measurement metric is consisted of data dimensions and performance evaluation criteria. Data dimensions

are classified into different levels from 10 to 50 data dimensions. In the meantime, Accuracy, F-Measure, Precision, and Recall are the performance evaluation criteria. The Accuracy of each feature selection technique for dimensionality reduction will be calculated after Precision, Recall, and F-Measure have already been calculated respectively (represented in equations 2 to 8).

$$\text{Precision(TP)} = \frac{\text{TP}}{\text{TP+FP}} \quad (1)$$

$$\text{Precision(TN)} = \frac{\text{TN}}{\text{TP+FP}} \quad (2)$$

$$\text{Precision} = \frac{\text{Precision(TP)+Precision(TN)}}{2} \quad (3)$$

$$\text{Recall(TP)} = \frac{\text{TP}}{\text{TP+FN}} \quad (4)$$

$$\text{Recall(TN)} = \frac{\text{TN}}{\text{TP+FN}} \quad (5)$$

$$\text{Recall} = \frac{\text{Recall(TP)+Recall(TN)}}{2} \quad (6)$$

$$\text{F-measure} = \frac{\text{2 x (Recall x Precision)}}{\text{Recall+Precision}} \quad (7)$$

$$\text{Accuracy} = \frac{\text{TP+TN}}{\text{TP+TN+FP+FN}} \quad (8)$$

Where

TP : Number of True positives
TN : Number of True negatives
FP : Number of False positives
FN : Number of False negatives

4 Results

For the performance of Linear SVM Weights Ranking Method with the SVM classifier, we found that all four key criteria of performance evaluation reached 100% at 20 data dimensions for the Leukemia dataset. Nevertheless, precision was the only one performance evaluation criterion reaching 100% at 40 data dimensions for the DLBCL dataset. Table 2 represents the performance of using the Linear SVM Weights ranking method for dimensionality reduction.

For the performance of the ReliefF Ranking Method with the SVM classifier, we found that the accuracy of the ReliefF ranking method was 96.90% at 30, 40, and 50 data dimensions. In the meantime, other performance evaluation criteria were also the same at 30, 40, and 50 data dimensions, respectively (97.60% for F-Measure, 98.92% for Precision, and 96.32% for Recall) for the Leukemia dataset. When the same technique was applied to the DLBCL dataset, we discovered that the accuracy and precision of the ReliefF feature selection technique were below 90% at all data dimensions. Table 3 shows the performance of using ReliefF ranking method for dimensionality reduction.

Table 2. The performance of the Linear SVM Weights algorithm for dimensionality reduction

Data Dimensions	Performance of the Linear SVM Weights Ranking Method (%)							
	Leukemia				DLBCL			
	Accuracy	F-Measure	Precision	Recall	Accuracy	F-Measure	Precision	Recall
50	99.31	99.48	98.96	100.00	95.81	97.12	99.10	95.22
40	99.31	99.48	98.96	100.00	95.16	96.63	100	93.48
30	98.62	98.95	98.95	98.95	94.19	95.96	99.07	93.04
20	100.00	100.00	100.00	100.00	95.81	97.10	99.54	94.78
10	97.24	97.91	97.40	98.42	95.16	96.66	99.09	94.35

Table 3. The performance of the ReliefF algorithm for dimensionality reduction

Data Dimensions	Performance of ReliefF Ranking Method (%)							
	Leukemia				DLBCL			
	Accuracy	F-Measure	Precision	Recall	Accuracy	F-Measure	Precision	Recall
50	96.90	97.60	98.92	96.32	85.81	90.83	87.20	94.78
40	96.90	97.60	98.92	96.32	87.42	91.86	88.35	95.65
30	96.90	97.60	98.92	96.32	84.84	90.39	85.33	96.09
20	95.52	96.57	96.83	96.32	82.26	88.84	83.27	95.22
10	95.52	96.57	96.83	96.32	78.81	87.01	79.50	96.09

For the performance of the combined Linear SVM Weights with ReliefF Ranking feature selections for the SVM classifier, we found that all four performance evaluation criteria of the combined Linear SVM Weights with ReliefF Ranking Method reached 100% for 20 and 40 dimensions of data. In the meantime, the combined Linear SVM Weights and ReliefF method was also efficient at 50 data dimensions where the accuracy was 99.66%. When the combined method was experimented on the DLBCL dataset, there was no single key evaluation criterion reaching 100% at all data dimensions. Particularly, all four evaluation criteria at 10 data dimensions were significantly lower than those at other data dimensions (92.26% for accuracy, 94.76% for f-measure, 95.18% for precision, and 94.35% for recall). Table 4 shows the performance of the combined Linear SVM Weights and ReliefF ranking methods.

Table 4. The performance of the combined Linear SVM Weight and ReliefF algorithm for dimensionality reduction

Data Dimensions	Performance of the Linear SVM Weights + ReliefF Ranking Method (%)							
	Leukemia				DLBCL			
	Accuracy	F-Measure	Precision	Recall	Accuracy	F-Measure	Precision	Recall
50	99.66	99.74	99.48	100.00	96.77	97.80	99.11	96.52
40	100.00	100.00	100.00	100.00	97.74	98.47	98.69	98.26
30	98.62	98.95	98.95	100.00	98.06	98.69	99.12	98.26
20	100.00	100.00	100.00	100.00	98.39	98.91	99.56	98.26
10	98.28	98.70	97.44	100.00	92.26	94.76	95.18	94.35

After all three feature selection methods had been measured; we discovered that the combined Linear SVM Weight and ReliefF was the most efficient feature selection technique in comparison with others for all data dimensions. Table 5 shows the accuracy comparison results.

Table 5. The accuracy comparison of feature selection techniques for dimensionality reduction

Feature Selection	The Accuracy of each Data Dimensions (%)									
	Leukemia					DLBCL				
	50	40	30	20	10	50	40	30	20	10
Linear SVM Weights	99.31	99.31	98.62	100	97.24	95.81	95.16	94.19	95.81	95.16
ReliefF	96.90	96.90	96.90	95.52	95.52	85.81	87.42	84.84	82.26	78.71
Linear SVM Weights + ReliefF	99.66	100	98.62	100	98.28	96.77	97.74	98.06	98.39	92.26

When the accuracy performance of each feature selection technique applied to the Leukemia dataset was graphed, there were significant differences between the combined Linear SVM with ReliefF technique and the ReliefF technique alone. However, there were no significant differences between the combined Linear SVM with ReliefF technique and the Linear SVM technique alone. Figure 3 shows the accuracy differences of three ranking methods.

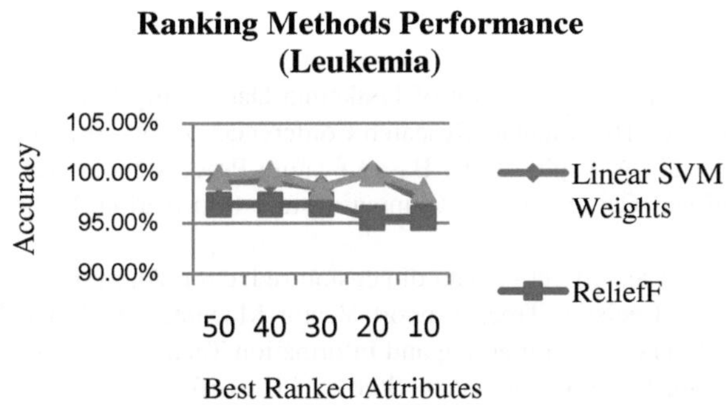

Fig. 3. The accuracy performance of ranking methods (Leukemia)

For the accuracy performance graphed from the DLBCL dataset, there were also significant differences between the combined Linear SVM with ReliefF technique and the ReliefF technique alone. Differences between the combined method and the Linear SVM alone were noticeable from 10 to 50 data dimensions. Although the combined method was more accurate than the Linear SVM technique alone at 20, 30, and 40 data dimensions, it was not comparable to the Linear SVM technique alone at 10 data dimensions.

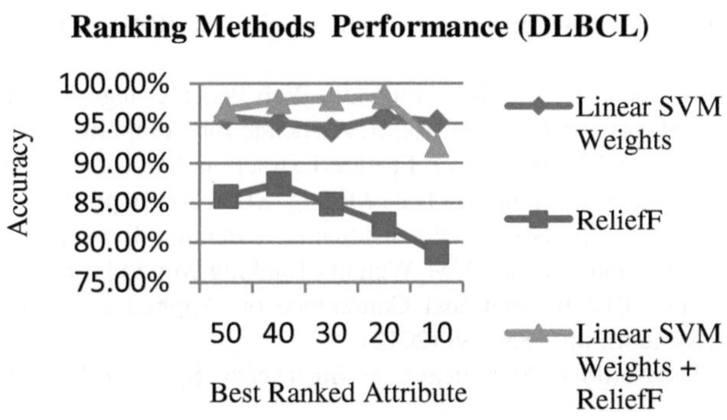

Fig. 4. The accuracy performance of ranking methods (DLBCL)

5 Conclusions

Based on our experimental findings, data dimensions could be downsized to 20 data dimensions with all performance evaluation criteria at 100% for the Leukemia dataset. At the same data dimensions for the DLBCL dataset, the combined method also showed satisfactory results for all performance evaluation criteria, which were higher than those in other feature selection techniques. Although we can summarise that using the combined Linear SVM Weight with ReliefF ranking method is more efficient than using Linear SVM Weight or ReliefF alone, the accuracy performance of the combined method at 10 data dimensions is something not ignorable.

References

1. Saengsiri, P., et al.: Classification of Leukemia Data Using Ranking and Support Vector Machine. In: The 11th Graduate Research Conference, pp. 30–38 (2010)
2. Sriurai, W., et al.: A Topic-Model Based Feature Processing for Text Categorization. In: The 5th National Conference on Computing and Information Technology, pp. 146–151 (2009)
3. Buathong, W., et al.: Effect of data dimension reduction Analysis and Data Classification Performance of Decision Tree. Support Vector Machine and Naïve Bayes. In: The 8th National Conference on Computing and Information Technology, pp. 568–573 (2012)
4. Pongpatharakan, P.: A Comparative Study of Classification Properties between CART. SVM, C5.0 and Hybrid methods. In: The 5th National Conference on Computing and Information Technology, pp. 1102–1106 (2009)
5. Zang, Y., Ding, C., Li, T.: A Two-Stage Gene Selection Algorithm by Combining ReliefF and mRMR. In: 7th IEEE International Symposium on Bioinformatics and Bioengineering, pp. 164–171. IEEE Press, New York (2007)
6. Chang, Y.N., Lin, C.J.: Feature Ranking Using Linear SVM. In: Workshop and Conference Proceedings, pp. 53–64 (2008)
7. Afizi, M., Shukram, M., Zakaria, O., Wahid, N., Mujahid, A.: A Classification Method for Data Mining Using Svm-Weight And Euclidean Distance. Australian Journal of Basic and Applied Sciences, 2053–2059 (2011)
8. Sarrafrzadeh, A., Atabay, H.A., Pedram, M.M., Shanbehzadeh, J.: ReliefF Based Feature Selection In Content-Based Image Retrieval. In: Proceeding of International MultiConference of Engineers and Computer Scientists (IMEC), Hong Kong, pp. 19–22 (2012)
9. Jin, X., Li, R., Shen, X., Bie, R.: Automatic Web Pages Categorization with ReliefF and Hidden Naïve Bayes, pp. 617–621. ACM: Associate for Computing Machinery (2007)
10. Sun, Y., Wu, D.: A RELIEF Based Feature Extraction Algorithm. In: Proceedings of the SIAM International Conference on Data Mining (SDM 2008), pp. 188–195 (2008)
11. Buathong, W.: A comparison of Dimensionality Reduction Techniques Using Information Gain. Gain Ratio and Linear SVM Weights Ranking Methods. In: 5th ACTIS National Conference and 2012 International Conference on Applied Computer Technology and Information Systems, pp. 185–189 (2012)
12. Vapnik, V.: The Nature of Statistical Learning Theory. Springer, New York (1995)

Dhaka Stock Exchange Trend Analysis Using Support Vector Regression

Phayung Meesad and Risul Islam Rasel

Faculty of Information Technology
King Mongkut's University of Technology North Bangkok, Thailand
pym@kmutnb.ac.th, rasel.kmutnb@gmail.com

Abstract. In this study we combine support vector machine (SVM) and windowing operator in order to predict share market trend as well as the share price. The instability of the time series data is one of the main reasons to lead to decrease of prediction accuracy in this analysis. On the other hand, some special SVM parameters such as c, ε, g should be carefully determined to gain high accuracy. In order to solve this problem mentioned above we use windowing operator as preprocess in order to feed the highly reliable input to SVM model. And train the model in iterative process such that we can find out the best combination of SVM parameters. This study is done on some listed company of Dhaka stock exchange (DSE), Bangladesh. And the training and testing data sets are real time values are collected from DSE. Four years historical data (2009-2012) are used in this analysis. And finally, we compare the output with the real time trend from DSE.

Keywords: SVM, Windowing operator, Stock market, Time series data.

1 Introduction

Stock market is the emerging sector in any country of the world now. Many people are directly related to this sector. It is important for the people who are directly related to the market to gain insight about the market trend. So, along with the development of stock market, forecasting stock has become an important topic among the people. Trend forecasting becomes an essential topic for stockholders, investors and the authority that are related to the stock market business.

Predicting stock price is regarded as a challenging task [1]. Stock market is essentially a non-linear, non-parametric, noisy and deterministically chaotic system [2][3][4]. Trend of a market depends on many things like liquid money, stocks, Human behavior, news related to stock market etc. All this together control the trend of a stock market. The goal of predicting market trend is to make assumption about the price of assets in stock market. The behavior of trend can be analyzed by using technical tools, parametric pricing methods or combination of these methods [5].

Neural Networks (NNs) and Support Vector Machines (SVMs) [6][7][8][9] are both standard, mature machine learning approaches with applications in prediction based on times series data. Many research works have been done before using these

P. Meesad et al. (Eds.): *IC²IT2013*, AISC 209, pp. 135–143.
DOI: 10.1007/978-3-642-37371-8_17

two techniques. In some recent researches, researchers have found that SVM can produce more accurate stock prediction than the NNs can do. Support Vector Regression (SVR) is the most common application form of SVMs [10][11]. One of the main characteristics of Support Vector Regression (SVR) is that instead of minimizing the observed training error, SVR attempts to minimize the generalized error bound so as to achieve generalized performance. This generalization error bound is the combination of the training error and a regularization term that controls the complexity of the hypothesis space [12][13]. Kim [1], Lai and Liu [3] showed in their research that SVM can be more accurate in producing result if we can chose the best combination of SVM parameters. Ince & Trafalis [2] showed the way how Kernel Principal Component can be analyzed to get better result. Lai and Liu [3] compare the prediction performances of NN and SVM in predicting exact stock prices on the Hang Seng Index (HSI) over 5 days and a 22 days horizon. As preprocessing or input selection techniques for SVR and NN, they used 15 days Exponential Moving Average (EMA15) and relative difference in percentage of price (RDP) RDP-5, RDP-10, RDP-15, and RDP-20. The best MAPE result was 0.8 for the year of 2008 short term forecast (5 days). For long term prediction (22 days), the best result was 4.33 also for the 2008 dataset. The average result was 5.02.

This paper consists of five sections. Section 2 introduces the basic concept of SVM and windowing operator. Section 3 is about the research design. Section 4 is the experiment and analysis. Section 5 is the conclusions and limitations of this study.

2 Basic Concept of Proposed Model

2.1 SVM Regression

SVM regression perform linear regression in the high dimension feature space using ε – insensitivity loss and, at the same time tries to reduce model complexity by minimizing $||\omega||^2$. This can be described by introducing slack variables ξ_i and ξ_i^* where$i = 1, \dots, n$, to measure the deviation of training sample outside ε- sensitive zone [3][9][13].

$$\frac{1}{2}||\omega||^2 + C\sum_{i=1}^{n}(\xi_i + \xi_i^*) \tag{1}$$

$$\text{Min}\begin{cases} y_i - f(x_{i,\omega}) \le \varepsilon + \xi_i^* \\ f(x_i,\omega) - y_i \le \varepsilon + \xi_i \\ \xi_i, \xi_i^* \ge 0, i = 1, \dots, n \end{cases} \tag{2}$$

This optimization problem can transform into the dual problem and solution is given by

$$f(x) = \sum_{i=1}^{n_{sv}}(\alpha_i - \alpha_i^*)K(x_i, x) \tag{3}$$

$$\text{Subject to, } 0 \le \alpha_i^* \le C, 0 \le \alpha_i \le C,$$

Where n_{sv} is the number of support vector (SVs) and the kernel function

$$K(x, x_i) = \sum_{j=1}^{m} g_j(x) g_j(x_i) \tag{4}$$

SVM generalization performance depends on a good setting of kernel parameters C, ε and kernel parameters [9][13].

The following formula is the evaluation of the predicted value [3].

$$MAPE = 100 \frac{\sum_{i=1}^{n} |\frac{A-P}{A}|}{n} \tag{5}$$

Mean Absolute Percentage Error (MAPE) which is the measure of accuracy in a fitted time series value in statistics, specifically trending. A and P are the real close value and the predicted close value respectively and n is the time frame or number of days.

3 Research Design

3.1 Research Data

To conduct the study and verify the effectiveness of our proposed model we collect the data set from Dhaka stock exchange (DSE), Bangladesh. We separate our data set into two groups. Training data set contains data from year 2009 to 2011 (700 data) and testing data set contains data from year 2012 (124 data). There are 4 attributes of this data set. They are open price, high price, low price and close price. One special attribute is date field used as ID field for this study. Figure 1 shows the actual close

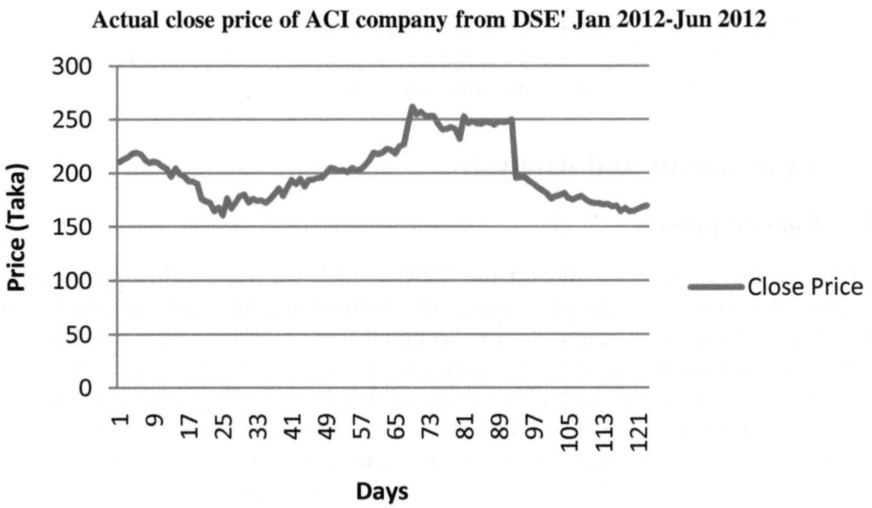

Fig. 1. Actual Share price from Jan 1, 2012 to Jun 28, 2012

price of ACI Company from DSE for the month of January 2012 to June 2012. Though we collect all listed company share price data from DSE but for the convenience of research analysis, we filter only one well-known company, named ACI group of company Limited's share price data in daily basis. But, the study can be apply to any kind of stock price data set which should contain 5 attributes, like date, open, high, low and close price value. Date is a special attribute used as id in this research and other 4 other attributes are used as regular attributes.

3.2 Analysis Steps

This study is conducted in two phases. Training phase and testing phase. The steps are given below:

Training Phase

Step 1: Read the training data set from local repository.
Step 2: Apply windowing operator to transform the time series data into a generic data set. This step will convert the last row of a window within the time series into a label or target variable. Last variable is treated as label.
Step 3: Accomplish a cross validation process of the produced label from windowing operator in order to feed them as input to the SVM model.
Step 4: Select kernel types and select special parameters of SVM (c, ε, g etc).
Step 5: Run the model and observe performance (accuracy).
Step 6: If performance accuracy is good than go to step 6, otherwise go to step 4.
Step 7: Exit from the training phase

Testing Phase

Step 1: Read the testing data set from local repository.
Step 2: Apply the training model to test the testing data set for price prediction.
Step 3: Produce the predicted trends and stock price.

4 Experiments and Analysis

4.1 Data Preprocess

To produce the optimized input for the SVM model, we use windowing operator to the time series data set as preprocessing step. Windowing allows us to take any time series data and transform it into a cross-sectional format. But for that, we should find out the proper windowing size and step size in order to produce the label. In our study, we use date as ID and close value as label. A Sliding Window Validation process is also applied to evaluate the output from the preprocess step. After doing the analysis, we get some proper combination of the windowing component to produce optimized input for the SVM model. Those are given in Table 1.

Table 1. Windowing Component

Window size	Step size	Training window width (TWW)	Training step size (TSS)	Testing window width (tww)
3	1	30	1	30

4.2 Kernel Component Analysis

In our research we use RBF kernel to predict the stock price and trend. Because the output from RBF kernel is good enough and it also takes short processing time. The most important components of RBF kernel are C value, g value, and epsilon (ε). In this research we try to find out the best combination of these values. And finally, we got some combination that produced good prediction result.

Table 2. RBF kernel component

SVM Models	Kernel	C	g	ε	$\varepsilon+$	$\varepsilon-$	Avg MAPE
Model-1	RBF	10000	1	2	1	1	0.42
Model-2	RBF	10000	1	2	1	1	0.27
Model-3	RBF	10000	1	2	1	1	0.22

In Table 2, there are three models. Model-1 is for predicting 1 day horizon stock price and trend. Model-2 and Model-3 are respectively for predicting 5 days horizon and 22 days horizon stock price and trend.

Table 3 shows support vector numbers (SV), bias values (b) and weights (w) for respective regression model for deferent horizons.

Table 3. Proposed SVM models for 1 day, 5 days and 22 days ahead prediction

Models	Horizon	Support Vector	Bias (offset)	Weight (w)		
	h	SV	b	w1 [close-2]	w1 [close-1]	w1 [close-0]
Model-1	1	696	400.686	1358.881	627.029	501.037
Model-2	5	692	381.482	825.014	734.139	-297.092
Model-3	22	675	421.296	1719.578	1631.468	805.925

4.3 Error Calculation

To evaluate the predicted stock price from our model, we apply Mean Absolute Percentage Error (MAPE). We compare our predicted stock price with real time DSE stock price for the month of January 2012 to May 2012. Table 4 shows the error calculation for 3 regression models. Model 1 is one day a-head prediction model, Model 2 is 5days a-head prediction model and Model 3 is 22 days a-head prediction model. And for evaluating prediction result and calculating MAPE we use 100 actual price data from 2nd January 2012 to 30th May 2012.

Table 4. MAPE calculation results for testing dataset (Jan-2012 to May-2012)

Models	Jan-12	Feb-12	Mar-12	Apr-12	May-12
Model-1	0.16	1.42	0.1	0.07	0.35
Model-2	0.18	0.7	0.04	0.07	0.33
Model-3	0.18	0.32	0.16	0.28	0.14

4.4 Graph Analysis

Figure 2 shows the actual and predicted share price of ACI group from DSE, for the month of January 2012 to May 2012. This is the outcome of the 1 day a-head prediction model. In this model the average error rate of predicting price value is 0.42.

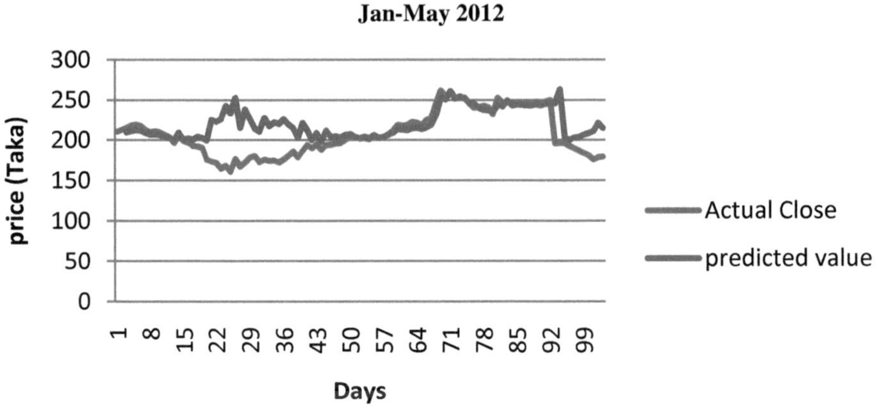

Fig. 2. Model-1 (1 day horizon)

Figure 3 shows the actual and predicted share price of ACI group from DSE, for the month of January 2012 to May 2012. This is the outcome of the model-2, which is based on 5 day a-head prediction. In this model the average error rate of predicting price value is 0.27.

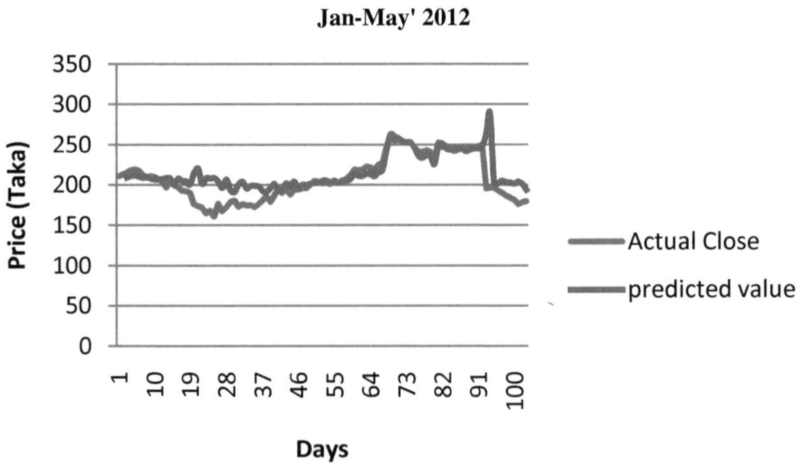

Fig. 3. Model-2 (5 days horizon)

Figure 4 shows the actual and predicted share price of ACI group from DSE, for the month of January 2012 to May 2012. This is the outcome of the model-3, which is based on 22 days advanced prediction. In this model the average error rate of predicting price value is 0.22.

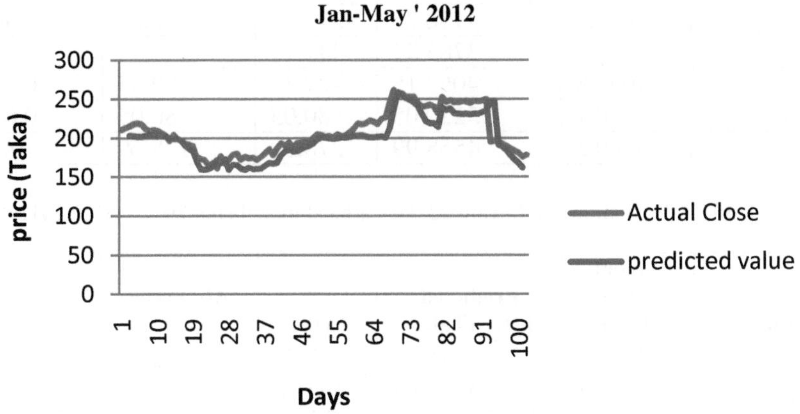

Fig. 4. Model-3 (22 days horizon)

4.5 Prediction Results

Table 5, 6, & 7 show predicted result and the error rate from 1 day, 5 days & 22 days a-head prediction model respectively, for the month of January 2012 to May 2012. In those tables, the actual price (Taka) is taken from current market trend of DSE in Bangladesh currency. And this share price is for ACI group of Company Limited, BD.

Figure 5 shows the MAPE values for the month of January 2012 to May 2012. From the figure, we see that 5days a-head and 22 days a-head regression model produce the less erroneous price value for the stock market.

Table 5. Prediction result from 1 day a-head model (Jan 2012 – May 2012)

Month	Actual price (Taka)	Predicted price (Taka)	Error	Abs error value	MAPE
January	3844.4	4147.85	-303.45	303.45	0.394665
February	3341.4	4241.33	-899.93	899.93	1.417512
March	4032.8	4110.07	-77.27	77.27	0.095802
April	5300.1	5218.71	81.39	81.39	0.069801
May	4280.3	4578.76	-298.46	298.46	0.348644

Table 6. Prediction result from 5 days a-head model (Jan 2012 – May 2012)

Month	Actual price (Taka)	Predicted price (Taka)	Error	Abs error value	MAPE
January	3844.4	4156.26	-311.86	311.86	0.405603
February	3341.4	3783.34	-441.94	441.94	0.696115
March	4032.8	4062.15	-29.35	29.35	0.036389
April	5300.1	5220.05	80.05	80.05	0.068652
May	4280.3	4558.99	-278.69	278.69	0.32555

Table 7. Prediction result from 22 days a-head model (Jan 2012 – May 2012)

Month	Actual price (Taka)	Predicted price (Taka)	Error	Abs error value	MAPE
January	3844.4	3876.92	-32.52	32.52	0.042295
February	3341.4	3138.04	203.36	203.36	0.32032
March	4032.8	3906.77	126.03	126.03	0.156256
April	5300.1	4979.68	320.42	320.42	0.274798
May	3922.4	3825.77	96.63	96.63	0.136863

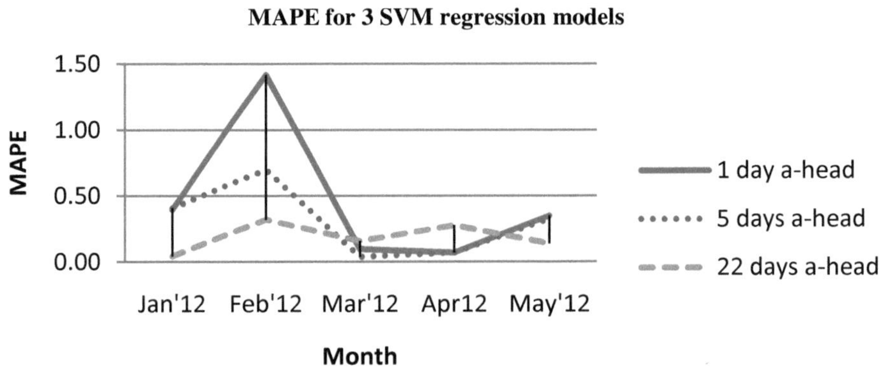

Fig. 5. MAPE for the month of January 2012 to May 2012

5 Conclusion and Limitation

5.1 Conclusion

Our motivation is to predict a good trend for the stock market. So, we focused on accurate stock price prediction and as well as trend prediction. After doing this study we get that, 5 days a-head and 22 days a-head prediction model produce less erroneous stock price for DSE. But, all these three models can produce a good prediction results.

5.2 Limitations and Future Work

In this work we only use data set from DSE and also evaluate our result comparing with DSE. In future, we will apply our model to other stock market data set and will also compare our research result with other types of data mining techniques.

Acknowledgement. The Authors are very grateful to Mr. Ashraf Uddin, Sr. Executive at LankaBangla Securities Ltd, for providing us the real time stock price data from Dhaka stock exchange (DSE), Bangladesh.

References

1. Kim, K.: Financial time series forecasting using support vector machines. Neurocomputing 55, 307–319 (2003)
2. Ince, H., Trafalis, T.B.: Kernel Principal Component Analysis and Support Vector Machines for Stock Price Prediction, pp. 2053–2058 (2004) 0-7803-8359-1/04/2004 IEEE
3. Lucas, K., Lai, C., James, N., Liu, K.: Stock Forecasting Using Support Vector Machine. In: Proceedings of the Ninth International Conference on Machine Learning and Cybernetics, pp. 1607–1614 (2010)
4. Lu, C., Chang, C., Chen, C., Chiu, C., Lee, T.: Stock Index Prediction: A Comparison of MARS, BPN and SVR in an Emerging Market. In: Proceedings of the IEEE IEEM, pp. 2343–2347 (2009)
5. Kannan, K.S., Sekar, P.S., Sathik, M.M., Arumugam, P.: Financial Stock Market Forecast using Data Mining Techniques. In: Proceedings of the International Multiconference of Engineers and Computer Scientists, pp. 555–559 (2010)
6. Hu, Y., Pang, J.: Financial crisis early warning based on support vector machine. In: International Joint Conference on Neural Networks, pp. 2435–2440 (2008)
7. Chen, K., Ho, C.: An Improved Support Vector Regression Modeling for Taiwan Stock Exchange Market Weighted Index Forecasting. In: The IEEE International Conference on Neural Networks and Brain, pp. 1633–1638 (2005)
8. Xue-Shen, S., Zhong-Ying, Q., Da-Ren, Y., Qing-Hua, H., Hui, Z.: A Novel Feature Selection Approach Using Classification Complexity for SVM of Stock Market Trend Prediction. In: 14th International Conference on Management Science & Engineering, pp. 1654–1659 (2007)
9. Debasish, B., Srimanta, P., Dipak, C.P.: Support Vector Regression. Neural Information Processing – Letters and Reviews 11(10), 203–224 (2007)
10. Hsu, C., Chang, C., Lin, C.: A Practical Guide to Support Vector Classification. Initial version (2003) Last updated version (2010)
11. Thissena, U., Brakela, R., Weijerb, A.P., Melssena, W.J., Buydensa, L.M.C.: Using support vector machines for time series prediction. Chemometrics and Intelligent Laboratory Systems 69, 35–49 (2003)
12. Cao, L.: Support vector machines experts for time series forecasting. Neurocomputing 51, 321–339 (2003)
13. Alex, J., Bernhard, S.: A tutorial on support vector regression. Statistics and Computing 14, 199–222 (2004)

Ontology-Driven Automatic Generation of Questions from Competency Models

Onjira Sitthisak[1], Lester Gilbert[2], and Dietrich Albert[3,4]

[1] School of Computer and Information Technology,
Faculty of Science, Thaksin University, Thailand
onjira.sitthisak@gmail.com
[2] School of Electronics and Computer Science, University of Southampton,
Highfield, Southampton, SO17 1BJ, United Kingdom
lg3@ecs.soton.ac.uk
[3] Knowledge Management Institute, Graz University of Technology, Austria
[4] Department of Psychology, University of Graz, Austria
dietrich.albert@tugraz.at,
dietrich.albert@uni-graz.at

Abstract. The paper explores some pedagogical affordances of machine-processable competency models. Self-assessment is a crucial component of learning. However, creating effective questions is time-consuming because it may require considerable resources and the skill of critical thinking. There are very few systems currently available which generate questions automatically, and these are confined to specific domains. Using ontologies and Semantic Web technologies certain limitations in automation, integration, and reuse of data across diverse applications can be overcome. This paper presents a system for automatically generating questions from a competency framework. This novel design and implementation involves an ontological database that represents the intended learning outcome to be assessed across a number of dimensions, including the level of cognitive ability and the structure of the subject matter. This makes it possible to guide learners in developing questions for themselves, and to provide authoring templates which speed the creation of new questions for self-assessment. The system generates a list of all the questions that are possible from a given learning outcome. Such learning outcomes were collected from the INFO1013 'IT Modeling' course at the University of Southampton. The way in which the system has been designed and evaluated is discussed, along with its educational benefits.

Keywords: assessment, knowledge representation, adaptivity, competency modelling, ontology-driven.

1 Introduction

In recent years, a variety of technology-based tools and learning environments have been created and installed in schools, universities, and organisations to support learning. Mostly these tools have been created to support e-learning content delivery and collaborative learning activities, much like a virtual classroom [1]. However,

P. Meesad et al. (Eds.): *IC²IT2013*, AISC 209, pp. 145–154.
DOI: 10.1007/978-3-642-37371-8_18

e-learning suggests not only new technologies for instruction but also new pedagogical approaches to enhance learning.

Self-assessment is a crucial component of learning. Summative evaluation is needed to demonstrate the current learner's knowledge [2]. Dewey observed that learners can learn from asking themselves questions and attempting to answer them [3]. Creating effective questions is time consuming because it may require considerable resources and skill in critical thinking. The questions have to be carefully defined in order to accurately represent the intended learning outcome and the subject matter content involved. Questions should be appropriate to the learner's level of knowledge based on the concept of hierarchies of content structure and cognitive abilities in order to use questioning more effectively as a pedagogical strategy [4].

There are currently many systems available to generate questions automatically; these are, however, confined to specific domains. A number of pioneering systems such as QuizPACK [5], Jeliot 3 [6], iClass [7] and A Web-based English CAT prototype system [8] have explored the use of automatic generation of questions using parameterised templates. Currently, such systems offer remarkable automatic generation of questions, but only for specific domains, and lack integration, interoperability, portability, and reusability with other systems and environments [9].

For identifying the knowledge or competence level of a learner, prerequisite structures on intended learning outcomes and assessment problems are extremely useful [0]. This involves with investigating whether the knowledge and competence structures assumed for a knowledge domain adequately reflect the real world [11]. The emerging Semantic Web technologies present new possibilities for learning. The Semantic Web can add meaning to resources available in a domain by enabling the creation of metadata for each resource; metadata can express meaning for a resource by using terms defined formally and unambiguously in an Ontology. An ontology is an explicit and formal specification for the description of the main concepts of a domain and their relationships, thus providing a shared understanding of a domain [12]. In this study, Web Ontology Language (OWL) is used to express competences composed of subject matter, cognitive ability, and other objects such as contexts, situations, and tools. These ontologies are domain, not structure, ontologies using a controlled vocabulary from Simple Knowledge Organisation System (SKOS) [13] .

This paper reports our provision of interoperable, portable, and reusable automatic generation of questions for self-assessment. We describe a specific approach for the automatic generation of questions in any domain by using a particular model of competencies. The model combines the intended learning outcomes with the subject matter to be assessed, using ontologies. The generated questions are expressed in the IMS Question and Test Interoperability specification (IMS QTI) to enable interoperability. We consider an implementation of the proposed competency framework, named COMpetence-Based learner knowledge for personalized Assessment (COMBA), present an experiment to test its outputs against the criteria of clarity, usefulness, challenge, and match with the learning outcomes, discuss the results, and draw some conclusions.

2 An Ontology for Competency Modelling

The concept of competency is increasingly important since it conceptualizes intended learning outcomes within the process of acquiring and updating knowledge throughout a learner's life [14].

Competency is defined as the integrated application of knowledge, skills, values, experience, contacts, external knowledge resources and tools to solve a problem, to perform an activity, or to handle a situation [15, 16]. In order to support lifelong learning, assessment systems have to focus on representation and updating a variety of knowledge domains, rules, assessments and learner's competency profiles.

The adoption of machine-processable competency records and their interoperability may be enhanced by adherence to emerging standards for competency definition. Existing e-learning competency standards (IMS RDCEO, HR-XML), however, are not able to accommodate complicated competencies, link competencies adequately, support comparisons of competency data between different communities, or support tracking of the knowledge state of the learner [17].

We proposed an improved competency model named COMpetence-Based learner knowledge for personalized Assessment (COMBA) [18]. The COMBA model involves a capability, its association with subject matter content, any attitudinal components, a proficiency level, evidence, any required tools, and a definition of the situation or context of the competency. Each competency, proficiency level, capability, attitude, and subject matter content has a source, an ontology or taxonomy. Drawing on the proposed competency model, we derive an ontology for competency modelling that combines the concepts of subject matter, cognitive ability, and other objects such as contexts, situations and tools. Such models provide ways to define competencies of individual students, prerequisites and goals for resource content, the student's knowledge state, and personalization capabilities for e-Learning.

Representation of knowledge and competency is a crucial area in e-Learning [19]. Without a good representation of the knowledge and competency to be processed, a delivery system will be unable to help its users according to their present and expected competency state[11]. The association between learning resources (documents, tools, actors, activities) and the knowledge and competencies they posses, contain, or process is a key challenge that Semantic Web technologies can address.

In this study, the knowledge domain was structured in form of a "domain ontology". The domain subject matter content, capability taxonomy, and competence were represented based on SKOS. A domain expert expressed the domain content, the capability taxonomy, and competences in an English-like form. A knowledge engineer represented these elements in the form of a semantic network, and then transformed them into an ontology. In this study, the ontology was based on OWL-Lite which was sufficiently expressive to describe the subject matter hierarchy and provides for higher performance reasoning. The ontology of the proposed system is shown in Fig 1. The definitions of the elements in the competence ontology are shown in Table 1.

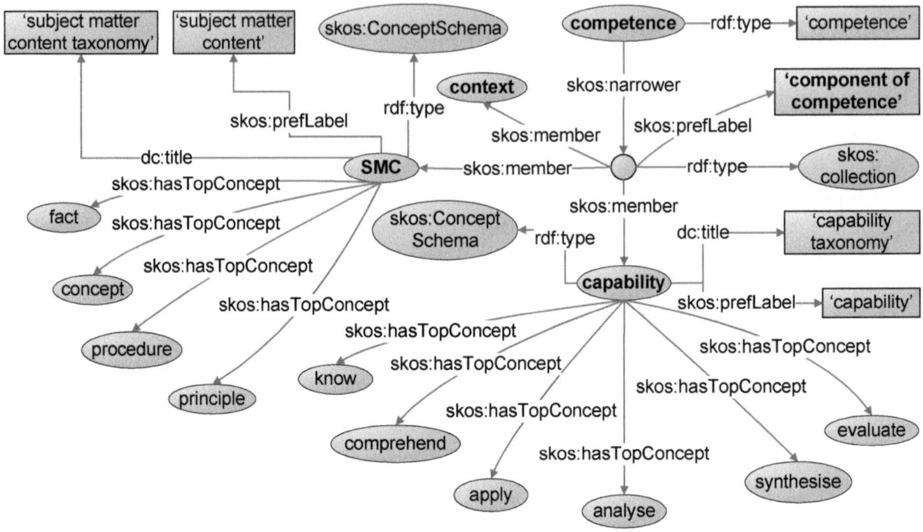

Fig. 1. Ontology of the proposed competency model

Table 1. The definitions of each element in the competence ontology

Class	Definition
Competence	Defines a capability associated with subject matter content (SMC), a proficiency level, evidence, any required tools, and definition of the situation which contextualises the competency.
SMC	Defines the subject domain of what the learner can do by the end of the unit of teaching and learning.
Capability	Defines behaviour that can be observed, based on a taxonomy of learning such as Bloom's, Gagné's nine areas of skill, or Merrill's cognitive domain.
Context	Defines the particular context and conditions of the competency, such as tools and situations.
Fact	Defines statements, or factual information which consists of an attribute and a value.
Concept	Defines a group of objects or ideas which are designated by a single word or term. A concept has a number of attributes which are used to classify or categorise objects according to their values on those attributes.
Procedure	Defines a sequential set of steps to accomplish a task or make a decision.
Principle	Defines the cause-effect relationships describing the behaviour of a system. It can usually be expressed as an equation if the system is in the scientific or engineering domain.
Know	Cognitive domain capabilities according to Bloom
Comprehend	
Apply	
Analyse	
Synthesise	
Evaluate	

3 Using the Competency Ontology in the COMBA System

COMBA aims to provide a system which is able to accommodate complicated competencies, link competencies adequately, and support tracking of the knowledge state of the learner. The system focuses on the identification and integration of appropriate subject matter content (represented by a content taxonomy) and cognitive ability (represented by a capability taxonomy) into a hierarchy of competencies. This makes identification of the assessment that would demonstrate successful teaching and learning straightforward. OWL-Lite is used to express competences composed of subject matter, cognitive ability, and other objects such as contexts, situations, and tools. Using the competency model and ontologies make it possible to overcome limitations in interoperability, portability, and reusability.

The system was built upon an ontological database that describes all resources and the relationships between them. The advantage of ontological schemas over database schemas is that ontological schemas define explicit formal specifications and include machine-interpretable definitions to share common understanding of the structure of the information among people or software agents [12]. An ontological database is flexible and extensible, allowing Semantic Web descriptions of the system resources, interoperability between different systems, and reasoning about the described resources.

COMBA consists of a number of modules [9]: competence navigator, subject matter navigator, capability navigator, question assembler, question to QTI schema converter, and sequencing manipulator. The framework of the ontologies is implemented in Protégé 3.31. The Protégé tool supports knowledge acquisition and knowledge base development [20]. Protégé includes an ontology editor and a system for generating and custom-tailoring forms for data entry by domain specialists. In this research, Protégé stores OWL ontologies in tables. Eight tables were implemented in the COMBA system, comprising three tables for the capability ontology, four tables for four categories (fact, concept, procedure, principle) for the subject matter ontology, and one table for the competency ontology. Three tables of the capability ontology were implemented as 'capability category', 'capability key verbs', and 'capability ordering'. The 'capability category' table referred to the six capability categories in the capability taxonomy. These were 'know', 'comprehend', 'apply', 'analyse', 'synthesise', and 'evaluate'. The 'capability key verbs' table referred to the key verbs in each capability category such as 'explain', 'calculate', and 'define'. The 'capability ordering' table linked two capabilities such that the first capability must be mastered before the next one. For example, the 'comprehend' capability must be mastered before the 'apply' capability.

The ontology repositories for the COMBA system were native stores. Native stores are directly built on the file system, thereby contributing positively to load reduction and minimising update latency [21]. In order to populate the OWL models, store them in a native store, and query them programmatically, a solution based on the Jena Semantic Web Framework[2] was implemented. Jena is an inference engine which can

---

[1] `http://protege.stanford.edu/`

[2] `http://jena.sourceforge.net/`

use the SPARQL query language [22] to query ontology-based models and descriptions in OWL. SPARQL follows SQL-style syntax such as using the SELECT query to process the student query; an example is shown in Fig. 2.

In this study, the 'SELECT' SPARQL query was used to extract data described in OWL, returning it as a tabular result set. The 'UNION' SPARQL query was used to combine result sets into a larger result set. SPARQL result sets were serialised into XML format to allow their direct manipulation.

```
SELECT ?capability ?SMC ?imply ?context ?relatedSMC
WHERE {?competence Comp:SMC ?SMC
          ?competence Comp:capability ?capability
         ?competence Comp:Context ?context
         ?capability Comp:imply ?imply
         {?SMC Comp:the_input ?relatedSMC} UNION
         {?SMC Comp:RelatedConcept ?relatedSMC}};
```

Fig. 2. An example of the SPARQL query language

Question generation begins with the competency of interest submitted to the system. The competence navigator module retrieves subject matter and capability nodes relevant to the competency using the competency ontological database.

Given the subject matter and capability of the submitted competency, the related topics in the four subject matter category tables and capability in the 'capability ordering' table are retrieved as well. For example, if the requested subject matter is 'confidence intervals', the retrieved related subject matter includes 'critical z score' and 'standard error'. For the 'calculate' capability, 'explain' and 'define' capabilities were retrieved as well. Question templates are used to assemble the retrieved subject matter and capability nodes into questions. For example, given the 'confidence interval' competency, the related subject matter and capabilities are inserted into the question templates to yield questions such as, 'Explain the importance of the critical z score'. The process of traversing competencies, retrieving the relevant nodes, and converting these to questions is recursive. The generated questions are transformed for conformance to the IMS QTI by a conversion process using the QTI schema.

Question templates are used to assemble the retrieved subject matter and capability nodes into questions. For example, given the 'confidence interval' competency, the related subject matter and capabilities are inserted into the question templates to yield questions such as, 'Explain the importance of the critical z score'. The process of traversing competencies, retrieving the relevant nodes, and converting these to questions is recursive. The generated questions are transformed for conformance to the IMS QTI by a conversion process using the QTI schema.

The relatively unsophisticated method of generating questions, in particular the use of simple question templates, yields some questions which are inappropriate, do not make good sense, or show poor grammar and syntax, such as, 'Calculate ER Diagram'. The generated questions needed to be filtered by a domain expert.

4 Experimentation, Results and Discussion

The objective of the experimentation was to determine the opinions of students about the quality of the generated questions and to explore how well the generated questions were rated on the criteria of 'clarity', 'usefulness', 'challenge', and 'match with the learning outcomes'. The independent variable was the type of the generated question: 'generic' or 'specific'. Generic questions were free of context, while specific questions included a specific problem context. The dependent variables were the student ratings on the criteria of clarity, usefulness, challenge, and match with the learning outcomes.

A prototype was developed to generate assessment questions from a competency data model. The competencies were collected from the INFO1013 'IT Modeling' course at the University of Southampton. The topics involved confidence intervals and associated content involving: critical z score, Alpha value, standard error, measure of dispersion, and sample size. We may note that representing competencies based on COMBA could be implemented for any domain whose competencies can be expressed in terms of subject matter and capability taxonomies.

The system generated 42 questions within the confidence interval topic. These questions were filtered to 25 questions based on two domain experts' selection of the questions which would most appropriately address the experimental questions. In this system, we used the question template file. Examples of generated questions are: "Define the meaning of the confidence interval", "Calculate the confidence interval for this situation", and "Explain the importance of the standard error in this situation". The questions involved three distinct capabilities: Define, Explain, and Calculate. The three 'specific' questions involved one question for each capability, while the two 'general' questions involved one question for each of the Define and Explain capabilities. There are some questions that the experts would have expected, such as, "What is the effect of sample size on the width of a confidence interval?" and, "In computing a confidence interval, when do you use t and when do you use z?" The content of these questions is not directly represented in the intended learning outcome and the subject matter involved, and these questions may be considered meta-questions. The use of a question template approach did not allow the generation of such questions. A questionnaire asked the students to rate a sample of five of the generated questions against the four criteria on a 3-point Likert scale ('Yes', 'No opinion', and 'No' coded as 1, 2, and 3 respectively). Different versions of the questionnaire presented different questions, such that there were 5 questionnaire versions in total. The questionnaires were randomly distributed to all attending students at the end of a lecture. The study gathered data from 27 students.

SPSS reports four test statistics for the multivariate test of differences in mean ratings of questions; Wilks's Lambda, Hotelling's Trace, the Pillai-Bartlett trace, and Roy's largest root. In this experiment, Wilks's Lambda and Hotelling's Trace are the best for our purpose because group differences are concentrated on the variate of rating classification. As can be seen in Table 2, the multivariate tests for differences in rating according to question type, capability type, and the question by capability type interaction showed significance only for differences between question types (for Wilks' Lambda, $p = 0.004$ and Hotelling's Trace, $p = 0.004$).

Table 2. Multivariate Test

Effect	The statistic method	Value	F	Hypoth df	Error df	Sig.
Question type	Wilks' Lambda	.888	4.023	4	127	.004
	Hotelling's Trace	.127	4.023	4	127	.004
Capability type	Wilks' Lambda	.940	.992	8	254	.443
	Hotelling's Trace	.063	.993	8	252	.442
Question type * Capability type (interaction)	Wilks' Lambda	.996	.134	4	127	.970
	Hotelling's Trace	.004	.134	4	127	.970

Table 3. Estimated Marginal Means for Question Type

Dependent Variable	Question type	Mean	Std. Error	95% Confidence Interval	
				Lower Bound	Upper Bound
Clear	Specific Q	1.975	.071	1.834	2.116
	Generic Q	1.630	.087	1.457	1.802
Useful	Specific Q	1.630	.071	1.490	1.769
	Generic Q	1.759	.086	1.588	1.930
Match to learning outcomes	Specific Q	1.877	.070	1.738	2.015
	Generic Q	1.778	.086	1.608	1.948
Challenging	Specific Q	1.346	.057	1.233	1.459
	Generic Q	1.500	.070	1.361	1.639

Table 4. Tests of Between-Subjects Effects

Source	Dependent Variable	Type III Sum of Squares	df	Mean Square	F	Sig.
Question type	Clear	4.48	1	4.48	10.87	.001
	Useful	.33	1	.33	.83	.37
	Match to learning outcomes	.15	1	.15	.37	.54
	Challenging	1.33	1	1.33	5.03	.03
Error	Clear	53.56	130	.41		
	Useful	52.52	130	.40		
	Match to learning outcomes	51.93	130	.40		
	Challenging	34.44	130	.27		

Table 3 provides the estimated marginal means for the four ratings according to question type. Table 4 provides the tests of between-subject effects for question type, where it may be seen that there were significant differences in mean ratings of ‘Clear’ and ‘Challenging’, but there were no significant differences in mean ratings of ‘Useful’ and ‘Match to learning outcomes’. An inspection of the profile graphs shows that the students rated the clarity of generic questions significantly higher than that of specific questions, while rating the challenge of the specific questions significantly higher than that of the generic questions. The students rated the specific and the generic questions as not significantly different with regard to mean ratings of ‘Useful’ and ‘Match to learning outcome’.

The results indicate that the generated questions were of acceptable value to the students. Six out of eight of the 95% confidence intervals were below 2, indicating a tendency to rate "Yes" rather than "No opinion" or worse. That they found the specific questions more useful, and the generic questions more challenging, is not an unexpected finding, and neither is the finding that both types of question did not differ significantly on the two other criteria, of their clarity and whether they matched the intended learning outcomes. Interestingly, there was no effect of capability type, and no interaction between capability type (define, explain, and calculate) and question type (specific and generic), indicating that ratings were similar for the three capability types. Overall, the results gave support to the research and suggest that further work would be useful. The authoring question template used as the starting point in formulating the format of questions exhibited a rather low efficiency of 59.52% (the number of the generated questions, 42, in relation to the number of selected questions, 25). It may be possible to use some natural language processing for developing the format of questions, and this point will be examined in future work.

5 Conclusion

This study presented one possibility of using ontologies for automation of generating questions. Using OWL to express competences addresses successfully many of the problems of extending and combining structured content of different formats from different schemas. This allows for creative uses of content in novel ways. From this study, the use of ontologies can enable an implementation of intelligent software agents helping the student to find and use globally distributed learning resources, collaborative and distributed authoring and course construction, and reuse of learning material for future study. In addition, using the competency ontology provides interoperable, portable, and reusable resources to define and update knowledge throughout a learner's life for e-learning and knowledge management applications.

While this research successfully demonstrates a data model and a method of automatically generating acceptable and useful questions, representing competencies and the subject matter was the critical challenge. The results indicate that the generated questions were of acceptable value to the students. Successful deployment of the system was required for the development of a detailed and systematic database comprising all the competencies involved in the particular domain of interest required the definition of a competency ontology.

A major challenge in the construction of a competency ontology was to ensure that existing competencies in the course syllabus could be properly represented in the new model. In addition, more effective algorithms are needed for generating questions. The template approach was unable to generate meta-questions, and more advanced methods would be required to accommodate such generation. Furthermore, any generating mechanism must ensure a high standard of English grammar in the resulting questions.

The construction of a sequence or series of questions, that is the construction of an adaptive examination or assessment, is a topic for future work. We believe that a competency model is critical to successfully managing assessment and achieving the goals of resource sharing, collaboration, and automation to support lifelong learning.

References

1. Koper, R., Specht, M.: TenCompetence Lifelong Competence Development and Learning. In: Sicilia, M.-A. (ed.) Competencies in Organizational E-Learning, Concepts and Tools. Idea Group (2007)
2. Paramythis, A., Weibelzahl, S., Masthoff, J.: Layered evaluation of interactive adaptive systems framework and formative methods. User Model User-Adap. Inter. 20, 383–453 (2010)
3. Dewey, J.: The significance of the Problem of knowledge Logic. The Theory of Inquiry. Holt and Co., New York (1938)
4. Gilbert, L., Gale, V.: Principles of eLearning Systems Engineering. Chandos (2007)
5. Brusilovsky, P., Sosnovsky, S.: Individualized exercises for self-assessment of programming knowledge. An evaluation of Quiz PACK. Educational Resources in Computing (JERIC) 5 (2005)
6. Myller, N.: Automatic Generation of Prediction Questions during Program Visualization. Electronic Notes in Theoretical Computer Science (ENTCS) 178, 43–49 (2007)
7. Albert, D., Nussbaumer, A., Steiner, C.: Using Visual Guidance and Feedback Based on Competence Structures for Personalising E-Learning Experience. In: The 16th International Conference on Computers in Education (ICCE 2008), pp. 3–10 (2008)
8. Yu-Huo, T., Yu-Lung, W., Hsin-Yi, C.: A Practical Computer Adaptive Testing Model for Small-Scale Scenarios. Educational Technology & Society 11, 259–274 (2008)
9. Sitthisak, O., Gilbert, L., Davis, H.C.: An evaluation of pedagogically informed parameterised questions for self assessment. Learning, Media and Technology, 33 (2008)
10. Steiner, C.M., Albert, D.: Personalising Learning through Prerequisite Structures Derived from Concept Maps. In: Leung, H., Li, F., Lau, R., Li, Q. (eds.) ICWL 2007. LNCS, vol. 4823, pp. 43–54. Springer, Heidelberg (2008)
11. Falmagne, J.-C., Doignon, J.-P.: Interdisciplinary Applied Mathematics. Springer, Berlin (2011)
12. Antoniou, G., van Harmelen, F.: A Semantic Web Primer. The MIT Press (2004)
13. SKOS Specification, `http://www.w3.org/TR/swbp-skos-core-guide`
14. Albert, D., Nussbaumer, A., Steiner, C.M.: Towards Generic Visualisation Tools and Techniques for Adaptive E-Learning. In: The 18th International Conference on Computers in Education (ICCE 2010). Asia-Pacific Society for Computers in Education (2010)
15. Sandberg, R.: Competence-the Basis for a Smart Workforce. In: Gerber, R., Lankshear, C. (eds.) Training for a Smart Workforce. Routledge, London (2000)
16. Friensen, N., Anderson, T.: Interaction for lifelong learning. British Journal of Educational Technology 35, 679–687 (2004)
17. Sitthisak, O., Gilbert, L., Davis, H.C., Gobbi, M.: Adapting health care competencies to a formal competency model. In: The ICALT. IEEE Computer Society Press (2007)
18. Sitthisak, O., Gilbert, L., Davis, H.C.: Deriving e-assessment from a competency model. In: The 8th IEEE International Conference on Advanced Learning Technologies (2008)
19. Paquette, G.: An Ontology and a Software Framework for Competency Modeling and Management. Educational Technology & Society 10, 1–21 (2007)
20. Gennari, J.H., Musen, M.A., Fergerson, R.W., Grosso, W.E., Crubézy, M., Eriksson, H., Noy, N.F., Tu, S.W.: The Evolution of Protégé. An Environment for Knowledge-Based Systems Development. International Journal of Human-Computer Studies 58, 89–123 (2003)
21. Kiryakov, A., Ognyanov, D., Manov, D.: OWLIM – A pragmatic semantic repository for OWL. In: Dean, M., Guo, Y., Jun, W., Kaschek, R., Krishnaswamy, S., Pan, Z., Sheng, Q.Z. (eds.) WISE 2005 Workshops. LNCS, vol. 3807, pp. 182–192. Springer, Heidelberg (2005)
22. SPARQL for RDF, `http://www.w3.org/TR/rdf-sparql-query`

A Comparative Study on Handwriting Digit Recognition Classifier Using Neural Network, Support Vector Machine and K-Nearest Neighbor

Chayaporn Kaensar

Department of Mathematic, Statistic and Computer Science, Faculty of Science,
Ubon Ratchathani University, Thailand
scchayka@ubu.ac.th

Abstract. The aim of this paper is to analyze efficiency of three classifiers which will be experimented and compared to find out the best techniques. They were experimented on a standard database of handwritten digit. However, not only recognition rate is considered, but also other issues (ex. error rate, misclassified image rate and computing time) will be analyzed. The presented results show that SVM is the best classifier to recognize handwritten digits. That is, the highest recognition rates (96.93%) are obtained. But the computing time of training is the main problem for them. Conversely, other methods, like neural networks, give insignificantly worse results, but their training is much quicker. However, all of the techniques also represent an error rate of 1–4% because of confusion with digits 1 and 7 or 3, 5 and 8 respectively.

Keywords: Handwritten Digit Recognition, Artificial Neural Network, Support Vector Machines, K-Nearest Neighbor and Recognition Rate.

1 Introduction

Handwritten digit recognition has attracted a great deal of research and analysis because there are many places where handwritten digit documents still exist, for example, automatic letter sorting at the post office, the cheque processing in the bank, or historical documents. [1][2]. Unfortunately, such writing documents are very hard to recognize even by people. Thus, a system which could aid an automatic recognition of handwritten documents would be very desirable.

Handwritten digit recognition is an important filed of Optical Character Recognition (OCR) and can be seen as a sub problem of OCR. It is the ability of a computer to receive and interpret intelligible handwritten digit input from sources such as documents, image and other devices.

Although more number of proposed system and classification techniques have been developed for this area such as [3][4][5], proper accuracy of predicting the pattern is still questionable.

So the comparison of proper techniques became a challenge and seems difficult to determine the best one because their performance is data-dependent. It also depends

P. Meesad et al. (Eds.): *IC²IT2013*, AISC 209, pp. 155–163.
DOI: 10.1007/978-3-642-37371-8_19

on many factors including high accuracy, low run time, low memory requirement and reasonable training time [6]

Thus, we aimed to study and compare three classifier techniques, Neural Network with BP Algorithm, K-Nearest Neighbor and SVM respectively. The performance evolution (i.e. training time,) is done to analyze the various classification algorithms to select the proper classifier and other issues for handwritten digit recognition.

To do so, we took Optical Recognition of Handwritten Digits Data Set from UCI Machine Learning Repository [7]. It is input data for analyzing the various classification techniques which are normalized from 32x32 bitmap to 8x8 bitmap already.

This paper is organized as follows: In next section, we present the related works on handwritten digit classification. In Section 3, we have discussed the three classification techniques (i.e. ANNs, SVM and K-Nearest Neighbor Classifier) used for recognizing handwritten digits. We also compare the classification accuracy among them and analyze the experimental result on Section 4. Finally, Section 5 contains conclusion and future work discussions.

2 Literature Survey

Handwritten digit recognition is an important problem in optical character recognition and it has been used as a test case for theories of pattern recognition and machine learning algorithms for many years.

It can be classified into two categories: online recognition and offline recognition. [8] The on-line recognition technology, which emerges in recent years, uses the geometry and temporal dynamics information of the users' input. The methods for online recognition relatively pose low resource and processing requirement, and may effectively use many kinds of clues to capture users' input customs. They are effective with good user adaptation. [9]

Inversely, Offline Recognition mainly processes and recognizes the user input handwritten digit based on images (the scanned images of handwritten digit, or the digital images transformed from the real time handwritten). A lot of methods have been proposed to solve offline recognition. [10].

In this paper, we focus on Offline Recognition. We have found that a number of researches have been concerned with the offline recognition of handwritten digits. For example:

In [1] the author addressed some issues in designing high reliability system for hand-written digit recognition using SVM classifiers. However, the presented result shows that it is difficult to achieve the good recognition rates by using only one selection method.

In [11] the author proposed a decision tree learning to classify different writing styles of the identical digits. Several direction features were employed to implement the classification. That is, when the stroke direction of some digits is similar, the decision tree learning can classify them properly. However, it is difficult to manage sets of possibilities as more features.

In [12] the author analyzed the learning rate using BP algorithm of Artificial Neural Network for handwritten digit recognition application. That is, they used various parameters such as learning rate, number of hidden neurons in hidden layer, and momentum term to analyze the learning rate which shows its impact for the performance of application.

The performance comparison is also applied and studied in different techniques ex. [6] [13].

Moreover, the comparison of classification Handwritten Digit Recognition have been published continuously. For example in [14] the authors have performed the experiments by extracting structural features from the handwritten digits by using SVM and tree classifier. The recognition rate of SVM classifier is more than the Tree classifier.

And in [15] they use three stage classifiers for hand written digit recognition, at stage 1 and 2 neural network is used, and at stage 3 support vector machine is used. The recognition rate obtained is among the best on MNIST database. The results were also better than single SVM using the same feature set.

Although there are a number of published research provide high recognition rate, little work has been done on analysis processing time consuming and comparison of three techniques, which provide good result, especially among efficiency classifier like ANNs, K-Nearest Neighbor and SVM. [6]

3 The Classification Method

3.1 BackPropagation Artificial Neural Network

Backpropagation Neural Network (BPN), which was developed by Rumelhart, et al. in 1986, is the most common neural network learning algorithm. It should be noted that input signals propagate forwards through the network, and error signals propagate backwards. Weight adjustments are made to reduce error. [16]

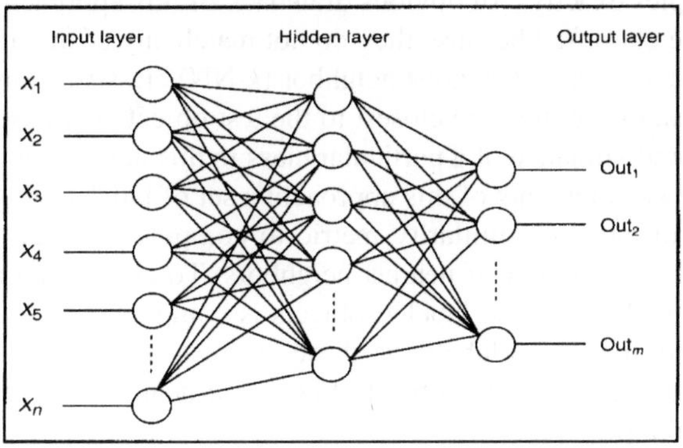

Fig. 1. Structure of backpropagation neural network. (Source: [17], p. 273)

In a BP Neural Network, the learning algorithm has two phases as follows [19][20]:

- Propagation; A training input pattern is presented to the network input layer. The network propagates the input pattern from layer to layer until the output pattern is generated by the output layer.
- Weight Update; If this pattern is different from the desired output, an error is calculated and then propagated backwards through the network from the output layer to the input layer. The weights are modified as the error is propagated.

In this paper, to solve a problem, we have used throughout one hidden layer BP-ANN and designed with three layers, namely an input layer, a hidden layer, and an output layer (Figure 1). Moreover, we used Rule-of-Thumb methods in [18][19] and the result set from [12] approximately define 45 nodes in the hidden layer because Rule-of-Thumb methods is used to calculate the proper node in the hidden layers.

The signal of input from outside spread to the output layer and gives the result through processing layer for layer of neurons in input layer and hidden layer. If the expected output cannot be obtained in output layer, it shifts to the conversed spreading processing and the true value. The error outputted by network will return along the coupled access formerly. The error is reduced by modifying contacted weight value of neurons in every layer and then it shifts to the positive spreading processing and revolves iteration until the error is smaller than the given value. There are no changes of weights in the recognition processing, except the data of the input or output layers. [18] [20]

3.2 K-Nearest Neighbor (KNN)

K-Nearest Neighbor (KNN) is one of the simple methods, which memorize the entire training data and perform classification. The attributes of the test object match one of the training examples exactly. An obvious drawback of this approach is that many test records will not be classified because they do not match any of the training records. A more sophisticated approach, *k*-nearest neighbor (*k*-NN) classification finds a group of *k* objects in the training set that are closest to the test object, and bases the assignment of a label on the predominance of a particular class in this neighborhood. [19][20]

There are three key elements of this approach: a set of labeled objects, e.g., a set of stored records, a distance or similarity metric to compute distance between objects, and the value of *k*, the number of nearest neighbors. To classify an unlabeled object, the distance of this object to the labeled objects is computed, its *k*-nearest neighbors are identified, and the class labels of these nearest neighbors are then used to determine the class label of the object. For the *k*-nearest neighbor classification algorithm is given in [6][19].

3.3 Support Vector Machines (SVMs)

Support Vector Machines (SVM) were introduced as a machine learning method by Cortes and Vapnik (1995). Since SVM is a binary classifier. Thus, the aim of SVM is to find the best classification function to distinguish between members of the two classes in the training data. The metric for the concept of the "best" classification function can be realized geometrically.

In the case of classification, an SVM constructs an optimal separating hyperplane in a high-dimensional feature space. The computation of this hyperplane relies on the maximization of the margin. In this case, we select Non-linear mapping and the kernel function, called Gaussian basis functions (RBF kernel). [6] [19]

This technique defines that the input vectors are only involved through their inner product. Thus, to map the data in a feature space, one does not need to consider the feature space in explicit form. One only has to calculate the inner products of the vectors in the feature space via the kernel function. This is the kernel trick that allows the construction of a decision function that is nonlinear in the input space but equivalent to a linear decision function in the feature space. The equation is described in [6][19].

4 Experimental Results and Evaluation

In this section, we will show the experiments which were performed in order to test our approach. The aim of this experiment is to compare three classifier techniques, Artificial Neural Network using Back Propagation (BP ANNs), K-Nearest Neighbor with Euclidean and SVM with Gaussian kernel, which were experimented and concluded to find out the best techniques. Moreover, the different values, such as error rate, misclassified image rate and processing time for computing step, were also analyzed as well.

To define the classifier for this experiment, we employed three techniques because they are all effective established techniques for pattern recognition and classification area. For example, BPN is one of the most known methods used for character recognition. It is able to segment non-linear separable classes. So, it is a common choice for the digit recognition task.

Another widely accepted technique is the use of SVM. It is beneficial from its generalization power because it is capable of moving the entire problem into a representation of greater dimension, enabling it to separate more complex problems. This is achieved by the use of kernel function, such as the RBF function.

The other one is KNN, which is also commonly used because it is simple and it also produce high recognition rate [13][15][19].

We first download data set, Optical Recognition of Handwritten Digits Data Set, from UCI machine learning repository (http://archive.ics.uci.edu/ml/datasets/) in the form of text file. The data was collected it from a total of 43 people, 30 people contributed for the training set and different 13 people for the test set. Each Digit was written in the form of matrix of 8x8 and contains 64 input attributes of continuous format, which was normalized size from 32x32 bitmap. That is, E. Alpaydin and C. Kaynak, who created this data, used preprocessing programs made available by NIST (The US National Institute of Standards and Technology) to extract normalized bitmaps of handwritten digits from a preprinted form. Some samples data from the UCI database [3] are shown in Figure 2.

In Figure 2, the values of the pixels are normalized and the target values are 16 gray-scale images of size 8x8. That is, they generates an input matrix of 8x8 where each element is an integer in the range 0..16 and the last digit is class code from 0 to 9.

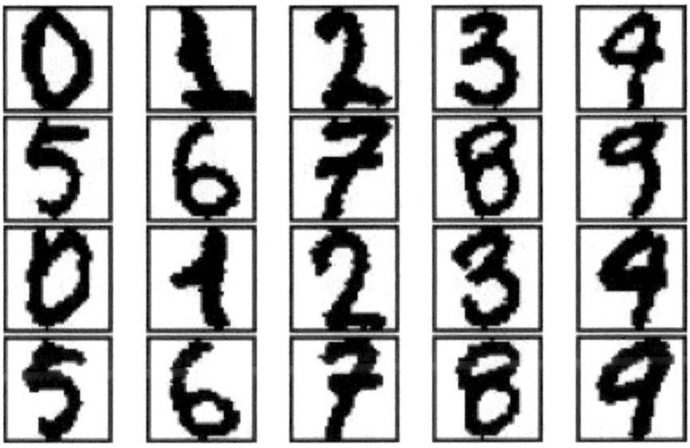

Fig. 2. Example of handwritten digit in the UCI database (Source: [21], p. 2765)

Moreover, the database provides number of items in each class code differently, 389 items in the largest category for class 1 and 3. And 376 items in the smallest category for class 0 and 5 respectively.

The total number of handwritten digits used is 5,620 instances. In the training phase, 3,823 hand written digits (68 percentages) are used as training pattern, 1,797 patterns of different digits (32 percentages) are used as a validation pattern to check. The Classification Techniques such as KK-Nearest with Euclidean, SVM and Artificial Neural Network are applied for Handwritten Digit Recognition.

Three classifiers were tested using the open source Wega Tool Kit [22]. All tests were performed on 1.60GHz Intel CPU T2050 processor under Windows 7.

To do so, we started with experimenting BPN. We set algorithm with one hidden layer and set 45 hidden nodes parameter in Weka because it was proved to be a fast according to Q. Abbas et. al. [12]. Then, the WLSVM software toolbox (i.e. Weka LibSVM - Integrating LibSVM into Weka) was employed for SVM which was seen as a form of implemented LibSVM working in Weka [23]. Moreover, all the running parameters were set as software defaults. Moreover, we also use the implementation of KNN in Weka. That is, the value k in all k-related algorithms is set to 1 because G. Daqi. and L. Jie [24] defined the classifier which is approximately equal to the 1-*NN* classifier in classification accuracy.

Table 1. Rates of Different Methods On UCI Data Set

Classifier	Recognition. Rate (%)	Misclassified Image (%)	Error Rate (%)	Recognition. Time. (s)
BP ANN 64-45-10	95.10	0.73	4.17	0.658
SVM with Gaussian kernel	96.93	1.34	1.73	76.112
K Nearest Neighbor	95.66	0.89	3.45	1.034

The result of all the classifier used for hand written digits is shown in the form of Table 1. That is, the individual recognition rate (Rec. Rate), misclassified image (Misc Image), error rates and recognition times (Rec. Time) of the three NNs, SVM and K-NN are presented.

It is obvious that the SVM has a superior recognition rate but it is a level of magnitude slower than the NNs and K-NN. However, an ANNs provide a low misclassification rate. It showed us that SVM is the best classifier but the time is still the main problem. Other methods like K-NN and ANNs give insignificantly worse results, but their time is much quicker.

Table 2. Recognition Rate (%) Per Digit

Digit	*Train (Number of Digits)*	*K-NN (%Recognition Rate)*	*ANNs (%) (% Recognition Rate)*
0	376	97.19	97.12
1	389	97.32	94.32
2	380	96.38	94.65
3	389	95.01	94.16
4	387	94.82	96.18
5	376	95.38	95.02
6	377	93.45	95.94
7	387	95.65	95.32
8	380	94.41	93.13
9	382	96.94	94.74
	Average	95.66	95.10

Moreover, the second experiment is to look at the recognition rate obtained for each of the individual ten digits as shown in Table 2. We found that the highest recognition rate was reached just over 97%. And the highest recognition rate obtained was with digit 0 and 1, whereas the lowest recognition rate obtained is for digit 9 and 6. Besides some digits were mainly confused with digits 1 and 7, like digit 3, 5 and 8 due to the similarity in writing these digits when it comes to different handwriting styles.

5 Conclusion and Future Work

This paper concludes that different classifier affects the recognition rate for handwritten digit recognition. To do so, we used three techniques from different good classifier and also used the opensource Wega tool kit for training and testing the dataset, which was from the UCI repository.

The presented results show that SVM is the best classifier to recognize handwritten digits. The highest recognition rates (96.93%) are obtained. But the time of training is the main problem for their use. Conversely, other methods like neural networks give insignificantly worse results, but their training is much quicker. However, to fully evaluate it, further study is necessary.

For future work, we will try to apply some techniques for character recognition in Thai Printed Characters. Moreover, we will consider reasonable factor which affects the recognition rate such as run time, memory requirement, proper parameter and any.

References

1. Gorgevik, D., Cakmakov, D.: Combining SVM Classifiers for Handwritten Digit Recognition. In: 16th International Conference on Pattern Recognition, vol. 247, pp. 529–551. IEEE Press (2002)
2. Oliveria, A.L.I., Mello, C.A.B., Silva, E.R.: Optical Digit Recognition for Images of Handwritten Historical Documents. In: Proceedings of the Ninth Brazilian Symposium on Neural Networks, pp. 166–171. IEEE Press (2006)
3. Khalighi, S.: A Novel OCR System for Calculating Handwritten Persian Arithmetic Expressions. In: International Conference on Machine Learning and Applications, pp. 755–758. IEEE Press (2009)
4. Khalighi, S., Tirdad, P., Rabiee, H.R., et al.: Handwritten Digit Recognition Using State-of-the-Art Techniques. In: International Conference on Machine Learning and Applications, pp. 320–325 (2009)
5. Deng, L.: The MNIST Database of Handwritten Digit Images for Machine Learning Research. IEEE Signal Processing Magazine, 141–142 (2012)
6. LeCun, Y., et al.: Learning Algorithm for classification: A comparison on handwritten digit recognition. CiteSeerX (1995)
7. UCI Machine Learning Repository – Optical Recognition of Handwritten Digits Data Set, `http://archive.ics.uci.edu/ml/datasets/Optical+Recognition+of+Handwritten+Digits`
8. Cheng, L.L., Nakashima, K., Hiroshi, S., et al.: Handwritten digit recognition: benchmarking of state of the art techniques. Pattern Recognition, 2271–2285 (2003)
9. Artieres, T., Marchand, J.M., Gallinari, P., et al.: Stroke Level Modeling of On-Line Handwriting Through Multi-Modal Segmental Models. In: The 10th International Workshop on Frontier in Handwriting Recognition, pp. 1–6 (2000)
10. Khorsheed, M.S.: Off-Line Arabic Character Recognition, A Review. Pattern Analysis & Application 5, 31–45 (2002)
11. Jiang, W.L., Sun, Z.X., et al.: User-Independent Online Handwritten Digit Recognition. In: Proceedings of the Fifth International Conference on Machine Learning and Cybernetics, pp. 3359–3364. IEEE Press (2006)

12. Abbas, Q., Ahmad, J., Bangyal, W.H.: Analysis of Learning Rate using BP Algorithm for Hand Written Digit Recognition Application. In: The 2nd International Conference on Information Engineering and Computer Science, pp. 1–4. IEEE Press (2010)
13. Gorgevik, D., Cakmakov, D., Radevski, V.: Handwritten Digit Recognition Using Statistical and Rule-Based Decision Fusion. In: The 11th Mediterranean Electrotechnical Conference, pp. 131–135 (2002)
14. Garg, N.K., Jindal, S.: An Efficient Feature Set For Handwritten Digit Recognition. In: The 15th International Conference on Advanced Computing and Communications, pp. 540–544. IEEE Press (2007)
15. Gorgevik, D., Cakmakov, D.: An Efficient Three-Stage Classifier for Handwritten Digit Recognition. In: Proceedings of the 17th International Conference on Pattern Recognition, pp. 507–510. IEEE Press (2004)
16. Ngu, S.C., Cheung, C., Leung, S.H.: Fast Convergence for Back Propagation Network with Magnified Gradient Function. In: International Joint Conference on Neural Networks, pp. 903–908. IEEE Press (2003)
17. Changa, T.C., Chaob, R.J.: Application of back-propagation networks in debris flow prediction. Engineering Geology 85, 270–280 (2006)
18. Zhang, C., Huang, X.: Off-line Handwritten Digit Recognition Based on Improved BP Artificial Neural Network. In: The IEEE International Conference on Service Operations and Logistics, and Informatics, vol. 1, pp. 626–629. IEEE Press (2008)
19. Tan, P.N., Steinbach, M., Kumar, V.: Introduction to Data Mining. Pearson Addison-Wesley (2006)
20. Russell, S., Norvig, P.: Artificial Intelligence: A Modern Approach, 2nd edn. Prentice-Hall (2002)
21. Kim, H.C., et al.: Constructing support vector machine ensemble. Pattern Recognition 36, 2757–2767 (2003)
22. Weka home page, `http://www.cs.waikato.ac.nz/-mJ/weka/index.html`
23. Integrating LibSVM into Weka Environment, `http://www.cs.iastate.edu/~yasser/wlsvm`
24. Daqi, G., Jie, L.: Kernel Fisher Discriminants and Kernel Nearest Neighbor Classification: A Comparative Study for Large-Scale Learning Problems. In: The 2006th International Joint Conference on Neural Networks, pp. 1333–1338. IEEE Press (2006)

Printed Thai Character Recognition Using Standard Descriptor

Kuntpong Woraratpanya and Taravichet Titijaroonrog

Faculty of Information Technology,
King Mongkut's Institute of Technology Ladkrabang, Thailand
{kuntpong,bank2533}@gmail.com

Abstract. The various font-types, font-sizes, and font-styles have a great impact on recognition performance of optical character recognition (OCR) systems. This becomes a grand challenge for recognition improvement. In order to enhance the performance, this paper proposes the printed Thai character recognition using a standard descriptor. The descriptor construction consists of two principal phases—preprocessing and feature extraction. In the former phase, the preprocessing provides a standard form for each character image. In the latter phase, the singular value decomposition (SVD) is applied to all font-type, font-size, and font-style character images to extract features. Then the standard descriptor is constructed from the suitable order selection of the SVD feature decomposition. Finally, the projection matrix technique is applied to the recognition phase in order to measure the cosine similarity between the standard descriptor and test set. The experimental results show that the proposed method achieves a high recognition rate and is invariant to font-types, font-sizes, and font-styles.

Keywords: optical character recognition (OCR), standard descriptor, singular value decomposition (SVD), feature extraction.

1 Introduction

An evolution of a variety of printed Thai fonts leads to the difficulty of retaining a high recognition rate of optical character recognition (OCR) systems. This becomes a grand challenge for recognition improvement. Over the last 20 years, printed Thai-character recognitions have been continually researched. Most of the approaches focus on an improvement of recognition rate. For instances, Kimpan et al. [1] introduced to fine classification of printed Thai character recognition using the Karhunen-Loeve expansion. This work uses a two-step approach, i.e., the lowest order of eigenvectors for rough classification and the higher order of eigenvectors for fine classification. It achieves the high recognition rate in testing with a standard font. Duang-phasuk et al. [2] presented printed Thai character recognition using feature matching and adaptive resonance theory I (ART I). This approach also requires two steps for feature extraction and recognition, i.e., global features for rough classification and local features for fine classification. It provides the high recognition rate

P. Meesad et al. (Eds.): *IC²IT2013*, AISC 209, pp. 165–173.
DOI: 10.1007/978-3-642-37371-8_20

when experimented on various fonts and a few font-sizes, but takes more time-consuming. Tangsurakit et al. [3] proposed printed Thai consonant recognition based on character density and strip features. This approach is evaluated with only five basic fonts of consonance characters. In addition, Kruatrachue et al. [4] presented automatic state machine induction for string recognition that focuses on an enhancement of character classifications. The result of recognition is rather high efficiency, but in case of very similar characters the recognition rate is low.

However, as discussed previously, none of these papers point out the problems of a variety of printed Thai font-types, font-sizes, and font-styles. A few papers studied on font styles and types. For examples, Tanprasert et al [5] proposed Thai type style recognition, and Thammano et al. [6] presented hierarchical cross-correlation ARTMAP neural network for recognizing printed Thai characters of no-head fonts.

Therefore, this paper proposes a standard descriptor, which is a modularity method, to improve the recognition performance of OCR systems. The standard descriptor is made from all font-type, font-size, and font-style character features extracted by using SVD. It is believed that the proposed descriptor helps improve the recognition performance of OCR systems without reengineering software.

2 Standard Descriptor Construction

A standard descriptor plays an important role in representative of various font-types, font-sizes and font-styles, thus leading to the performance improvement of OCR. This section describes the effective standard descriptor construction consisting of two principal procedures, preprocessing and feature extraction. In addition, projection matrix technique applied to a recognition procedure is presented in the last subsection.

2.1 Preprocessing Procedure

Each character segmented from a document image is usually in different sizes. In order to provide a standard form for character segmentation images to construct the standard descriptor, image complementation, zero padding, and resizing procedures are applied to each character image as shown in Fig. 1(a). As a result, the character segmentation image is in a standard form of 32×32 pixels and zero backgrounds. Fig. 1(b) and (c) show a matrix of different font-types and font-sizes and a matrix of a standard form, respectively.

2.2 Feature Extraction

All matrices of a standard form in Fig. 1(c) are transformed into image vectors as depicted in Fig. 2. Then a training matrix is formed from such image vectors, which are contained features of all font-types, font-sizes, and font-styles. The procedure to construct a standard descriptor can be demonstrated in Fig. 3.

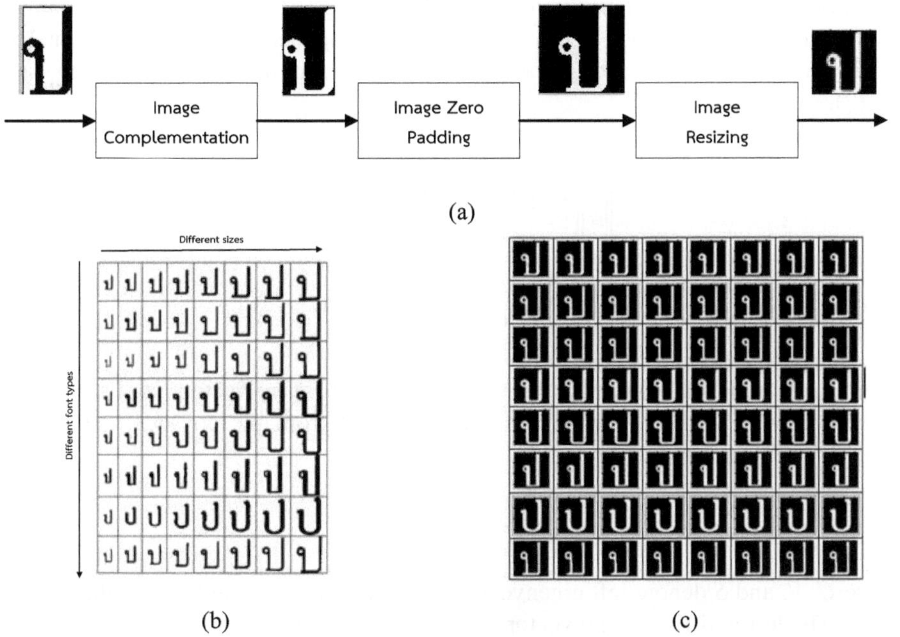

Fig. 1. Preprocessing procedures: (a) image complementation, zero padding, and resizing procedures, (b) matrices of different font-types and font-sizes, and (c) matrices of standard forms

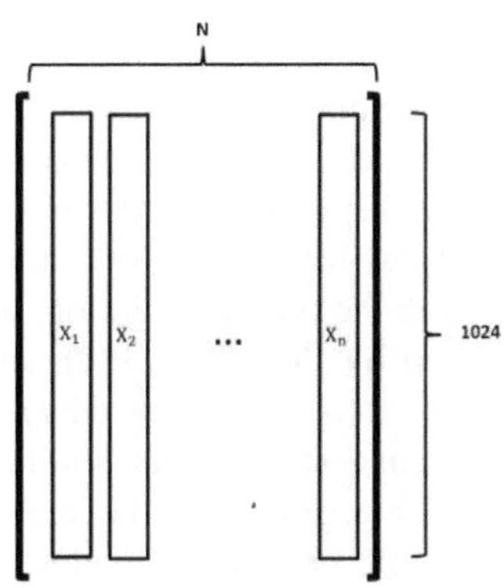

Fig. 2. A training matrix formed from image vectors

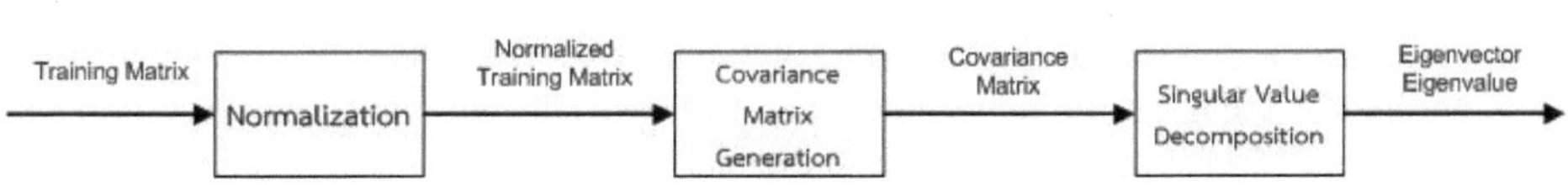

Fig. 3. Standard descriptor construction procedures

Initially, the training matrix is normalized by using Eq. (1). The result is a normalized training matrix. Then a covariance matrix is generated from the normalized training matrix by subtracting mean and determining matrix elements by using Eq. (2).

$$\vec{u} = \frac{u}{\|u\|}; \quad \|u\| = \sqrt{u_1^2 + u_2^2 + \cdots + u_N^2} \tag{1}$$

$$C_x = \frac{1}{N}\sum_{i=1}^{N}(X_i - M)(X_i - M)^T, \tag{2}$$

where X_i, N, and M are an image vector, a number of image vectors, and a mean value, respectively. Finally, the features are extracted from the covariance matrix by using SVD defined as Eq (3).

$$A = U \times S \times V^T, \tag{3}$$

where U, V, and S denote left eigenvector, right eigenvectors, and eigenvalue, respectively. The left and right eigenvectors are represented as $U=AA^T$ and $V=A^TA$, respectively. Finally, the standard descriptor is constructed by means of the suitable range of the higher order eigenvectors.

2.3 Recognition Procedure

The projection matrix technique is applied to the recognition phase in order to measure the cosine similarity between the standard descriptor and test set. In general, the misclassification occurs when the shape of characters is similar. It can be demonstrated with Fig. 4. This example shows features of four similar characters—ข, ฃ, ช, ซ—in a vector space. As mentioned in the previous section, the lower order eigenvectors of such characters are represented by P(ข), P(ฃ), P(ช), and P(ซ). On the other hand, the higher order eigenvectors are represented by H(ข), H(ฃ), H(ช), and H(ซ). It is evident that the lower order eigenvectors of four similar characters have the same directions, while the higher order eigenvectors have different directions. In other words, the higher order eigenvectors provide an evident classification. Suppose that there is a test vector y(ข) in a vector space as depicted in Fig. 4 and the cosine similarity is applied to measure the angle between P and y. In this case the cosine function gives a maximum value, since the angle is very small. This leads to the misclassification; that is, ข can be recognized as ฃ, ช, or ซ. Conversely, when the cosine similarity is applied to measure the angle between H and y, and the classification criteria is minimal; it is a largest angle. In this case the test vector can be recognized properly. It is summarized that the higher order eigenvectors provide the good features to classify the similar character shapes. Therefore, this paper proposes the construction of the efficient standard descriptor based on the higher order eigenvectors and the use of the projection matrix technique with minimal cosine similarity measure for recognition.

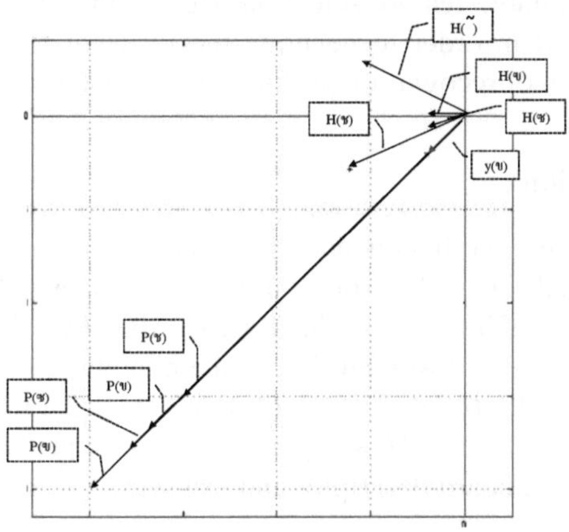

Fig. 4. A plot of the lower order eigenvectors versus the higher order eigenvectors

The overall recognition procedure is illustrated in Fig. 5. The manipulation begins with a test vector, y, is normalized with standard descriptor norms—$\|sd_1\|$, $\|sd_2\|,\ldots,\|sd_n\|$. This step generates n test unit vectors—$y_1, y_2,\ldots,y_n$. Then the projection matrix technique is applied to obtain cosine similarity values—$s_1, s_2,\ldots, s_n$—calculated from inner product of test unit vectors and standard descriptors. Finally, the minimum value of a set s_n is selected for recognition.

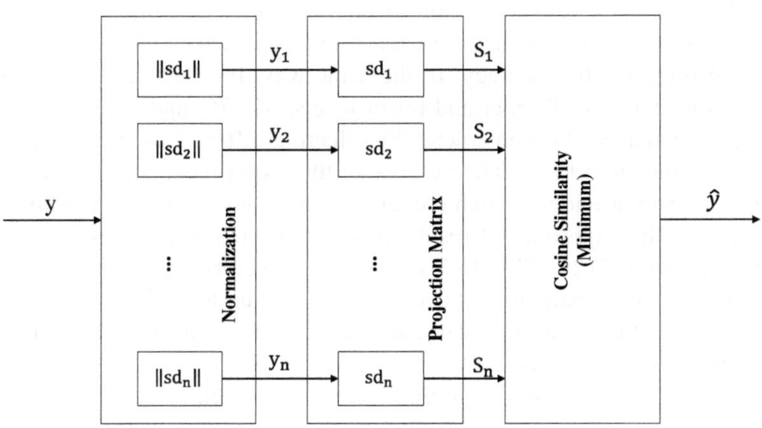

Fig. 5. A recognition procedure by using projection matrix technique

3 Experimental Results

As mentioned in the subsection 2.2, the standard descriptor constructed by SVD method depends on the suitable feature selection. Hence, the feature selection based on experiments is analyzed in order to generate the efficient standard descriptor. Furthermore, the evaluation of recognition performance is also illustrated.

3.1 Data Preparation

The test sets used in all experiments are Thai character image corpus consisting of consonants, vowels, and tones. A resolution of such images is a 400 dpi. Font types are composed of AngsanaUPC, BrowalliaUPC, CordiaUPC, DilleniaUPC, EucrosiaUPC, FreesiaUPC, IrisUPC, JasmineUPC, and unknown fonts, and font sizes are composed of 8, 10, 12, 14, 16, 18, 20, and 22. Font styles are regular, bold, italic, and bold-italic. There are totally 58,656 samples which are equally divided into training set for constructing the standard descriptor and test set.

3.2 Suitable Feature Selection and Recognition Accuracy

In this subsection, the appropriate feature selection is tested with two experiments. Then the results are analyzed. Hence, two standard descriptors, namely SD1 and SD2, are constructed by selecting the lower-order eigenvector and the higher-order eigenvector of SVD, respectively. Both standard descriptors are tested with Thai character images divided into three levels, upper, middle, and lower levels. In the first experiment, SD1 provides the lower recognition rate as illustrated in Fig. 6. It yields the good recognition accuracy, on average 87.58%, for all font styles, when seven components in lower-order eigenvectors are selected as depicted in Fig. 6(b). In addition, this experiment also reveals that the increase of a selected number of eigenvectors does not help improve the accuracy. In the same way, Fig. 6(a) and (c) exhibits the low recognition accuracy of upper and lower levels, 81.87% and 93.35%, respectively. On the other hand, SD2 is constructed by selecting different ranges of higher-order eigenvectors as shown in Fig.7. The purpose of this experiment is to obtain a suitable eigenvector order to provide the high recognition accuracy with the minimum number of eigenvectors. The experimental results prove that the suitable range and order of eigenvectors are from 512 to 704, 18% of total eigenvectors, and point out by a block as shown in Fig.7. The recognition accuracy of SD2 is up to 98.74% for middle level characters. The recognition accuracy in comparison of SD1 and SD2 is summarized in Table 1. It is obvious that SD2 outperforms SD1 in terms of recognition accuracy; however, the high accuracy of recognition rate trades off the increase of a number of components.

Furthermore, no-head fonts of printed Thai characters cause a serious problem in decreasing the recognition rate [6]. However, based on the experiment, SD2 can also provide the higher recognition accuracy, 94.05%, when compared to the method in [6], 83.77%.

As reviewed in Introduction section, the methods proposed in [1-2] need two steps to extract features, whereas SD2 uses only single-step to extract features. Fewer steps lead to the reduction of computing time. Moreover, SD2 outperforms the method in [4] in terms of misclassification, when the shape of characters is similar.

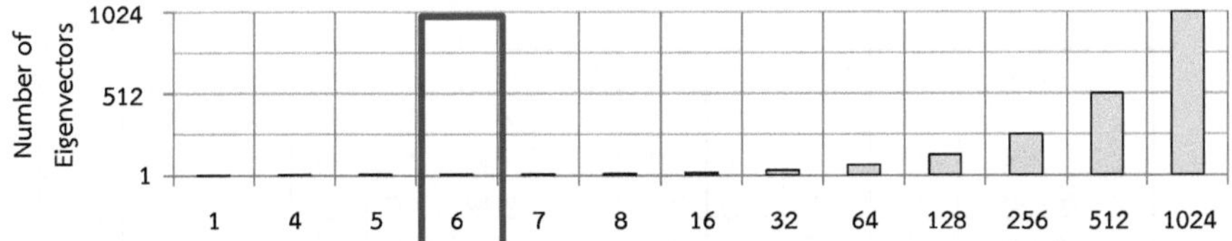

Accuracy(%)	Normal	79.88	82.80	82.00	81.47	81.03	80.67	75.27	68.71	59.49	52.57	39.72	26.77	08.33
	Bold	80.94	79.61	80.32	79.96	79.79	79.26	76.24	69.77	63.56	54.08	42.11	20.83	08.33
	Italic	80.94	81.65	81.65	83.24	83.16	83.24	76.77	69.68	61.79	56.03	46.81	22.96	08.33
	Bold-Italic	80.31	82.18	82.62	82.80	83.07	83.07	79.96	74.82	68.71	62.32	48.14	22.96	08.33
	Average	80.52	81.56	81.65	81.87	81.76	81.56	77.06	70.75	63.39	56.25	44.20	23.38	08.33

(a)

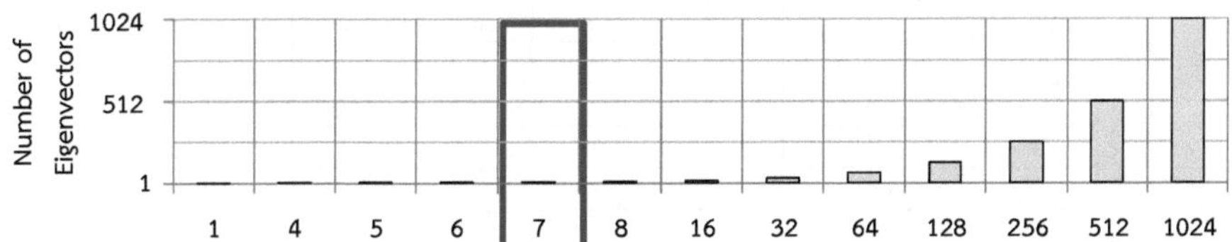

Accuracy(%)	Normal	72.96	82.47	83.49	83.76	83.86	83.98	80.88	75.42	67.71	54.12	17.77	03.58	01.54
	Bold	79.51	88.43	88.66	89.13	88.87	89.28	87.18	82.85	75.24	56.84	19.49	03.75	01.54
	Italic	74.99	84.82	85.79	85.86	86.69	86.09	84.16	79.92	72.73	59.87	21.54	04.47	01.54
	Bold-Italic	80.69	89.00	86.17	90.56	90.88	90.74	89.21	85.91	81.41	68.31	29.51	05.34	01.54
	Average	77.04	86.18	86.03	87.33	87.58	87.52	85.36	81.03	74.27	59.79	22.08	04.29	01.54

(b)

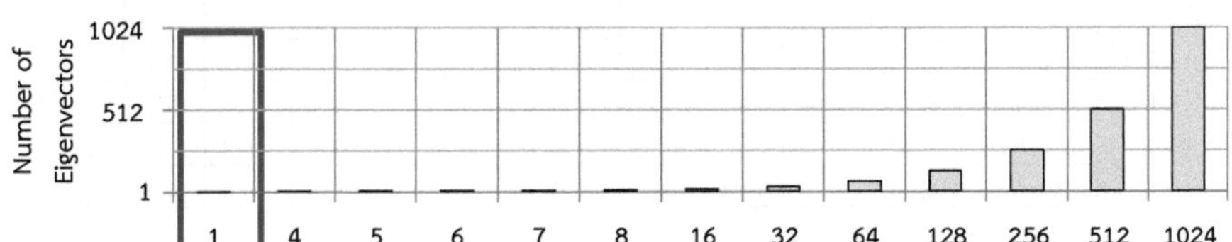

Accuracy(%)	Normal	95.04	93.97	93.26	93.26	92.55	91.84	89.01	87.23	82.98	80.85	76.70	33.33	33.33
	Bold	92.20	89.36	89.36	89.01	89.01	88.30	84.75	82.27	79.08	75.18	66.67	33.33	33.33
	Italic	93.62	89.36	89.72	90.70	90.70	90.07	83.69	80.50	77.66	73.40	64.18	33.33	33.33
	Bold-Italic	92.55	89.72	88.30	85.82	86.17	84.75	77.31	75.53	72.70	68.09	59.93	33.33	33.33
	Average	93.35	90.60	90.16	89.70	89.61	88.74	83.69	81.38	78.11	74.38	66.87	33.33	33.33

(c)

Fig. 6. Accuracy of the standard descriptor using the lower-order eigenvector of SVD in (a) upper level, (b) middle level, and (c) lower level

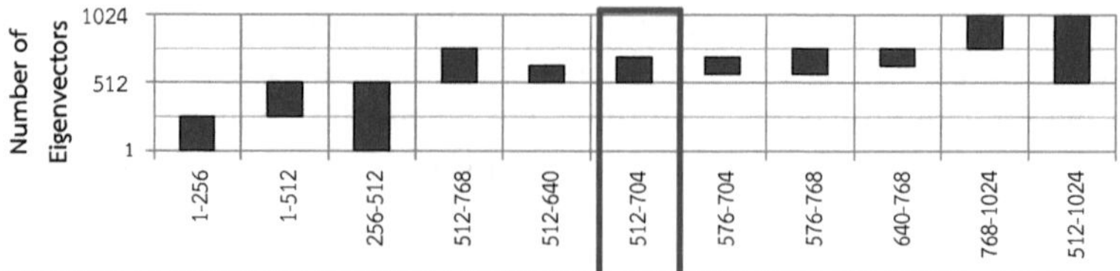

Accuracy(%)	Normal	05.23	95.04	08.33	96.54	95.74	96.19	95.74	96.01	95.65	95.30	96.45
	Bold	04.17	97.61	08.33	98.23	97.52	98.05	98.05	98.14	97.51	97.70	98.04
	Italic	07.36	96.10	08.33	97.43	97.16	97.96	97.16	96.90	96.54	96.45	97.34
	Bold-Italic	06.74	96.72	08.33	96.63	96.45	96.80	96.01	96.54	95.57	95.12	96.36
	Average	5.88	96.37	08.33	97.21	96.72	97.25	96.74	96.90	96.32	96.14	97.05

(a)

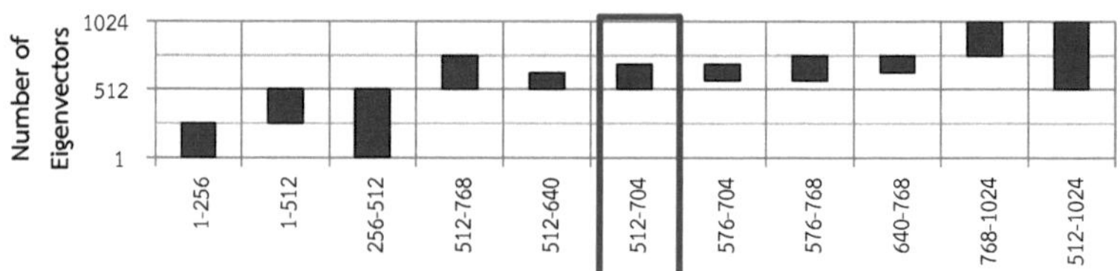

Accuracy(%)	Normal	00.00	98.23	00.00	98.37	97.90	98.34	97.93	98.23	97.82	98.18	98.50
	Bold	00.00	99.19	00.00	99.23	99.02	99.15	99.07	99.20	99.08	99.36	99.39
	Italic	00.00	97.76	00.00	98.41	97.81	98.17	97.82	98.36	97.86	98.35	98.77
	Bold-Italic	00.00	99.12	00.00	99.40	99.12	99.31	99.15	99.38	99.10	99.20	99.50
	Average	00.00	98.68	00.00	98.67	98.46	98.74	98.49	98.79	98.47	98.77	99.04

(b)

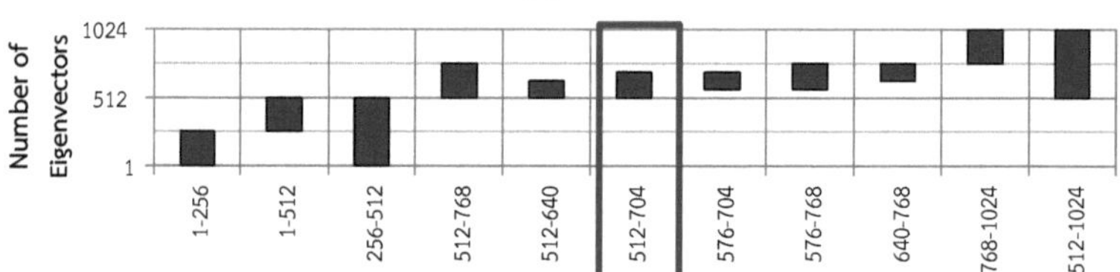

Accuracy(%)	Normal	00.00	98.58	02.13	99.65	99.65	99.65	99.65	99.65	99.65	99.65	99.65
	Bold	00.71	99.65	01.77	99.29	98.94	99.29	98.94	99.29	99.65	99.29	99.65
	Italic	00.00	98.58	00.00	99.29	99.29	99.29	99.29	99.29	99.29	99.65	99.65
	Bold-Italic	02.84	99.65	16.67	99.29	99.29	99.29	99.65	99.65	99.65	98.58	99.65
	Average	00.24	99.12	05.14	99.38	99.29	99.38	99.38	99.47	99.56	99.29	99.65

(c)

Fig. 7. Accuracy of the standard descriptor using the higher-order eigenvector of SVD in (a) upper level, (b) middle level, and (c) lower level

Table 1. A comparison of accuracy of SD1 and SD2

Level	Accuracy (%)							
	SD1				SD2			
	Normal	Bold	Italic	Bold-Italic	Normal	Bold	Italic	Bold-Italic
Upper	81.03	79.79	83.16	83.07	96.19	98.05	97.96	96.80
Middle	83.98	89.28	86.09	90.74	98.34	99.15	98.17	99.31
Lower	92.55	89.01	90.70	86.17	99.65	99.29	99.29	99.29
Average	85.85	86.03	86.65	86.66	**98.06**	**98.83**	**98.47**	**98.47**

4 Conclusions

In this paper, an efficient standard descriptor for character recognition is proposed. The main contribution of this paper is constructing the standard descriptor based on the higher order eigenvectors and recognizing the character with the projection matrix technique. The experiment results show that the proposed method, SD2, evidently outperforms the traditional method, SD1, in terms of a recognition rate. It is concluded that the proposed standard descriptor helps improve the performance of OCR systems without reengineering software.

References

1. Kimpan, C., Itoh, A., Kawanishi, K.: Fine Classification of Printed Thai Character Recognition Using the Karhunen-Loeve Expansion. In: IEEE Proceedings, vol. 134, pp. 257–264 (1987)
2. Duangphasuk, S.: Thai Printed Character Recognition Using Feature Matching Method and ART1. M.S. Thesis, Dept. of Information Technology, King Mongkut's Institute of Technology Ladkrabang, Bangkok (2002) (in Thai)
3. Tangsurakit, N.: Thai Printed Consonant Recognition Based on Character Density and Strip Features. M.S. Thesis, Dept. of Telecommunication Engineering, King Mongkut's Institute of Technology Ladkrabang, Bangkok (2005) (in Thai)
4. Kruatrachue, B., Pantrakarn, N., Siriboon, K.: Automatic State Machine Induction for String Recognition. In: Proceedings of World Congress on Engineering, pp. 153–158 (2007)
5. Tanprasert, C., Sae-Tang, S.: Thai type style recognition. In: Proceedings of the 1999 IEEE International Conference on Circuits and Systems, pp. 336–339 (1999)
6. Thammano, A., Duangphasuk, P.: Printed Thai Character Recognition Using the Hierarchical Cross-correlation ARTMAP. In: Proceedings of the 17th IEEE International Conference on Tools with Artificial Intelligence, pp. 695–698 (2005)

Development of an Image Processing System in Splendid Squid Grading

Nootcharee Thammachot[1], Supapan Chaiprapat[1], and Kriangkrai Waiyakan[2]

[1] Department of Industrial Engineering, Faculty of Engineering,
Prince of Songkla University, Thailand
nootcharee.t@gmail.com, supapan.s@psu.ac.th
[2] Department of Industrial Management Technology,
Faculty of Agro-Industry, Prince of Songkla University, Thailand
kriangkrai.w@psu.ac.th

Abstract. Quality inspection of commercial squids is a labor intensive process. This study proposes an approach to develop a computer vision system for size and specie classification of squids. Both species were differentiated by distinction of mantle shapes. A Multi Layer Perceptron (MLP) with a back propagation algorithm was used to sort squid samples to pre-defined sizes based on a standard of National Bureau of Agricultural Commodity and Food Standards (ACFS). Features extracted from squid images including area, perimeter and length of the squid mantle were used as parameters into the network. Differences between species could be distinguished by using a ratio of length and width of the squid mantle. Results showed that approximately 90% of size and specie classification accuracy could be achieved from the approach proposed in this study.

Keywords: sorting process, computer vision system, splendid squid, neural network.

1 Introduction

About 70% of splendid squids commercially caught in the fishing ground of the Gulf of Thailand and the Andaman Sea are Loligo duvauceli, where the rest are Loligo chinensis[1] as shown in Fig.1. Both are the most demanded cephalopods in the market. On average the first specie, L.duvauceli has a wider and shorter mantle as compared to L.chinensis. L.duvauceli is more preferable by some seafood processing companies as a requirement from their customers. If it is requested, specie classification must be handled before the squids are distributed to processing lines. In addition to specie classifying, sizing is a process carried out simultaneously at the beginning of the process line to guarantee customer satisfaction and prevent yield loss. Size is categorized according to National Bureau of Agricultural Commodity and Food Standards (ACFS) mainly based on the product's weight (Table 1), although occasionally length can also be used as a sizing criterion. The size code in Table 1 is basically named from the approximate number of squids in 1 kilogram. This process is very

P. Meesad et al. (Eds.): *IC²IT2013*, AISC 209, pp. 175–183.
DOI: 10.1007/978-3-642-37371-8_21

labor intensive. During these years, the seafood processing industry has been facing problems related to employing human operators in this skillful position. Sizing workers must be well trained to be able to distinguish squids of different sizes. If they are retired, a process line will be interrupted while a company is training a new worker for a replacement. As labor shortage increasingly becomes worse, finding a replacement could take months. Besides, products sized by human operators are often found to have uncontrollable variation. They use mainly visual sensory assessment and sometimes they simply estimate product weight using their hands. A digital scale is allowed only when it is really needed because its usage causes low productivity. As a result, more automatic systems are introduced to a manufacturing process line to alleviate impact related to human employment.

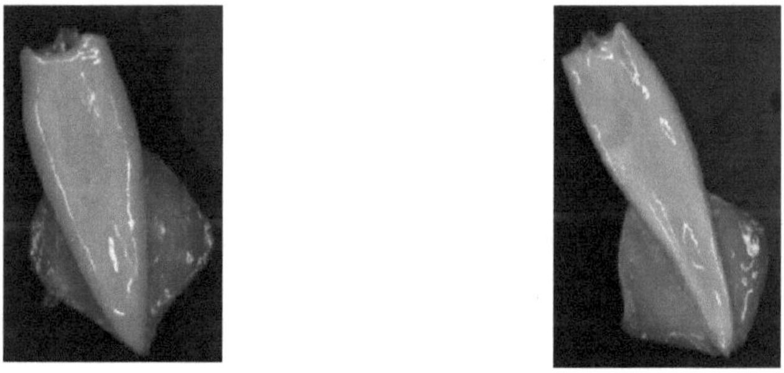

(a) (b)

Fig. 1. The splendid squid species: (a) *L.duvauceli* (b) *L.chinensis*

Table 1. Weight ranges of squid based on ACFS

Code	Weight (g./each)
U/5	> 200
U/10	100-200
U11/20	50-100
U21/40	25-50

Image processing is one of the most interesting and successful techniques for inspection and quality evaluation of agricultural products. The advantages of image processing are high accuracy, reproducibility, nondestructive testability and multi-parameter detections including shape, size, color and defection. There are quite a number of published articles related to applications of image processing in quality inspection and a grading system for food and agricultural industry. Nagata and Tallada[2] developed a strawberries sorting system using size and shape of the fruit. A three-layer Artificial Neural Network (ANN) was used to classify shape based on linear features extracted from strawberries images. Size was classified by a simple regression analysis and a linear relationship between the projected area and the measured weight of the fruit was formulated. The authors then improved an accuracy of strawberries classification by developing a simple grading system for size and shape of strawberries. Shape was classified based on the ratio of area, when the total projected area in strawberries image was used in size sorting. Other a strawberry grading system is developed by Liming and Z. Yanchao[3] using a set of three characteristics including size, shape and color. Size of strawberries was classified by its maximum diameter. The shape parameter was obtained from ratios of section lines in strawberry image before being classified using a K-means clustering method. The color classification used a dominant color method. S. Riyadi et al.[4] developed a computer vision system for papaya size grading using an analysis on shape characteristics (area, mean diameter and perimeter). These three characteristics were measured from papaya images before being fed to a Multi-Layer Perceptron (MLP) Artificial Neural Network (ANN) model for classification. Jarimopas B. and Jaisin N.[5] proposed a sweet tamarind sorting system for shape and size classification and defect detection. Shape was identified by curvature of the tamarind pod. Size was sorted based on length of the pod. Broken areas are an indicator for defects. Kavdır and Guyer[6] investigated the effects of using different feature sets and classifiers on classification performance. Among such classifiers as plug-in decision rule, 1-Nearest Neighbor, 2-Nearest Neighbor, 3-Nearest Neighbor, decision tree classifier and MLP, it was concluded that the last technique provided the highest classification results. In addition, MLP can solve problems which are not linearly and obtain result of classification over 90%. In using MLP, a Back Propagation (BP) algorithm is best suitable for classifying agriculture produces[7-8]. A BP neural network consists of an input layer, at least one hidden layer and one output layer. The layers are connected by weights among neurons. In a training cycle of the network, parameters of known samples are fed as input to the network and the output is computed. The output is compared with the desired result and the weight of the network is modified to reduce the error. The training cycle is repeated to modify the weights of the network until the error is optimized.

An application of image processing technique for squid classification has not been reported. Therefore, this paper aims to present a new approach using a Multi Layer Perceptron Neural Network with a Back Propagation algorithm for classifying specie and size of splendid squids.

2 Materials and Methods

2.1 Splendid Squid Samples

Splendid squids are available commercially fresh and processed in many forms. Generally, there are six types of fresh splendid squid product being manufactured in seafood processing industry, which are *whole round, whole cleaned, tube, fillet, head and wing*. Among these, *whole cleaned* and *tube* are the most demanded by the market. *Whole cleaned* squid is a squid with skin peeled off and having head and gut removed, leaving only its flattened body and fins. *Tube* squid is a *whole cleaned* with its fins off. In this study, samples were collected from a *whole cleaned* processing line.

As mentioned, to guarantee customer satisfaction, size and specie classification must be carried out before the squids will be delivered to subsequent processes. In this study, an image processing system was developed to sort the squids into pre-defined sizes as according to Table 1, as well as to distinguish between both species. The system included four steps; (1) data acquisition, (2) pre-processing, (3) feature extraction and (4) classification.

2.2 Data Acquisition

An image acquisition system was setup by a light control box with a dimension of 0.45 m x 0.56 m x 1m. Light sources were four 10 watt fluorescent lamps mounted at each edge of the box 0.4 m above target objects. The images were captured from the top using a CCD camera (scA640-70gc) with 659 x 490 pixels image resolution installed approximately 1 m above the object. Images were acquired using National Instruments Vision Development Module version AI 2011.

2.3 Pre-processing

Image pre-processing is a procedure for enhancing image data and transforming it into a format ready for further computational processing. The squid images were initially taken in an RGB format before being converted to a grayscale image (Fig. 2). Fig. 3 (b) shows a histogram of a pixel intensity distribution of the image. An automatic threshold suggested a threshold value at 130 to best segment the region of interest from the background. Every pixel having a gray level over such value would take on a new value of 1, otherwise 0 would be its new value, and a binary image was obtained. Morphology was then used to remove small objects (noises) and fill holes. Edge detection was performed to extract contour of the region of interest (Fig. 4).

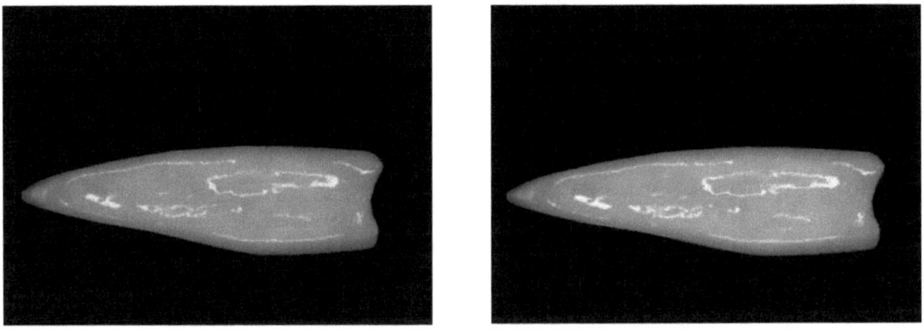

(a) (b)

Fig. 2. (a) Original RGB image and (b) Grayscale image

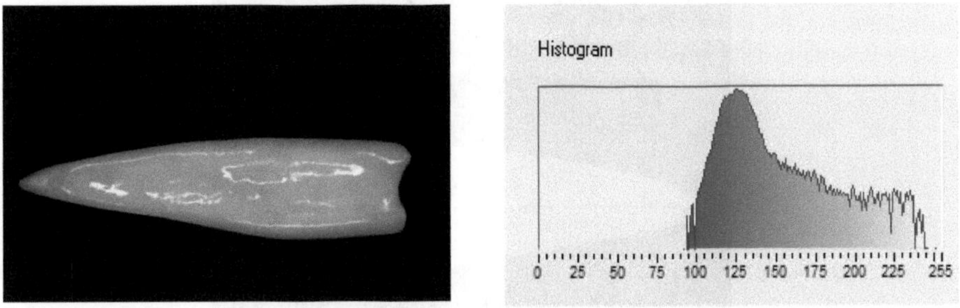

(a) (b)

Fig. 3. (a) Grayscale image and (b) Histogram graph (T=130)

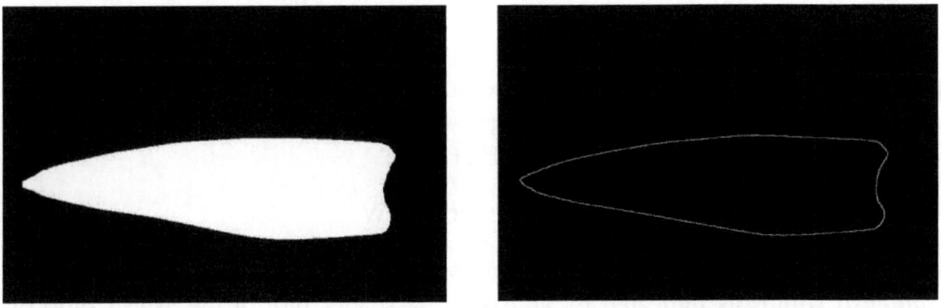

(a) (b)

Fig. 4. (a) Image segmentation and (b) Edge detection

2.4 Feature Extraction

The objective of the developed image processing system is to distinguish between the two species and sort squids into pre-defined sizes. From[1], it was reported that distinct features that differentiate these two species were characteristics of their mantle shapes. As mentioned earlier, *L.duvauceli* has a wider and shorter mantle compared to *L.chinensis*. Parameters used to describe their main characteristics of shape were therefore width and length of their mantles. The length (L) was measured horizontally from the end tip of the squid mantle toward the vertical side. The width (W) was measured vertically along the widest side of its mantle as shown in Fig. 5.

Squid sizes as according to Table 1 was categorized based on their weights. Although weights could not be retrieved directly from the image, other data such as diameter, area, perimeter, and length of the region of interest could be used as a weight estimate[4]. In this study, the authors used area, perimeter and length altogether as parameters into a neural network system. The area was obtained by a summation of pixels within the region of interest, when the perimeter was a distance around it. The image was spatially calibrated to obtain real world dimensional measurements. In this camera setting, pixel width and height were approximately 0.457 mm.

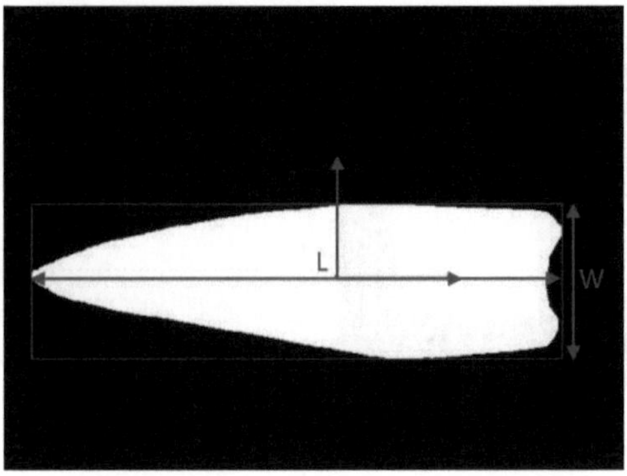

Fig. 5. Parameter determination of squid mantle

2.5 Specie and Size Classification

Specie Classification. Two species of squids were distinguished from each other by using differentiation in their shapes. Shape was characterized by an aspect ratio of the mantle length and width as described in Eq.(1)

$$\text{Shape aspect ratio} = L/W \tag{1}$$

Samples randomly collected were composed of 110 of *L.duvauceli* and 50 of *L.chinensis* squids as indicated by an expert. A scatter plot of the aspect ratio of all samples is as shown in Fig. 6. From the plot, it can be seen that the aspect ratio of 3.4 would be the best divider between *L.duvauceli* and *L.chinensis* as it is the point where minimum misclassification can be obtained.

If the aspect ratio < 3.4, then the squid is *L.duvauceli* , else the squid is *L.chinensis.*

Size Classification. In this study, size was classified by a Back Propagation Neural Network (BPNN) using a Multi Layer Perceptron (MLP) model. The network had an input layer with 3 neurons: area, perimeter and length, one hidden layer with 4 neurons and an output layer. The number of neurons in the hidden layer was calculated by Eq.(2)[7]. In general, determining the optimal number of hidden neurons is important for designing classier. If the neural network has too few or has too many hidden neurons, it may affect the system's generalization capability.

$$h = (m+n)^{(1/2)}+a \tag{2}$$

when

h is the number of hidden neurons,
m is the number of output neurons,
n is the number of input neurons,
a is the member of digit from 1 to 10.

In the output layer, neurons were pre-defined sizes: U21/40, U11/20, U/10 and U/5. The outputs were calculated out of the above inputs through a sigmoid transfer function (Fig. 7).

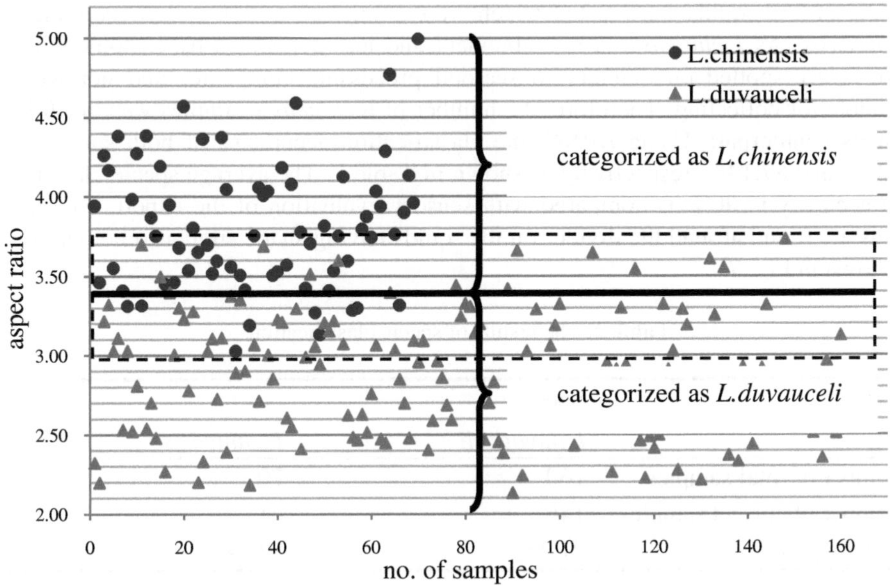

Fig. 6. Comparison of an aspect ratio of both species

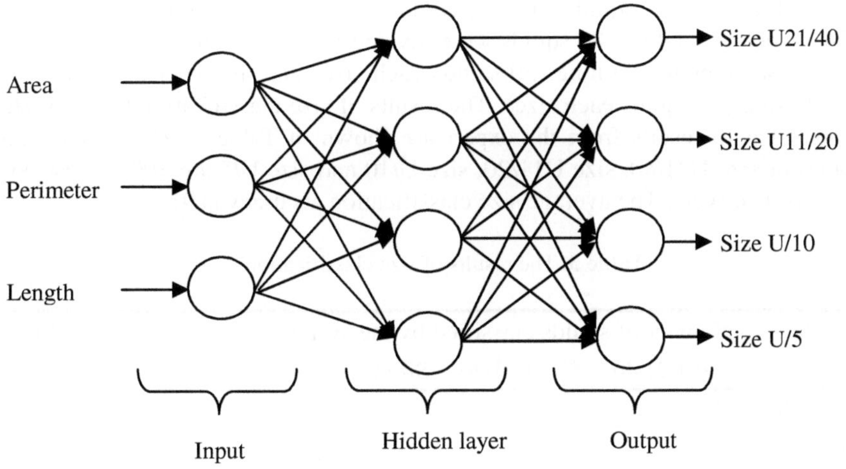

Fig. 7. Basic structure of Multi Layer Perceptron (MLP) for size classification

3 Results and Discussion

3.1 Squid Specie Classification Test

A total of 230 squids were tested which were composed of 160 *L.duvauceli* and 70 *L.chinensis*. As *L.chinensis* is less abundant and less demanded by customers, they were rarely spotted on delivery to seafood processing companies and not widely available for collecting. Therefore, the number of *L.chinensis* samples was much less than its counterpart. Using BPNN, the classification accuracies of both species are 91.25% and 87.14%, respectively as shown in Table 2. The average specie classification accuracy is 90% as compared with sensory evaluation of the expert. Disagreements are found around the divider on the squids having shape similar to both species as shown in Fig. 7.

Table 2. The results of specie classification

Item	Specie	
	L.duvauceli	*L.chinensis*
Number of total samples	160	70
Number of correct samples	146	61
Accuracy (%)	91.25	87.14

3.2 Squid Size Classification Test

Four pre-defined sizes of squid were clustered by a Back Propagation Neural Network (BPNN) model. A total of 280 squids were divided into two groups; 200 squids were used as the samples for training (50 squids each size) and the other 80 squids were used for testing (20 squids each size). The results of size classification by the system as compared with results from the expert are shown in Table 3. The classification accuracies of size U21/40, size U11/20, size U/10 and size U/5 are 100%, 80%, 90% and 95%, respectively. The average size classification accuracy is 91.25%.

Table 3. The results of size classification

By the expert	Number of squids classified by the system				Correct rate (%)
	Size U21/40	Size U11/20	Size U/10	Size U/5	
Size U21/40	20	-	-	-	100
Size U11/20	4	16	-	-	80
Size U/10	-	-	18	2	90
Size U/5	-	-	1	19	95
Average					91.25

4 Conclusion

This paper presents an algorithm for specie and size classification of splendid squids. The algorithm used mantle shape as a feature to classify the species. Both species were differentiated by an aspect ratio of length and width of mantle shape. As in accordance with National Bureau of Agricultural Commodity and Food Standards (ACFS) of Thailand, size was categorized mainly based on weight. Three parameters including area, perimeter and length extracted from the squid image. Combination of these parameters was fed into a BPNN model for size classification. The algorithm provided satisfactory classification results.

Acknowledgement. This work was supported by the Higher Education Research Promotion and National Research University Project of Thailand, Office of the Higher Education Commission.

References

1. Boonwanich, T., Thossapornpitakkul, S., Chotitummo, U.: Reproductive Biology of Squid Loligo duvauceli and L. chinensis in the Southern Gulf of Thailand, Technical paper, Southern marine Fisheries Development Center, Marine Fisheries Division, Department of Fisheries (1998)
2. Nagata, M., Tallada, J.G.: Quality Evaluation of Strawberries. Computer Vision Technology for Food Quality Evaluation, 265–287 (2008)
3. Liming, X., Yanchao, Z.: Automated strawberry grading system based on image processing. Computers and Electronics in Agriculture 71, S32–S39 (2009)
4. Riyadi, S., Rahni, A.A., Mustafa, M.M., Hussain, A.: Shape Characteristics Analysis for Papaya Size Classification. In: 5th Student Conference on Research and Development, pp. 1–5. IEEE (2007)
5. Jarimopas, B., Jaisin, N.: An experimental machine vision system for sorting sweet tamarind. Journal of Food Engineering 89(3), 291–297 (2008)
6. Kavdır, İ., Guyer, D.E.: Evaluation of different pattern recognition techniques for apple sorting. Biosystems Engineering 99(2), 211–219 (2007)
7. Yousef, A.O.: Computer vision based date fruit grading system, Design and implementation. Journal of King Saud University - Computer and Information Sciences 23(1), 29–36 (2010)
8. Chen, X., Xun, Y., Li, W., Zhang, J.: Combining discriminant analysis and neural networks for corn variety identification. Computers and Electronics in Agriculture 71, S48–S53 (2009)

Cardiac Auscultation with Hybrid GA/SVM

Sasin Banpavichit[1], Waree Kongprawechnon[1], and Kanokwate Tungpimolrut[2]

[1] School of Information, Computer and Communication Technology, Sirindhorn International Institute of Technology, Thammasat University, Thailand
sasin_rvd@yahoo.com, waree@siit.tu.ac.th
[2] National Electronics and Computer Technology Center, NSTDA, Thailand
kanokvate.tungpimolrut@nectec.co.th

Abstract. Cardiac Auscultation is the act of listening to a heart sound with the purpose to analyze the condition of a heart. This paper proposes an alternative screening system for patients using a hybrid GA/SVM. GA/SVM technique will allow the system to be able to classify the heart sound base on the heart condition with high accuracy by using GA in a feature selection part of the system. This method improves the training input samples of SVM resulting in a better trained SVM to classify the heart sound. GA in the system is used to generate the best set of weighing factor for the processed heart sound samples. The system will be low cost but has high accuracy.

Keywords: Cardiac Auscultation, Genetics Algorithm, Support Vector Machine, Wavelet packet, Shannon's Entropy, PCA.

1 Introduction

Since cardiac disease is one of the leading cause of death in the world, responsible for over 13% of death according to department of public health of Thailand [1]. Cardiac diseases can be prevented by regular check-up for treatment in time before the diseases become fatal [2]. In Thailand regular health check-up is not widely done by the people especially in the suburban area where availability of check-up are low. The availability of cardiac auscultation is even lower since traditionally cardiac auscultation requires a skilled doctor to use stethoscope to listen to heart sound of the patient directly. When the diagnostics tool for cardiac auscultation is introduced skilled doctor that can conduct cardiac auscultation became rare [3], but the tool itself is only available in major hospitals making cardiac auscultation inaccessible in suburban area.

This study is aimed to produce a reliable screening system to analyze heart sound of patients. Since the system is low cost, only equipment are computer and heart sound files, more people will have access to the screening system and increases their chance to detect any heart diseases. There has been many work done on the classification of heart sound using various methods such as [4], [5], [6], [7] and [8]. Some of these work relies heavily on segmentation of the heart sound signal to determine cycle of heart sound and classify them [5], but some heart sound signals can be difficult to be segmented. This proposed system do

P. Meesad et al. (Eds.): *IC²IT2013*, AISC 209, pp. 185–194.
DOI: 10.1007/978-3-642-37371-8_22

not requires heart sound signal to be segmented which eliminate any difficulty regarding segmentation. The accuracy of the classification on [7] and [8] are quite accurate but still can be improved especially in the False Negative(FN) area. The proposed system will not only overcome this problem but also provide high accuracy of classifying the heart condition. Since false negative area is where the classifier result indicates that the heart sample is healthy when it is actually not. False negative result is the most dangerous since it means that after patient received the diagnostic he will be mislead by the fact that he is healthy when he is actually not. If this false negative can be reduced the system will be improved significantly. The implementation of GA is considered since its implementation of of GA has been used as feature selection in [9], [10] and [11] and yield significant improvement in finding a more suitable set of training data.

The organization of this paper is divided into 5 sections the first one is this section the introduction. Section 2 starts with the methodology, describing all the techniques that will be used in the system and a description of overall system, Section 3 is simulation result of the proposed system. Conclusion is in section 4 and lastly future work is described in section 5.

2 Methodology

The overall block diagram of the proposed system is depicted in Figure 1.The system depicted consists of 4 main parts. These parts are Preprocessing, Feature Extraction, Genetic Algorithms and Classification. All of the processes and algorithms used are all done in MATLAB. The preprocessing, Wavelet packet based feature extraction and PCA are studied in [7], this method is implemented in the proposed system. The input of the system are heart sound files labeled by their conditions. There are total of 352 heart sound samples consisting of 144 normal heart sound and 208 pathological murmur comprised of Aortic Stenosis, Aortic Regurgitation, Mitral Stenosis and lastly Mitral Regurgitation. These heart sound first enter the first component of the system, the Preprocessing. In this first part since each heart sound comes from different source we need to resample, reduce noise and normalize these signals. Each sound signal then passes through Feature Extraction to extract significant features, this is particularly very important since dimension of the features used to train SVM has significant influence on determining the system performance time and accuracy. Here the Wavelet packet based feature extraction is used. The number of features will then further be reduced by using PCA and then enters Genetic Algorithms(GA) to find an optimum weighing factor for the features to be used to train SVM. The classification accuracy of the system is evaluated using a cross validation method. The final trained SVM can classify healthy heart sound from abnormal heart sound.

2.1 Data Collection and Preprocessing

During preprocessing from [7] length of each heart cycle is equalized and resampled at the rate of 4kHz. After that the resampled heart sound then enters

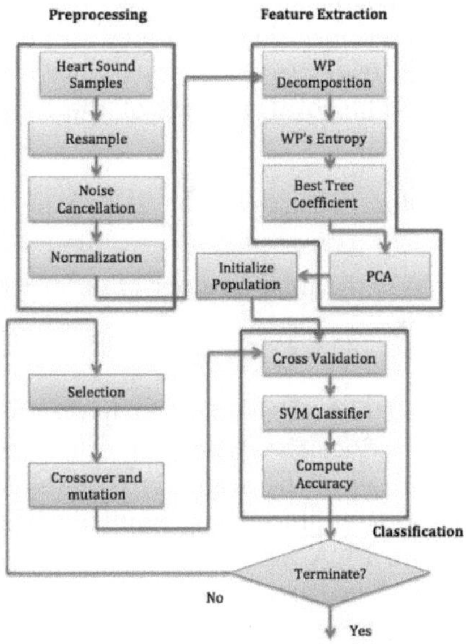

Fig. 1. Overall block diagram of the proposed system

the process of noise cancellation, here 5-level DWT via soft thresholding is used along with Daubechies-6 wavelet family for detail coefficients. Then lastly the physiological variations in intensity of each heart sound is discarded by normalization using (1).

$$\widehat{x} = \frac{x - \mu}{\sigma} \tag{1}$$

To have zero mean and unity variance where $\widehat{x}$ is a final heart sound signal and x is a signal before normalization.μ is the mean of the samples and σ is the variance of the samples.

2.2 Feature Extraction

Wavelet Packet(WP)-Based Feature Extraction. The feature extraction part of the proposed system is based on [7]. Since heart sound is a non-stationary signal in order to retain most of the time-frequency information in the signal, WP-based filter is used since it has high performance yield at characterizing these heart sound signals. Higher-order Daubechies family wavelets can be used to improve frequency resolution and reduce aliasing. After WP decomposition each subband energies is assessed by employing of nonnormalized Shannon's entropy criterion as (2)

$$E(t) = -\left|\widehat{x}(t)\right| \cdot log\left|\widehat{x}(t)\right| \tag{2}$$

$E(t)$ is Shannons's entropy and $\widehat{x}(t)$ is a heart sound signal, the logarithmic function allows a greater entropy weight measure to signal intensity while attenuating noise and disturbances. The WP that is suitable for the heart sound signal is 6-level decomposition on Daubechies- 3 mother wavelet function compiled to. The selected percentage of retained energy is 99.9%. This is to characterize the best-basis feature with 96% of compression rate approaching hierarchically on the decomposition indices. The total outputs of the 2-channel filter banks are obtained as (3).

$$Total\ Features = \sum_{i=1}^{n} 2^i, n = 6. \tag{3}$$

After the performance of WP the final number of different extracted features on time-frequency plane is 126.

Feature Selection Using PCA. PCA or Principal Component Analysis is the method commonly used to reduce the dimension of the given feature. The algorithm will try to locate a subset of the original extracted feature. The dimension of these features set is important since they significantly shape the performance and accuracy of the system. PCA uses subsets of eigenvector from the covariance matrix of the extracted feature. The feature that has the sum of 90% of eigenvalue to previous sum is retained as related features [7], meaning that these features are significant in identifying each sample. The result number of features remaining in this particular system is 12 features, which is considerably smaller that the previous 126 features.

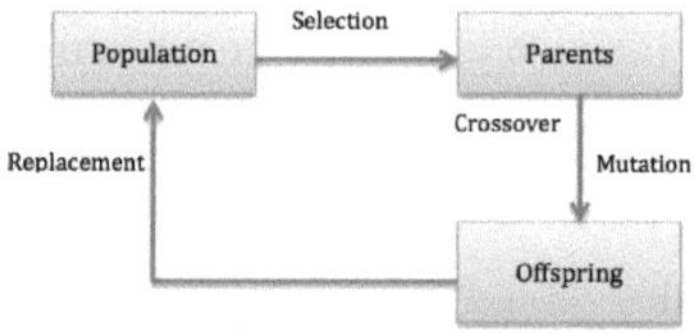

Fig. 2. Evolution diagram of GA

Genetic Algorithms. Genetic algorithm(GA) is a optimization technique based on Darwins theory of natural selection. GA starts with a first group of candidate solution called population consists of many members called chromosomes. Each chromosome has a set of data called genes, the setting of these genes can be configured to suit the application. Based on Darwins theory of survival of the fittest GA will find an optimal solution of chromosomes after some number of iteration of the algorithm. GA generates the solution by creating the next generation of the population that is represented by chromosomes until desired result is obtained. The quality of the solution chromosome of GA is determined by its fitness value. Fitness value is obtained through the fitness function determined to evaluate the quality of each chromosome. When GA generates the next generation of population it selects chromosomes base on their fitness value, the fitter the chromosomes the higher the probability of them to remain in the next generation through a reproduction process. The reproduction process involves two functions crossover and mutation. Crossover allows two chosen chromosomes to exchange gene between each other. Mutation is when any gene inside the chromosome is randomly altered. The offspring that is a result of crossover and mutation will replace the previous population. This process is called evolution and GA will evolve until specified condition is met [12]. The evolution diagram is illustrated in Figure 2.

In this proposed system GA acts as a feature selection to choose the best set of weighed features to be used to train SVM.

Since the final 12 features result from each sample are significant factors that determine the accuracy of the proposed system, selection of these features is very important. From comparing and observing these features with the same condition labels, it is found that most of the features set values of the same label resemble each other with some difference through out each samples. These observations indicate that each feature has different significance in determining the classification accuracy of the SVM classifier. Therefore if each feature is weighed differently with some value, the accuracy of the classifier may increase. This is the reason why a weighing factor is selected to be used in the system to enhance the accuracy of the system.

The method of GA that is used in the proposed system is based on a selection of the weighing factors that will be used to weigh each features of the heart sound [13].This means that importance of some features with large weighing factor will be maximize while importance on some features with small weighing factor will be minimized. GA first initializes a population of chromosomes and keep them in a matrix G. The matrix G is illustrated below with m rows and n columns.

$$G = \begin{bmatrix} g_{11} & g_{12} & \cdots & g_{1n} \\ g_{21} & g_{22} & \cdots & g_{2n} \\ \cdots & \cdots & \cdots & \cdots \\ g_{m1} & g_{m2} & \cdots & g_{mn} \end{bmatrix},$$

each row of the matrix represents each chromosome of the population. The m and n are the number of chromosomes and the number of features remaining from PCA, in this case the number of features is 12. Each gene of chromosome inside the matrix G is a weighing factor of value ranging from 0 to 10. These weighing factors are used to weigh features corresponding to the matrix G column.

In order to used matrix G each row of the matrix is put in to a diagonal matrix G_{diag} below.

$$G_{diag} = \begin{bmatrix} g_{11} & 0 & 0 & 0 \\ 0 & \ddots & 0 & 0 \\ 0 & 0 & \ddots & 0 \\ 0 & 0 & 0 & g_{1n} \end{bmatrix},$$

Each sample set features is defined to be a row vector x_f with n elements.

$$x_f = \begin{bmatrix} x_{f1} \\ x_{f2} \\ \vdots \\ x_{f11} \\ x_{fn} \end{bmatrix},$$

After the first set of population is generated, each chromosome of the weighing factor is used to weigh the heart sound by multiplying G_{diag} with x_f. The weighed heart sound samples are then evaluated using 10 fold cross validation method [14] to obtained the accuracy of the weighed data set when using SVM as classifier. 10 fold cross validation divides the samples set in to 10 subset and uses 9 of them to train SVM and 1 subset to test the SVM for 10 times, this means that every samples are used both to train and test the SVM classifier. SVM classifier will be further discuss in detail later on in the next section. The accuracy product of each weighing factor set is acquired, it is then used as a fitness value to indicate the fitness of each chromosome. Chromosome that has higher accuracy rate is fitter than that of lower fitness value.

The selection process of GA uses these accuracy to select how to process each chromosome. Elite chromosomes are the first 4 chromosomes that has the highest accuracy among population, the elite chromosomes are kept unchanged into the next generation. The last 4 chromosomes with lowest accuracy however will be eliminated. The rest of the chromosomes of the next generation are generated by crossover and mutation. Crossover that is used in the system is a one point crossover method as illustrated in Figure 3, where random gene in the chromosome is selected to be the point of crossover that two chromosomes swapped their genes. Each crossover process produces 2 offspring chromosomes.

Fig. 3. One point crossover

In this system the determination of which chromosomes to be crossover to each other is simply done by randomly pair up every chromosomes in the population after elimination of the last 4 least accurate chromosomes. After crossover there is 30% chance that the offspring of crossover will mutate in the first generation. The mutation rate keeps decreasing as GA generates the next generation, this setting is due to the fact that GA approaches the solution as GA generates next population. The mutation in the proposed system is done by randomly alter 4 genes inside the chromosome. When chromosome mutates, those weighing factor in each gene will be regenerated. After the population goes through selection, crossover and mutation the final product is a new population that will replace the previous one. This process is continued until a specified number of generations is reached.

2.3 SVM Classification

In order to classify all the heart sound samples into two categories, normal and unhealthy the Support Vector Machine method is chosen [7]. Training of SVM classification using RBF kernel is proven to be an effective classification method for higher dimension classification such as heart sound signal in the proposed system, which has 12 features per sample. The classification accuracy is evaluated using a 10 fold cross validation method as stated earlier. The accuracy is then used to determine the fitness value of each set of weighing factors for features set generated from GA.

SVM classifier with RBF-Kernel function finds the solution of the model by mapping the nonlinear input data into a higher dimensional feature space. This is done to make the feature data separable. These separable features are in a form of data point z_i. The hyperplane that is the best at separating data are constructed from many possible hyperplanes by the following equation (4) and (5) where y_i is a possible hyperplanes

$$y_i(w \cdot z_i + b) \geq 1 - \xi, \forall i \tag{4}$$

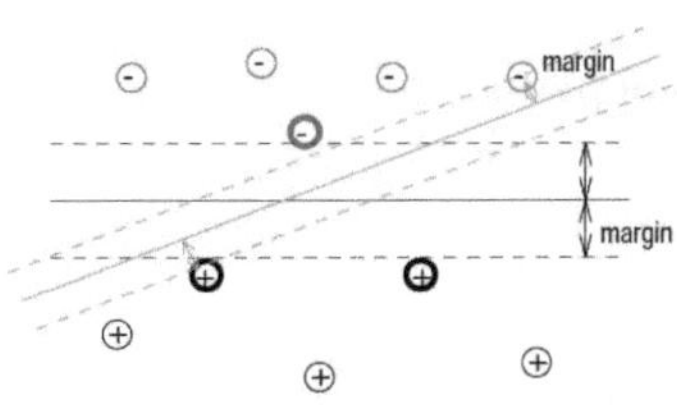

Fig. 4. SVM plane

The maximum margin in Figure 4 is attained by (5) and (6)

$$minimizing \quad \frac{1}{2} w \cdot w + C \sum_{i=1}^{L} \xi_i \tag{5}$$

C is regularization parameter determined by a user and ξ_i is a non negative slack variable.

$$subject\ to\ constraint\ y_i(w \cdot z_i + b_) \geq 1 - \xi_i \text{ and } \xi \geq 0, \forall i \tag{6}$$

And lastly the minimum training error by (7) and(8)

$$maximizing \quad W(\alpha) = \sum_{i=1}^{L} \alpha_i - \frac{1}{2} \sum_{i=1}^{L} \sum_{j=1}^{L} \alpha_i \alpha_j y_i y_j K(x_i x_j) \tag{7}$$

$$subject\ to\ constraint \sum_{i=1}^{n} \alpha_i = 0, 0 \leq \alpha_i \leq C;\ i = 1, ..., n \quad (8)$$

In these equations variables are as followed, i is number of input data, x_i is a nonlinear input, w is the weight vector, b is scalar threshold value, C is regularization parameter, ξ_i is a nonnegative slack variable and the coefficient and lastly α_i is a nonnegative Lagrangian multipliers found from solving the Quadratic Programming problem. RBF kernel function which is chosen as kernel function is (9)

$$K(x, x_i) = exp(-\frac{\parallel x - x_i \parallel^2}{2\phi^2}), i = 1, ..., n \quad (9)$$

Here the variable ϕ is a kernel width which is selected by cross-validation process to define the optimum separate group of the classified support vectors.

3 Simulation Result

The result depicted here is from an initialized population of 60 chromosomes, meaning that GA starts with 60 random weighing factors sets. The reason that a small population is chosen is that the number of features is 12, which is considerably higher than most of the work done using GA, training and testing SVM classifier with them takes a significant amount of computational time. The simulation result of SVM classifier from simulation with no GA and a result from GA with 60 generations is depicted in Table 1.

Table 1. False Negative(FN) , false positive , true negative(TN), true positive(TP) false positive(FP)

Setting	FN	FP	TN	TP	Sensitivity	Specificity	Accuracy
No GA	27	0	144	181	87.02%	100%	92.33%
GA	23	0	144	185	88.94%	100%	93.47%

The result in Table 1 is produced from the highest accuracy set of weighing factors from GA. The accuracy is improved as expected. To be precise when SVM is trained while implementing GA of 60 initialized population size and 60 generations the accuracy is improved by 1.14%. Table 1 here also shows false negative, false positive, true negative, true positive, sensitivity and specificity of the system. Positive is a abnormal classification and negative is a normal classification. Sensitivity is the percentage of true positive divided by total real positive. Specificity is the percentage of true negative divided by total real negative.

One significant drawback of using GA here is that the GA method in this system evaluate its chromosomes by using the classification accuracy of SVM as fitness function. Training and testing SVM classifier can take a significant amount of computational time especially when the dimension of the data is high.

For the simulations in Table 1 the average time the simulation takes for each generation of GA is approximately 300 seconds per generation of GA responsible for the execution time of over 5 hours. One thing to be noted is that if the system either increases number of samples or number of features this performance time can increase significantly.

4 Conclusion

This paper aims to propose a new alternative screening diagnostics tool at low cost. The system uses Wavelet packet based feature extraction to extract the significant features from the heart sound samples to form 126 features set then the dimension of the features is then further reduced using PCA. To obtain the best solution of the weighing factors for the features to be used to train SVM classifier GA is used. GA will generate a random population of weighing factors as chromosomes. These chromosomes are then evolve with the next generation determined by the fitness of each chromosomes where the chromosomes that have high classification accuracy will be chosen to be parents of the next generation and enter the process of crossover and mutation. The fittest chromosome of the last generation will be used to train the final SVM classifier. The result classifier accuracy is improved comparing to not using weighing factors.

5 Future Work

The accuracy of the system may be improved simply by adjusting the population size and number of generations of GA or adjusting the weighing factors setting and the crossover method to, this require some more experiments to determine the best setting. Another thing that may increase the accuracy as well is the number of features SVM uses, but if the number of features is increased computational time also increases significantly.

The proposed system yields higher accuracy but requires significantly longer computational time, this increase in computational time is mainly from the SVM cross validation. The problem could be solved by using a new evaluation method with less computational time.

The proposed system should be tested on real raw heart sound collected on site from the patients using digital stethoscope. Graphic User Interface(GUI) can also be developed for real-time usage of the algorithm.

Acknowledgment. This research is financially supported by Center of Excellence in Biomedical Engineering, Sirindhorn International Institute of Technology (SIIT), Thammasat University (TU) and the National Research University Project, Thailand Office of Higher Education Commission.

References

1. Public Health Organization of Thailand: Public Health Statistics, http://bps.ops.moph.go.th/Healthinformation/statistic50/statistic50.html
2. World Health Organization: Prevention of Cardiovascular Disease Pocket Guidelines for Assessment and Management of Cardiovascular Risk (2007)
3. Clark III, D.: An argument for reviving the disappearing skill of cardiac auscultation (2012), http://www.ccjm.org/content/79/8/536.full
4. Bunluechokchai, C., Ussawawongaraya, W.: A Wavelet-based Factor for Classication of Heart Sounds with Mitral Regurgitation. International Journal of Applied Biomedical Engineering 2(1) (2009)
5. Gupta, C.N., Palaniappan, R., Swaminathan, S., Krishnan, S.M.: Neural Network Classiícation of Homomorphic Segmented Heart Sounds. Biomedical Engineering Research Center, Nanyang Technological University, Singapore (2007)
6. Chebil, J., Al-Nabulsi, J.: Classification of Heart Sound Signal Using Discrete Wavelet Analysis. International Journal of Soft Computing 2(1), 37–41 (2007)
7. Chatunapalak, I., Phatiwuttipat, P., Kongprawechnon, W., Tungpimolrut, K.: Childhood Musical Murmur Classification with Support Vector Machine Technique. International Institute of Technology, Thammasat University, Pathum Thani, Thailand (2012)
8. Phatiwuttipat, P., Kongprawechnon, W., Tungpimolrut, K.: Cardiac Auscultation Analysis System with Neural Network and SVM Technique. In: ECTI Conference, Konkaen (2011)
9. Eads, D., Hillb, D., Davisa, S., Perkinsa, S., Maa, J., Portera, R., Theilera, J.: Genetic Algorithms and Support Vector Machines for Time Series Classification. Rochester Institute of Technology, New York (2012)
10. Huang, C.L., Wang, C.J.: A GA-based feature selection and parameters optimization for support vector machines. Department of Information Management, National Kaohsiung First University of Science and Technology, Taiwan (2006)
11. Dacal-Nieto, A., Vazquez-Fernandez, E., Formella, A., Martin, F., Torres-Guijarro, S., Gonzalez-Jorge, H.: Genetic Algorithms approach for feature selection in potatoes classification by computer vision. Computer Science Department, University of Vigo, Spain (2009)
12. Whitley, D.: A Genetic Algorithm Tutorial. Computer Science Department, Colorado State University, Colorado, USA (1993)
13. Pei, M., Goodman, E.D., Punch, W.F.: Feature Extraction Using Genetic Algorithms. Department of Computer Science Genetic Algorithms Research and Applications Group(GARAGe), Michigan State University, Michigan, USA (1997)
14. Kohavi, R.: A Study of Cross-Validation and Bootstrap for Accuracy Estimation and Model Selection. Computer Science Department, Stanford University, CA, USA (1995)

Real-Time FPGA-Based Human Iris Recognition Embedded System: Zero-Delay Human Iris Feature Extraction

Amirshahram Hematian[1], Suriayati Chuprat[1], Azizah Abdul Manaf[1], Sepideh Yazdani[2], and Nadia Parsazadeh[1]

[1] Advanced Informatics School (AIS),
[2] Center for Artificial Inteligence and Robotics (CAIRO),
Universiti Teknologi Malaysia. 54100 Kuala Lumpur, Malaysia
{hamirshahram2,ysepideh,pnadia4}@live.utm.my
{suria,azizah07}@ic.utm.my
http://www.utm.my

Abstract. Nowadays most of iris recognition algorithms are implemented based on sequential operations running on central processing units (CPUs). Conventional iris recognition systems use a frame grabber to capture a high quality image of an eye, and then system shall locate the pupil and iris boundaries, unwrap the iris image, and extract the iris image features. In this article we propose a prototype design based on pipeline architecture and combinational logic implemented on field-programmable gate array (FPGA). We achieved to speed up the iris recognition process by localizing the pupil and iris boundaries, unwrapping the iris image and extracting features of the iris image while image capturing was in progress. Consequently, live images from human eye can be processed continuously without any delay or lag. We conclude that iris recognition acceleration by pipeline architecture and combinational logic can be a complete success when it is implemented on low-cost FPGAs.

Keywords: Biometrics, FPGA, Human Identification, Iris Recognition, Pattern Matching, Zero-Delay.

1 Introduction

In biometric recognition systems, features are extracted from human fingerprints, voice, face, eye and iris. With the unique structure of the human iris, iris recognition technology is founded as a leading-edge technology in human identification. In the first proposed method by Daugman's algorithms [1], a Gabor wavelet transform was used for feature extraction to generate the iris code and in another algorithm [2] they used pattern matching. In Daugman's algorithms [1], amplitude information of the pixels was discarded and the 2048 bits was generated from phase information of the transform. Time-consuming operations like traversing gigantic number of pixels within an image to localize the pupil and

P. Meesad et al. (Eds.): *IC²IT2013*, AISC 209, pp. 195–204.
DOI: 10.1007/978-3-642-37371-8_23

iris boundaries, non-linear calculations like unwrapping the iris image into a fixed-size rectangle, iris image enhancement and normalization, hamming distance calculation and all other pre/post processes have major effects on the system reliability. Consequently, usage of multi-processor platforms is increasingly adopted. On the other hand, FPGAs can provide high performance, configurable and complex combinational functions or simple logic gates like AND and XOR that can be used in hamming distance calculations, as an example for iris recognition purpose. FPGAs propound the hardware implementation of functions to deliver dedicated hardware for each part of the process. Hardware dedicated processes can function in parallel or sequential operations, depending on the purpose of use. Pipelined architectures for hardware dedicated functions can provide high performance computing. Furthermore, use of combinational logic instead of sequential logic can play a major role to achieve the optimum computing performance. In comparison with conventional iris recognition algorithms, we achieved zero-delay iris and pupil localization, iris image unwrapping, and iris image feature extraction by the proposed design. In simple words, when a process takes time between two vertical syncs of two video frames, it is considered as zero-delay or zero-lag process. Accordingly, delta time of the process is smaller than duration of a video frame. The value of delta time depends directly on the pixel clock of the incoming video stream. In the proposed design, almost all processes work based on pixel clock to return the result of analysis before the end of each vertical sync in parallel. Consequently, all video frames can be processed without any delay. Our iris recognition system (IRIS1) is capable of processing up to 32 video frames per second at clock speed 50MHz. Fig. 1 shows IRIS1 prototype.

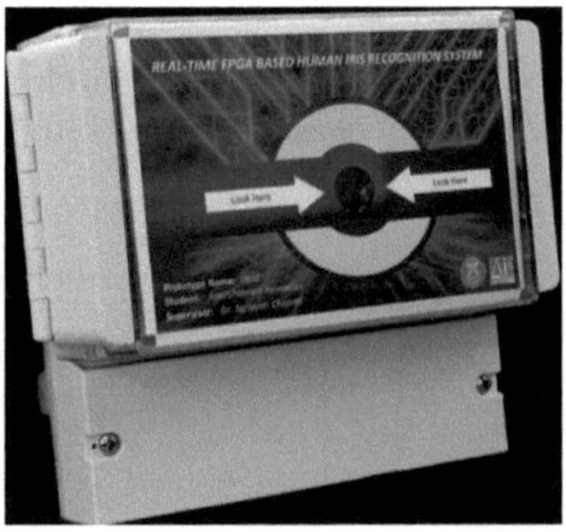

Fig. 1. IRIS1 prototype

2 Related Works

Iris recognition, as an essential method in biometric identification has been improved dramatically in the last decade. Various types of implementations have been done [1], [4–23] to achieve maximum accuracy and speed. In case of hardware implementation [24–30] they have proposed FPGA system models as a solution for real-time iris recognition system. In one case [24] they used a soft processor as the main processing unit of the system, but they did not implement image processing operations directly on the hardware. In other cases [25–30] they managed to get the advantage of FPGA in order to improve iris recognition performance using dedicated hardware resources. In order to optimize iris recognition performance, only some parts of image processing operations are implemented on the device, but they still have lags and delays in sequential parts. From the local feature extraction aspect, as demonstrated in [31], iris recognition methods can be categorized to three major types: phase-based methods [1], [6], [7], [22], zero-crossing representation methods [10], [15], and texture-analysis-based methods [5], [8], [11], [13], [16]. Daugman [1], [7], [22] used texture phase structure information of the iris image to generate a 2048 bit iris code. In this paper, we propose a new phase-based iris feature extraction method optimized for hardware implementation using VHDL. We also demonstrate experimental results of our implementation on IRIS1 prototype system based on Altera Cyclone-II EP2C20 (DE-1) Development Kit.

3 IRIS1 Embedded System Overview

We should highlight that all considerations in this paper refer to capturing video frames from human eye under Near Infra-Red lighting that only human eye is in the image, as shown in Fig. 2.

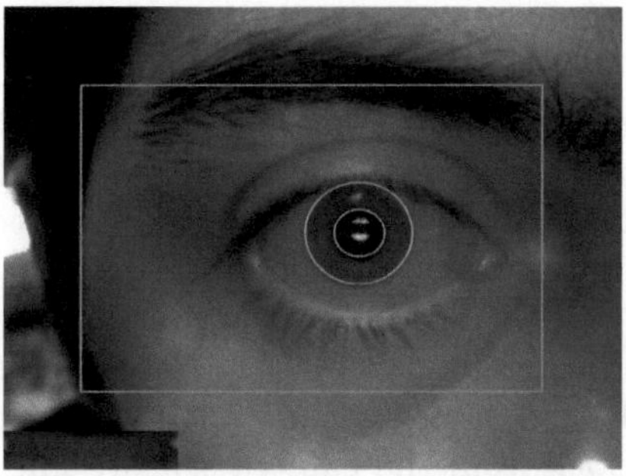

Fig. 2. IRIS1 prototype VGA output

We used OV7670 camera sensor in IRIS1 prototype to capture video frames with resolution 640x480 in YUV colour space and progressive scan mode. For image processing, we only used luma information of the image with bit depth 8-bit and stored the image data within 512KB SRAM on DE-1 board. Maximum video frame rate of OV7670 camera sensor is 32 frames per second. In this section, a novel method based on FPGA is presented and optimized for high performance iris recognition. Benchmark results of iris image enhancement, pupil and iris localization, iris image unwrapping, and iris image feature extraction are also included in this paper. Following is the overview of the proposed design in details as described below:

VHDL Design. In order to implement image processing functions in hardware language we used VHSIC Hardware Description Language (VHDL). At the earlier stage of development, we defined a system design diagram that all inputs, outputs and logic blocks were defined in generic, as shown in Fig. 3.

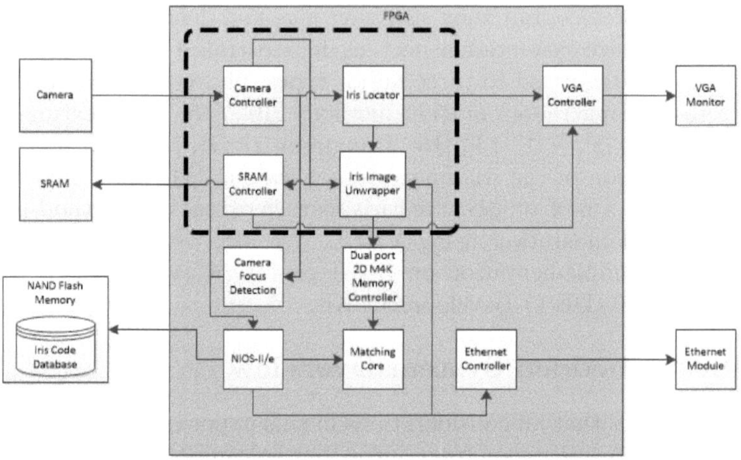

Fig. 3. IRIS1 System Design Diagram

SRAM Controller. The SRAM on DE-1 board (IS61LV25616-10T) delivers high speed access time up to 10ns. In order to cover address and data setup and hold time for each access, we used clock speed 50MHz (20ns) and a state machine with four states. First two states were dedicated to reading for VGA controller and the last two states were shared for read or write to SRAM by iris image unwrapper logic block. SRAM controller was designed to deliver read-only access for VGA controller at 640x480x60Hz with clock speed 25MHz driven from clock 50MHz, and read/write access for iris image unwrapper logic block which works based on pixel clock from the camera sensor. The SRAM controller has three separate ports for VGA controller and iris image unwrapper logic blocks, as shown in Fig. 4.

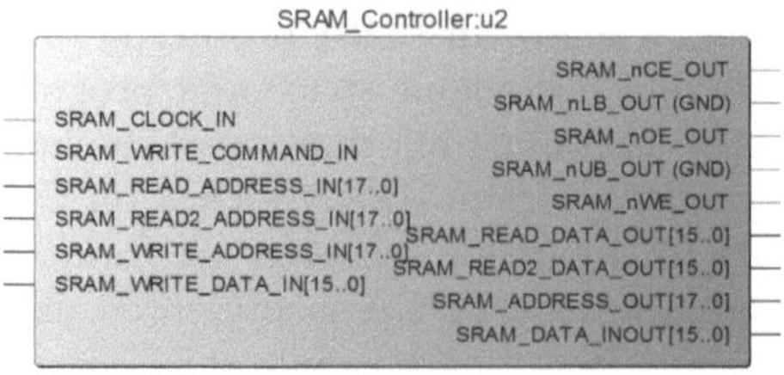

Fig. 4. IRIS1 SRAM Controller Logic Block

Camera Controller. The camera controller was designed to receive streaming digital video frames from the camera and extract luma information of the image data. Meanwhile, a digital filter normalized input luma information based on previous luma information in the Region Of Interest (ROI). In IRIS1 the margins of ROI were 80 pixels from every corner of video frame. The normalized image data was provided for the iris locator logic block which was responsible to localize the pupil and the iris. Luma normalization filter prevents low contrast images to fool the iris locator. This digital filter updates the luma scale factor at each vertical sync, and calculates the new luma scale factor if the incoming pixel data belongs to the ROI. Fig. 5 shows the timing of luma extraction and normalization analysis.

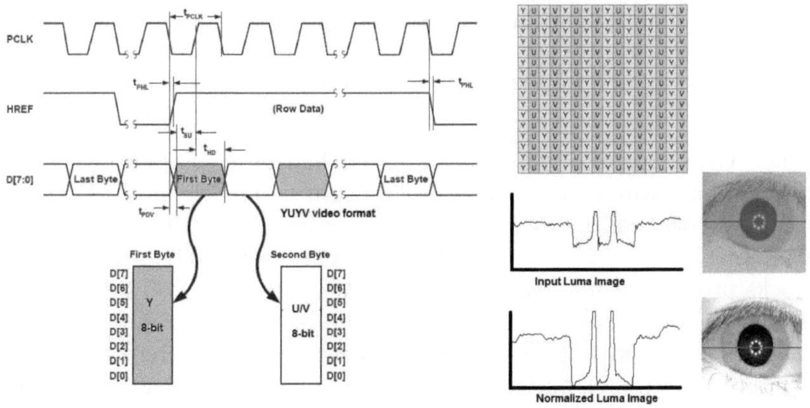

Fig. 5. (Left)IRIS1 Camera Controller Timing Diagram, (Right)Luma Normalization Analysis

Iris Locator. We should underline that we assumed that the pupil is the darkest part of the image of an eye, and our system receives pixels data in progressive-scan mode. In our proposed method, in order to reduce complexity of the operations, we decided to do the calculations based on concentric circles as pupil and iris boundaries. In the first step of iris localization, the pupil locator finds the pupil in the video frame by clustering video frame data via a fuzzy logic filter in order to convert gray-scale image data to binary image data. During this conversion, the pupil locator compares pairs of pixels in each line to find the maximum length of a continuous black line in the current line of the binary video frame. The longest black line within ROI is considered as pupil diameter, and pupil center point is calculated based on the center point of the longest black line. Moreover, diameter of the iris is bounded, and it cannot be more than four times bigger than the pupil diameter. Whilst the pupil localization is

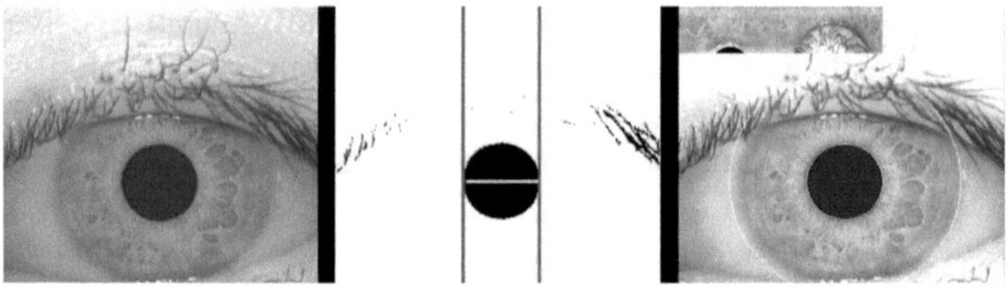

Fig. 6. (Left)Captured Image From CASIA Iris Database, (Middle)Monochrome Image For The Pupil Localization, (Right)The Iris and Pupil Localization

in progress, iris localization is also in progress based on pupil position and radius using histogram analysis of the horizontal line, crossing pupil center point to the right-side of the video fame until pixel intensity phase change value becomes negligible, and system reaches the sclera. Fig. 6 shows the result of the pupil and iris localization on an image of an eye from CASIA iris database[3].

Iris Image Unwrapper. One of the major challenges in complex process implementation on hardware is to use non-integral values in mathematical operations. We used fixed-point values for decimal arithmetics in the iris unwrapping logic block. Accuracy of the fraction part of the calculation is 0.001, sin(θ) and cos(θ) logic blocks provide decimal values in return when $0 \leq \theta \leq 360$ to convert pixel addresses from line scan type to circular type in the video memory (SRAM) where the iris image data is located. This process will be started when the pixel clock is out of ROI, then a fixed size rectangle (180x40) is used as a place that iris unwrapping logic block will store unwrapped iris image data in video memory (bottom-left of image in video memory). The reason that system stores the iris image data in this way is that image data format is in progressive-scan mode. Image data is started from top-left of image and will end up in bottom-right of the image. We should mention that any pixel clock out of ROI can be used for extra processes, in this case, after iris localization in ROI, system will unwrap the iris image into a fixed size rectangle. This logic block will drop any pixel data coming from the camera at this point and will replace them with the iris image data using sin(θ) and cos(θ) logic blocks. Meanwhile, each pair of pixels is encoded to one encoded bit as a feature, in parallel. Due to the light changes while capturing images from human eye, amplitude information of pixels is discarded and only phase information is encoded. In this case, slope direction between a pair of pixels is encoded as a feature. If the slope value is a positive value, the encoded bit is considered as 1, and if the slope value is a negative value, the encoded bit is considered as 0. Fig. 7 shows the result of iris image feature extraction.

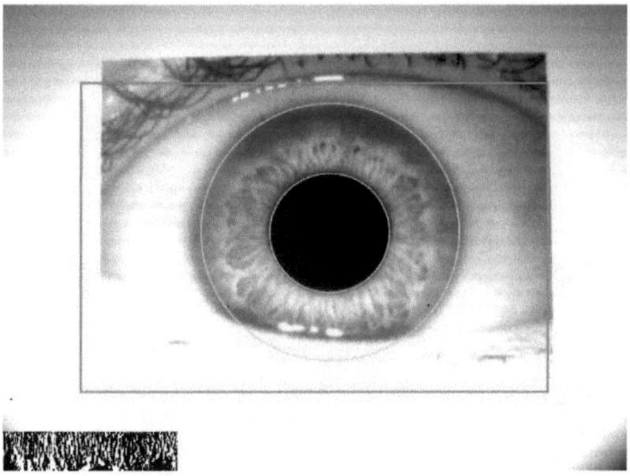

Fig. 7. Iris features are extracted from CASIA Iris Database

4 Experimental Results

Timing Considerations. As indicated in Fig. 3, this paper only demonstrates logic blocks related to the iris image feature extraction. Total number of clocks for an image with resolution 640x480 pixels with pixel format YUV (Y8U8Y8V8) is equal to 614,400 clocks. As we mentioned above, we only use luma (Y8) information for the described logic blocks. This will divide the number of total clocks in two which is equal to 307,200 clocks. If a process takes time between clock number 0 to 307,199 that process is considered as a Zero-Delay process. In IRIS1 prototype Zero-Delay processes do not cause any delay that may result in dropping video frames. For every video frame, the iris locator logic block finds the pupil and iris boundaries in the ROI and sends the information to the iris image unwrapper logic block. When the incoming pixel addresses refer to the bottom-left corner of the image (Out of ROI), the iris image unwrapper logic block replaces the incoming pixels with the iris image pixels which are stored recently in the video memory. In parallel, the iris image unwrapper logic block stores the extracted features of the iris image into a 2D on-chip M4K memory of the FPGA for identification and enrolment purposes. Fig. 8 shows the processing areas in the image. Thus before end of transfer of the image from camera to the FPGA, all operations are completed. Consequently, there is no delay or lag between video frames and all frames are processed. Moreover, Fig. 8 shows the comparison between conventional iris recognition method and the proposed method with Zero-Delay feature running on IRIS1 prototype.

Comparison. As indicated in [34] VeriEye from Neurotechnology judged one of the fastest and most accurate iris recognition algorithms in NIST IREX III Evaluation (April 16, 2012). In the best case with template size 2,328 bytes, the iris template extraction time is 80 milliseconds running on Intel Core i7-2600 3.4GHz (Quad Core). In comparison with IRIS1 prototype with template

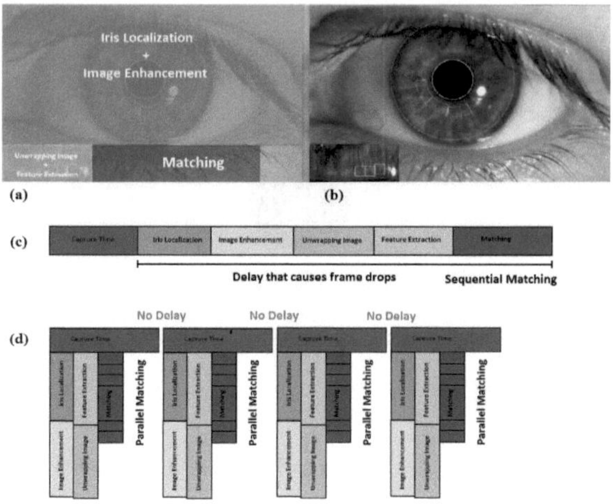

Fig. 8. (a)Processing Areas of The Image, (b)Real Image, (c)Timing of Conventional Iris Recognition Method, (d)Timing of IRIS1 Iris Recognition Embedded System

Table 1. Iris localization processing time comparison table

Method	Time of Iris Localization(ms)	Accuracy(Percent)
Daugman[1]	213	98
Tisse[21]	153	99
Wildes[23]	66	99.5
Proposed	Zero	94

size 900 bytes, the iris template extraction time is Zero in the proposed design. IRIS1 is capable of processing 32 video frames per second at memory clock 50MHz and memory size 512KB. Table. 1 shows the timing comparison for the iris localization between the proposed design on IRIS1 prototype and elder iris recognition methods that have been tested on Intel Celeron CPU 1.8GHz and 1.0GB RAM [33].

5 Conclusion

In this paper, we have proposed a new design based on hardware acceleration for iris recognition using low-cost FPGAs in order to satisfy all timing considerations by parallel computing. The iris and pupil localization, iris image unwrapping, enhancement and feature extraction have been demonstrated in this article including experimental results. It has been proved that using dedicated hardware resources for time consuming operations can end in high performance computing. Performance comparison of existing methods for iris localization and feature extraction have been conducted on CASIA iris database. Such experiments prove that dedicated hardware architectures can gigantically increase the performance of existing iris recognition methods. Moreover, Cyclone-II FPGAs provide a low-cost alternative for parallel computing systems like iris recognition that may need to compare millions of irises in a very short time at low-cost.

References

1. Daugman, J.G.: High Confidence Visual Recognition of Persons By a Test of Statistical Independence. IEEE Trans. Pattern Anal. Mach. Intell. 15, 1148–1161 (1993)
2. Wildes, R.P., Asmuth, J.C., Green, G.L., Hsu, S.C., Kolczynski, R.J., et al.: A System for Automated Iris Recognition. In: Second IEEE Workshop on Applications of Computer Vision, Sarasota, FL (1994)
3. CASIA iris database, `http://www.cbsr.ia.ac.cn`
4. Wildes, R.P., Asmuth, J.C., Hanna, K.J., Hsu, S.C., Kolczynski, R.J., et al.: Automated, Non-Invasive Iris Recognition System and Method. U.S. Patent 5572596 (1996)
5. Zhu, Y., Tan, T., Wang, Y.: Biometric Personal Identification Based on Iris Patterns. In: 15th International Conference on Pattern Recognition, Barcelona (2000)
6. Daugman, J.G.: Biometric Personal Identification System Based on Iris Analysis. U.S. Patent 5291560 (1994)
7. Daugman, J.G.: Demodulation by Complex-Valued Wavelets for Stochastic Pattern Recognition. International Journal of Wavelets, Multi-resolution and Information Processing 1, 1–17 (2003)
8. Lee, K., Lim, S., Byeon, O., Kim, T.: Efficient Iris Recognition Through Improvement of Feature Vector and Classifier. ETRI 23, 61–70 (2001)
9. McHugh, J.T., Lee, J.H., Kuhla, C.B.: Handheld Iris Imaging Apparatus and Method. U.S. Patent 6289113 (1998)
10. Boles, W.W., Boashash, B.: A Human Identification Technique Using Images of the Iris and Wavelet Transform. IEEE Trans. Signal Process. 46, 1185–1188 (1998)
11. Ma, L., Wang, Y., Tan, T.: Iris Recognition Based on Multichannel Gabor Filtering. In: International Conference on Asian Conference on Computer Vision (2002)
12. Flom, L., Safir, A.: Iris Recognition System. U.S. Patent 4641394 (1987)
13. Ma, L., Wang, Y., Tan, T.: Iris Recognition Using Circular Symmetric Filters. In: 16th International Conference on Pattern Recognition (2002)
14. Sanchez-Reillo, R., Sanchez-Avila, C.: Iris Recognition with Low Template Size. In: International Conference of Audio and Video-Based Biometric Person Authentication (2001)
15. Sanchez-Avila, C., Sanchez-Reillo, R., de Martin-Roche, D.: Iris-Based Biometric Recognition Using Dyadic Wavelet Transform. IEEE Trans. Aerosp. Electron. Syst. 17, 3–6 (2002)
16. Wildes, R.P., Asmuth, J.C., Green, G.L., Hsu, S.C., Kolczynski, R.J., et al.: A Machine-Vision System for Iris Recognition. Machine Vision and Applications 9, 1–8
17. Rozmus, J.M., Salganicoff, M.: Method and Apparatus for Illuminating and Imaging Eyes Through Eyeglasses. U.S. Patent 6069967 (1997)
18. Camus, T.A., Salganicoff, M., Chmielewski, T.A., Hanna, J.K.J.: Method and Apparatus for Removal of Bright or Dark Spots by the Fusion of Multiple Images. U.S. Patent 6088470 (1998)
19. Zhang, G.H., Salganicoff, M.: Method of Measuring the Focus of Close-up Iages of Eyes. U.S. Patent 5953440 (1999)

20. Tan, T., Wang, Y., Ma, L.: A New Sensor for Live Iris Imaging. PR China Patent ZL 01278644.6 (2001)
21. Tisse, C., Martin, L., Torres, L., Robert, M.: Person Identification Technique Using Human Iris Recognition, pp. 294–299 (2002)
22. Daugman, J.G.: Statistical Richness of Visual Phase Information: Update on Recognizing Persons by Iris Patterns. International Journal of Computer Vision 45, 25–38 (2001)
23. Wildes, R.P.: Iris Recognition: An Emerging Biometric Technology. IEEE 85(9), 1348–1363 (1997)
24. Hentati, R., Bousselmi, M., Abid, M.: An Embedded System for Iris Recognition. In: 5th International Conference on Design and Technology of Integrated Systems in Nanoscale Era, Hammamet (2010)
25. Liu-Jimenez, J., Sanchez-Reillo, R., Lindoso, A., Miguel- Hurtado, O.: FPGA Implementation for an Iris Biometric Processor. In: IEEE International Conference on Field Programmable Technology, Bangkok (2006)
26. Yasin, F.M., Tan, A.L., Reaz, M.B.I.: The FPGA Prototyping of Iris Recognition for Biometric Identification Employing Neural Network. In: 16th International Conference on Microelectronics (2004)
27. Hu-lin, Z., Mei, X.: Iris Biometic Processor Enhanced Module FPGA-based Design. In: Second International Conference on Computer Modelling and Simulation, Sanya (2010)
28. Grabowski, K., Sankowski, W., Napieralska, M., Zubert, M., Napieralski, A.: Iris Recognition Algorithm Optimized for Hardware Implementation. In: IEEE Symposium on Computational Intelligence and Bioinformatics and Computational Biology, Toronto (2006)
29. Rakvic, R.N., Ulis, B.J., Broussard, R.P., Ives, R.W., Steiner, N.: Parallelizing Iris Recognition. IEEE Trans. Inf. Forens. Security 4, 812–823 (2009)
30. Reaz, M.B.I., Sulaiman, M.S., Yasin, F.M., Leng, T.A.: Iris Recognition Using Neural Network Based on VHDL Prototyping. In: International Conference on Information and Communication Technologies: From Theory to Applications (2004)
31. Ma, L., Tan, T., Wang, Y., Zhang, D.: Personal Identification Based on Iris Texture Analysis. IEEE Trans. Pattern Anal. Mach. Intell. 25, 1519–1533 (2003)
32. Altera Corporation: Cyclone II Device Family Data sheet, Cyclone II Device Handbook 1 (2007)
33. Alvarez-Betancourt, Y., Garcia-Silvente, M.: A Fast Iris Location Based on Aggregating Gradient Approximation Using QMA-OWA Operator. In: IEEE International Conference on Fuzzy Systems, Barcelona (2010)
34. Verieye from NeuroTechnology, `http://www.neurotechnology.com`

Reconstruction of Triple-wise Relationships in Biological Networks from Profiling Data

Quynh Diep Nguyen[1,⋆], Tho Hoan Pham[1,⋆], Tu Bao Ho[2], Van Hoang Nguyen[3], and Dang Hung Tran[1]

[1] The Center for Computational Science, Hanoi National University of Education, Vietnam
{diepnq,hoanpt,hungtd}@hnue.edu.vn

[2] Japan Advanced Institute of Science and Technology, Japan
bao@jaist.ac.jp

[3] Hanoi University of Agriculture, Vietnam
hoangnv@hua.edu.vn

Abstract. Reconstruction of biological networks from profiling data is one of the most challenges in systems biology. Methods that use some measures in information theory to reconstruct local relationships in biological networks are often preferred over others due to their simplicity and low computation cost. Present mutual information-based methods cannot detect as well as provide relationships that take into account more than two variables (called multivariate relationships or k-wise relationships). Some previous studies have tried to extend mutual information from two to multiple variables; however the interpretation of these extensions is not clear. We introduce a novel interpretation and visualization of mutual information between two variables. With the new interpretation, we then extend mutual information to multiple variables that can capture different categories of multivariate relationships. We illustrate the prediction performance of these multivariate mutual information measures in reconstructing three-variable relationships on different benchmark networks.

Keywords: Mutual information, information theory, biological networks.

1 Background

Thanks to advances in biotechnology, expression of thousands of biological substances (genes, proteins or metabolites) can be simultaneously measured. Inferring biological relationships, pathways and networks from these high-throughput profiling data is one of the most challenges in systems biology. There have been a growing number of computational methods to reconstruct biological networks. They can be broadly divided into three main approaches. The first one includes information theoretic methods that find statistical dependences between two variables/biological substances (dependence here refers to any situation in

⋆ Corresponding authors.

P. Meesad et al. (Eds.): *IC²IT2013*, AISC 209, pp. 205–215.
DOI: 10.1007/978-3-642-37371-8_24

which random variables do not satisfy a mathematical condition of probabilistic independence), typically [1,7,9]. These methods rely on two basic measures of information theory, mutual information and correlation coefficient, to capture local relationships between substances. The second approach consists of graphical learning methods that maximize a scoring function over some alternative network models to find the best one fitting the data [13,24]. Since the network model space is often very huge, these graphical methods usually use some searching heuristic strategies from specific initial networks. In consequence, they might get stuck in local optima. The third one includes mathematical models that try to find parameters of a system of mathematical differential and difference equations from the training data. Due to the computational complexity, only linear or simple functions are considered in the mathematical equation models. Some excellent reviews on different aspects of biological network reconstructing methods can be found in [11,15,21].

Among the three above approaches, graphical and mathematical models can provide multivariate relationships (relationships involving more than two substances at the same time) in their learned models. However, they cannot work well on large networks since the number of parameters of the models grows exponentially in the fan-in or fan-out of a node. On the contrary, the information theoretic approach has advantages of simplicity and low computational costs; it is superior to the graphical and mathematical approaches in reconstructing large networks [9,20]. However, the information theoretic approach until now does not provide and in some cases cannot detect relationships broader than pair-wise ones.

The reason why information theoretic methods have two above disadvantages is that they use mutual information (MI), which is firstly defined as a dependence measure between only two variables [6,14]. Usually, people think that pair-wise relationships can provide evidence for relationships of multiple variables. But it is not always true. For example, if there are all three pair-wise relationships (X, Y), (X, Z) and (Y, Z) of three variables, it is not able to infer if X, Y and Z are from the same multivariate relationship because those pair-wise relationships might come from different independent events at different time. In additionally, mutual information can detect local relationships between two variables; it may miss some more general relationships, i.e. multivariate relationships, such as synergy ones [10,22]. For example, the *xor* relationship among three binary variables $X \; xor \, Y \; xor \, Z = 0$ will not be discovered if we use only mutual information between any pair of variables [18].

Extension of mutual information from two to multiple variables is not trivial, even for the case of three variables [5,16]. All efforts so far, in our understanding, seem to formulate a single measure that characterizes specially a n-variable relationship/dependence (or k-wise relationship), which remains imprecisely defined. Total mutual information and interaction information are such a kind of measures [3,6,8,14]. While the first one integrates all dependences among multiple variables at different levels and it is not able to extract the k-wise dependence from the total mutual information; the second has attracted much controversy on the interpretation and the usage [4,16].

Table 1. Pair-wise and triple-wise relationship networks. Original relationships in the first column have sub-relationships (derivatives of the original relationships) at pair-wise and triple-wise levels in the second and the third columns respectively

Full original relationships	pair-wise relationships	three-variable relationships
$G_2\_synthesis : G_4$	(G_2, G_4)	none
$G_3\_synthesis : (\sim G_1).G_2$	(G_1, G_3), (G_2, G_3)	(G_1, G_2, G_3)
$G_1\_synthesis : G_2 + G_4.(\sim G_3)$	$(G_1, G_2), (G_1, G_3)$ and (G_1, G_4)	$(G_1, G_2, G_3), (G_1, G_2, G_4)$, and (G_1, G_3, G_4)
$X_1\ xor\ X_2\ xor\ X_3 = 0$	(X_1, X_2), (X_1, X_3) and (X_2, X_3)	(X_1, X_2, X_3)

In previous work [18], we provided a novel interpretation of MI between two variables based on the physical view of distribution entropy. We then extended MI for more than two variables based on the proposed interpretation. Different from the case of two variables where there is a unique MI measure, in the case of three or more variables there are a variety of MI measures that quantify different dependence categories existing among multiple variables. In this work, we will investigate the prediction performance of proposed multivariate mutual information measures in reconstructing triple-wise relationship networks.

2 Methods

2.1 Pair-wise and Triple-wise Relationships in Biological Networks

The ultimate objective of a biological network inference is to model real events in biological organisms as more exact as possible. In the simplest way, a biological network consists of a set of nodes which correspond to biological substances and a set of edges which correspond to some direct relationship or influence between substances. All computational studies on the network reconstruction just aim to discover a set of edges in networks that fitting the available data. In other words, they just try to reconstruct networks at pair-wise relationship level (edge connecting two nodes), which have some shortcomings in representing multivariate relationships - relationships involving more than two substances at the same time.

Pair-wise relationship networks are the simplest representation of multivariate relationships. We say that two variables (G_1, G_2) have a relationship if they involve in an event in a biological organism. If three variables involve in the same event, there is also a three-variable relationship that is represented by the triple (G_1, G_2, G_3) (called triple-wise relationship). An event involving multiple variables can be represented at different relationship level: pair-wise, triple-wise, etc. Table 1 shows some examples of representing the original full relationships (in the first column) at pair-wise and triple-wise relationship levels. The first row of the table corresponds to an original relationship between G_2 and G_4: the target gene G_2 is regulated by the transcriptional factor G_4. This original relationship has only one pair-wise relationship (itself) and has not any three-or-more-variable relationships. The second row corresponds to an original three-gene relationship: the target gene G_3 is co-ordinately regulated by G_1 and G_2. This original relationship has two sub-relationships at the pair-wise level, (G_1, G_3) and (G_2, G_3), and one triple-wise relationship itself (G_1, G_2, G_3). The third row corresponds to an original four-gene relationship: G_1 is combinatorially regulated by three transcriptional factors G_2, G_3 and G_4. This original four-gene relationship has 3 sub-relationships at pair-wise level: (G_1, G_2), (G_1, G_3) and (G_1, G_4); and 3 sub-relationships at the triple-wise level: (G_1, G_2, G_3), (G_1, G_2, G_4) and (G_1, G_3, G_4). The last row is an original three-variable relationship (X_1 *xor* X_2 *xor* $X_3 = 0$) among three binary variables. This relationship is a typical example of synergical relationships and will re-appeared in the rest of this paper. This *xor* relationship has 3 sub-relationships at the pair-wise level: (X_1, X_2), (X_1, X_3) and (X_2, X_3).

From Table 1, we can see that the triple-wise relationships provide more knowledge than pair-wise ones about the full original relationships that involve more than two variables. For example, with the original relationships with exact three variables like the second row of the table, if using only pair-wise relationships $(G_1, G_2), (G_2, G_3), (G_3, G_1)$, we do not know if G_1, G_2 and G_3 involve in the same event of the G_3 synthesis; but if we know that there is a triple-wise relationship (G_1, G_2, G_3), we can infer that all three genes involve in the same biological event. Clearly, triple-wise relationships describe more closely the original reality than pair-wise ones. Unfortunately, three-variable relationships ignore the original relationships that concern only two variables (the first row of Table1). Therefore, we should use relationships at different levels to understand the full original relationships.

This study will focus on the reconstruction of triple-wise relationships. Triple-wise and pair-wise relationships will complement each other to provide more information about the real events happened in biological organisms.

2.2 MI Measures for Reconstructing Specially Triple-wise Relationships

In this work, we focus on the reconstruction of triple-wise relationships. There are some previous studies addressing synergical relationships of three variables [10,23] since these synergical relationships cannot be detected by pair-wise MI. The key idea of the previous methods is to use the interaction information (some authors called it as three-way mutual information) that is a Venn diagram-like extension from the pair-wise MI as follows

$$InteractInfo(Z, X, Y) = H(X) + H(Y) + H(Z) - H(X, Y) - \\ - H(Y, Z) - H(Z, X) + H(X, Y, Z) \quad (1)$$

or an equivalence

$$InteractInfo(Z, X, Y) = MI(X, Y) + MI(X, Z) - MI(X, [Y, Z]) \quad (2)$$

$$= MI(Y, X) + MI(Y, Z) - MI(Y, [X, Z]) \quad (3)$$

$$= MI(Z, X) + MI(Z, Y) - MI(Z, [X, Y]) \quad (4)$$

The interpretation and the usage of this interaction information are still in controversy [4,10]. With the interpretation and visualization of $MI(Z,X)$, $MI(Z,Y)$, $MI(Z,[X,Y])$ as in [18], we confirm that *InteractInfo(X, Y,Z)* (Equ. 4) can get a positive or negative value and we cannot say what it represent for.

In the previous work [18], we have introduced four MI measures of three variables as following:

1. Total mutual information

$$MI(X,Y,Z) = H(p_X \times p_Y \times p_Z) - H(p_{X,Y,Z}) = H(X) + H(Y) + H(Z) - H(X,Y,Z) \tag{5}$$

2. Mutual information between X and [Y,Z]

$$MI(X,[Y,Z]) = H(p_X \times p_{Y,Z}) - H(p_{X,Y,Z}) = H(X) + H(Y,Z) - H(X,Y,Z) \tag{6}$$

3. Mutual information between Y and [Z,X]

$$MI(Y,[Z,X]) = H(p_Y \times p_{Z,X}) - H(p_{X,Y,Z}) = H(Y) + H(Z,X) - H(X,Y,Z) \tag{7}$$

4. Mutual information between Z and [X,Y]

$$MI(Z,[X,Y]) = H(p_Z \times p_{X,Y}) - H(p_{X,Y,Z}) = H(Z) + H(X,Y) - H(X,Y,Z) \tag{8}$$

Among our four MI measures of three variables, the total or cube MI is a single measure to capture the triple-wise relationships. However, the cube MI includes all dependences of three variables and it does not distinguish triple-wise relationships from the pair-wise ones. To detect specially triple-wise relationships, we should use three cylinder MI measures (Equations 6, 7 and 8). The smallest value of these three quantities will characterize specially triple-wise relationships; we call it as *RelationshipMI*. In other words, *RelationshipMI* is the total MI after removing the largest pair-wise MI:

$$\mathit{RelationshipMI}(Z,X,Y) = \min(MI(X,[Y,Z]), MI(Y,[Z,X]), MI(Z,[X,Y])) \tag{9}$$

$$= MI(X,Y,Z) - \max(MI(X,Y), MI(Y,Z), MI(Z,X)) \tag{10}$$

2.3 Evaluation of Triple-wise Relationship Reconstructing Measures

We consider triple-wise relationship reconstruction as a binary classification problem. From the training data (expression profiles of genes, proteins or metabolites), reconstructing measures (or methods) will predict triple-wise relationships with some confidence. These predicted triple-wise relationships will be matched in some ways (usually using area under the ROC curve criterion or AUC for short) with the true triple-wise relationships to report the performance of reconstructing methods.

Similar to the previous studies, we evaluate network reconstructing measures (or methods) on *in silico* benchmarks. But different from previous studies that address the pair-wise relationship inference, we test the reconstructing methods at the triple-wise relationship level, where true positive triple-wise relationships

Table 2. Information on *in silico* networks that generate training datasets

Net. size	Dataset	Distribution of original kinetic equations			# Extracted three-variable relationships
		# pair-wise equations	# three-variable equations	# four-or-more--variable equations	
10	Ecoli1	3	1	2	7
	Ecoli2	3	6	0	6
	Yeast1	2	3	0	3
	Yeast2	1	4	4	29
	Yeast3	3	2	3	36
50	Ecoli1	28	11	3	29
	Ecoli2	21	13	10	59
	Yeast1	29	9	7	73
	Yeast2	15	4	26	369
	Yeast3	8	9	29	369
100	Ecoli1	53	21	9	58
	Ecoli2	61	26	2	32
	Yeast1	39	25	21	140
	Yeast2	8	18	66	974
	Yeast3	18	10	64	2504

are extracted from the validated networks as described in Section 2.1. We use the same benchmark data sets that have been used to evaluate pair-wise relationship reconstructing methods (DREAM3 challenge 4 [19]). These datasets are *in silico* generated from some modules of curated networks so that they look like realistic as much as possible (DREAM3 were generated from different parts of Ecoli and Yeast networks) [17]. The data sets can be downloaded at the DREAM website. Since the website does not provide true triple-wise relationship networks, we have to build these target triple-wise relationships by ourselves from the SBML network models that the authors provided. There are totally 15 benchmark data sets (see Table 2 for the summary information). Five of them have 10 genes, five have 50 genes and five have 100 genes. The structure of networks that generated the data sets has also different connectivity complexities. Each benchmark dataset in turn consists of three types of data: knock-down, knock-out and trajectory data.

3 Results and Discussion

3.1 Reconstructing Specially Three-Variable Relationships

MI is first defined as a symmetric dependence measure between two variables. Extension of MI from two to multiple variables is not trivial [5,16]. All efforts have tried to formulate a single measure that characterizes specially a n-variable relationship/dependence, which remains imprecisely defined. The total mutual information (see the formulas of three variables in Equ. 5) and interaction information (see the formulas of three variables in Equ. 4) are such a kind of measures [6,8,10]. However, they are suitable or not for detecting n-wise relationships is still in question. The first, total mutual information or *Total MI* for short, is a measure that quantifies an overall relationship among multiple variables, it includes all relationship categories. It is not special for n-wise relationship/dependence.

Table 3. The prediction performance (AUC) of different MI-measures in reconstructing triple-wise relationships for benchmark networks in Table 2. MI-measures are estimated in all three types of data: knock-down, knock-out and trajectory.

MI Measure	Net. size	Ecoli1	Ecoli2	Yeast1	Yeast2	Yeast3	Average	overall
	Size 10	0.74	0.79	0.89	0.66	0.52	0.72	
RelationshipMI	Size 50	0.85	0.75	0.58	0.63	0.68	0.70	0.68
	Size 100	0.80	0.68	0.61	0.50	0.53	0.63	
	Size 10	0.66	0.73	0.83	0.69	0.56	0.69	
Total MI	Size 50	0.85	0.73	0.53	0.60	0.68	0.68	0.66
	Size 100	0.79	0.63	0.59	0.50	0.52	0.61	
	Size 10	0.20	0.56	0.34	0.62	0.51	0.45	
InteractInfo	Size 50	0.70	0.64	0.38	0.51	0.57	0.56	0.52
	Size 100	0.69	0.57	0.49	0.45	0.48	0.54	
	Size 10	0.59	0.81	0.85	0.60	0.50	0.67	
Pair-wise MI_1	Size 50	0.78	0.66	0.58	0.49	0.52	0.61	0.63
	Size 100	0.78	0.66	0.58	0.49	0.52	0.60	
	Size 10	0.66	0.74	0.83	0.69	0.58	0.70	
Pair-wise MI_2	Size 50	0.83	0.70	0.54	0.60	0.68	0.67	0.66
	Size 100	0.78	0.66	0.60	0.49	0.52	0.61	

The second measure, interaction information or *InteractInfo* for short, has attracted much controversy in interpretation and usage to reconstruct biological networks [4,10,22]. Under our view, from the formulas of *InteractInfo* (Equ. 4) and interpretation/visualization of $MI(Z,X), MI(Z,Y), MI(Z,[X,Y])$ like in [18], we can see that $MI(Z,X) + MI(Z,Y) - MI(Z,[X,Y])$, or *InteractInfo(X,Y,Z)*, may have a positive or negative value. It is so difficult to interpret *InteractInfo(X,Y,Z)*. Some previous works have provided evidences that *InteractInfo* can detect synergical triple-wise relationships that pair-wise MI cannot [10,23].

We have introduced an additional measure called *RelationshipMI* that is the smallest cylinder MI, or an equivalence, *Total MI* after removing the largest pair-wise MI (Equ. 10). It means that after excluding the maximum pair-wise MI, *Total MI* should characterize specially triple-wise relationships. In the first experiment, we want to evaluate the prediction performance of *RelationshipMI*, *Total MI* and *InteractInfo* on datasets described in [18] (in this experiment we used the integrated data of all three available types: knock-down, knock-out and trajectory). We estimate these measures for all 3-combinations of genes in the networks. In all experiments of this work, we use an estimation method of entropies and mutual information by using B-spline functions as described in [12]. The measure of a triple is considered as a confidence of the prediction that the triple is a triple-wise relationship or not. We used a non-threshold criterion, area under the ROC curve (AUC), to report the prediction performance.

Table 3 shows AUC of MI-based measures in reconstructing triple-wise relationships on different benchmark networks. Overall AUC average of *RelationshipMI*, *Total MI* and *InteractInfo* on fifteen data sets are 0.68, 0.66 and 0.52, respectively. As our expectation, *RelationshipMI* outperforms both *Total MI* and *InteractInfo* since *RelationshipMI* concentrates specially triple-wise relationships as the analysis in the previous paragraph. This work is the first to address the evaluation of reconstructing measures at the level of triple-wise relationships.

Our results showed that *InteractInfo* gave very low prediction performance, overall average on fifteen datasets is 0.52 (nearly the random prediction). It did not surprise to us although it does not agree with some previous studies saying that *InteractInfo* can detect synergical relationships [10,23]. As the analysis of Section 2.2, *InteractInfo* does not relate the triple-wise relationships. It might capture synergical relationships (a special kind of triple-wise relationships), but it is not clear. Conversely, *RelationshipMI* and *Total MI* can capture all categories of three-variable relationships, so they gave a better prediction performance.

To check the advantage of three-variable MI measures over pair-wise ones in reconstructing triple-wise relationship networks, we also evaluate two simple methods that use only pair-wise MIs. The first will predict (X,Y,Z) to be a triple-wise relationship if all three measures of pair-wise MIs: $MI(X,Y)$, $MI(Y,Z)$ and $MI(Z,X)$ are greater than a threshold. This method is equivalent with using a reconstructing measure of the minimum of three pair-wise MIs (called *Pair-wise* MI_1, see Equ. 11 below). The second pair-wise-MI-based measuse is lightened the first one, which will predict (X,Y,Z) to be a triple-wise relationship if only two of three pair-wise MIs $MI(X,Y)$, $MI(Y,Z)$ and $MI(Z,X)$ are greater than a threshol. This method is equivalent to a measure of the second minimum of three pair-wise MIs (called *Pair-wise* MI_2, see Equ. 12 below.)

$$Pair\text{-}wiseMI_1 = \min(MI(X,Y), MI(Y,Z), MI(Z,X)) \tag{11}$$

$$Pair\text{-}wiseMI_2 = Second\text{-}\min(MI(X,Y), MI(Y,Z), MI(Z,X)) \tag{12}$$

The last two rows of Table 3 show the prediction performance of these *Pair-wise MI* measures. We can see that *Pair-wise* MI_2 gave higher peformance than *Pair-wise* MI_1. The overall AUC of two *Pair-wise MI* methods on fifteen data sets are 0.66 and 0.63, respectively. They are all smaller considerably than that of *RelationshipMI* and *Total MI*. This might be from the situation that there are some triples (X,Y,Z), each two of them involves a pair-wise relationship, but these pair-wise relationships are independent with others. These triples will be predicted as a triple-wise relationship with *Pair-wise MI* measures. However, the prediction is false.

From the prediction results we can see that all measures (except *InteractInfo* that is not suitable for relationship reconstruction) exhibit the highest performance on the Yeast1-Size10 dataset: *RelationshipMI* (0.89), *Total MI* (0.83). The reason explains why Yeast1-Size10 dataset provides the best performance is that its network (the model generating the dataset) is very simple, the network structure includes only 3 triple-wise relationships (see Table 2). For some networks with very high connectivity complexity (for example Yeast2 and Yeast3 in all sizes), the prediction performance of all measures are quite low.

3.2 Reconstruction of Triple-wise Relationships from Different Types of Data

In the previous subsection, we evaluate reconstructing measures on the integrated data that includes all three available types: knock-down, knock-out and trajectory. In this subsection, we will show the evaluation results on each type

Table 4. The overall average AUC of MI-measures in reconstructing triple-wise relationships from different data sets

	Knock-out	Knock-down	Trajectory	Integrated data
RelationshipMI	0.73	0.54	0.58	0.68
Total MI	0.76	0.58	0.59	0.66
InteractInfo	0.55	0.55	0.54	0.52
Pair-wise MI_1	0.60	0.53	0.56	0.63
Pair-wise MI_2	0.74	0.54	0.58	0.66

of data independently. Table 4 shows the overall average AUC of different MI-measures in reconstructing three-variable relationships from three independent types of data and from the integrated data. As we can see, the knock-out data provides the best performance with all measures. It proves that the knock-out data contains much more information than knock-down and trajectory data in reconstructing networks.

There are two other interesting findings from this experiment. Firstly, the integrated data that collects all three types of data does not improve the network reconstruction performance when compared with the single knock-out data. Secondly, while *RelationshipMI* gave the best performance on the integrated data (*RelationshipMI*: 0.68, *Total MI*: 0.66); *TotalMI* gave the best performance on the knock-out data (*RelationshipMI*: 0.73, *Total MI*: 0.76). Therefore, each measure can work well on a specific type of data in reconstructing networks.

4 Conclusions

We have introduced a new interpretation and visualization of mutual information (MI) between two variables and then proposed MI measures for multiple variables based on the proposed interpretation. Multivariate MI measures can detect relationships broader than pair-wise relationships. Multivariate relationships at different levels will provide more information about the real biological networks than the pair-wise ones. Furthermore, multivariate MI measures can detect the synergical multivariate relationships that pair-wise MI cannot. We have evaluated prediction performance of three-variate MI measures in reconstructing biological networks at the level of triple-wise relationships on some benchmark data sets. Though network reconstruction at triple-wise relationship level is more difficult than pair-wise network reconstruction, our predicting results showed that multivariate MI measures provide a promising performance, the best overall AUC is 0.76 on the knock-out data set.

In this work, we only used three-variable MI measures to reconstruct triple-wise relationships in the networks. These three-variable MI measures can detect some triple-wise relationships (for example, synergical ones) that pair-wise MI cannot; in turn, three-variable MI measures might not be able to detect synergical relationships of more than three variables. Therefore, to detect full relationships of biological networks, we should use all multivariate MI measures at different levels. Unfortunately, that is a problem of combinatorial explosion; we must estimate multivariate MI measures for all combinations of variables.

Acknowledgment. This work was supported by Vietnam's National Foundation for Science and Technology Development (NAFOSTED Project No. 102.03.21.09).

References

1. Butte, A., Kohane, I.: Mutual Information Relevance Networks: Functional Genomic Clustering using Pairwise Entropy Measurements. In: Pacific Symposium on Biocomputing, pp. 418–429 (2000)
2. Cover, T.M., Thomas, J.A.: Elements of Information Theory, 2nd edn. Molecular Systems Biology. Wiley-Interscience, A John wiley & Sons, Inc. (2006)
3. Fano, R.M.: A Statistical Theory of Communication. MIT Press, Cambridge (1961)
4. Jakulin, A.: Machine Learning Based on Attribute Interactions. PhD Dissertation, University of Ljubljana (2005)
5. Jakulin, A., Bratko, I.: Quantifying and Visualizing Attribute Interactions: An Approach Based on Entropy. CoRR, cs.AI/0308002 (2004), `http://arxiv.org/abs/cs.AI/0308002`
6. McGill, W.J.: Multivariate information transmission. Psychometrika 19(2), 97–116 (1954)
7. Steuer, R., Kurths, J., Daub, C.O., Weise, J., Selbig, J.: The Mutual Information: Detecting and Evaluating Dependencies between Variables. Bioinformatics 18, 231–240 (2002)
8. Watanabe, S.: Information Theoretical Analysis of Multivariate Correlation. IBM Journal of Research and Development 4, 66–82 (1960)
9. Margolin, A., Nemenman, I., Basso, K., Wiggins, C., Stolovitzky, G., Favera, R.D., Califano, A.: ARACNE: an Algorithm for the Reconstruction of Gene Regulatory Networks in a Mammalian Cellular Context. BMC Bioinformatics 7(sppl.1) (2006)
10. Anastassiou, D.: Computational Analysis of the Synergy among Multiple Interacting Genes. Molecular Systems Biology 3(83) (2007)
11. Cho, K.H., Choo, S.M., Jung, S.H., Kim, J.R., Choi, H.S., Kim, J.: Reverse engineering of gene regulatory networks. IET Syst. Biol. 1(3), 149–163 (2007)
12. Daub, C.O., Steuer, R., Selbig, J., Kloska, S.: Estimating Mutual Information using B-spline Functions–an Improved Similarity Measure for Analysing Gene Expression Data. BMC Bioinformatics 5(118) (2004)
13. Friedman, N., Linial, M., Nachman, I., Peer, D.: Using Bayesian Networks to Analyze Expression Data. J. Comput. Biol. 7, 601–620 (2000)
14. Han, T.S.: Multiple mutual information and multiple interactions in frequency data. Information and Control 46, 26–45 (1980)
15. Heckera, M., Lambecka, S., Toepferb, S., Somerenc, E., Guthke, R.: Gene regulatory network inference: Data integration in dynamic models - A review (2009)
16. Leydesdorff, L.: Interaction Information: Linear and Nonlinear Interpretations. Int. J. General Systems 6(36), 681–685 (2009)
17. Marbach, D., Schaffter, T., Mattiussi, C., Floreano, D.: Generating realistic in silico gene networks for performance assessment of reverse engineering methods. J. Comput. Biol. 16(2), 229–239 (2009)

18. Pham, T.H., Ho, T.B., Nguyen, Q.D., Tran, D.H., Nguyen, H.V.: Multivariate Mutual Information Measures for Discovering Biological Networks. In: The 9th IEEE - RIVF International Conference on Computing and Comunication Technologies, pp. 103–108 (2012)
19. Prill, R.J., Marbach, D., Saez-Rodriguez, J., Sorger, P.K., Alexopoulos, L.G., Xue, X., Clarke, N.D., Altan-Bonnet, G., Stolovitzky, G.: Towards a rigorous assessment of systems biology models: the DREAM3 challenges 5(2) (2010)
20. Stolovitzky, G., Prill, R.J., Califano, A.: Lessons from the DREAM2 Challenges: A Community Effort to Assess Biological Network Inference. Ann. N.Y. Acad. Sci., 159–195 (2009)
21. Styczynski, M.P., Stephanopoulos, G.: Overview of computational methods for the inference of gene regulatory networks. Computers & Chemical Engineering 29(3), 519–534 (2005)
22. Walters-Williams, J., Li, Y.: Estimation of mutual information: A survey. In: Wen, P., Li, Y., Polkowski, L., Yao, Y., Tsumoto, S., Wang, G. (eds.) RSKT 2009. LNCS, vol. 5589, pp. 389–396. Springer, Heidelberg (2009)
23. Watkinson, J., Liang, K.C., Wang, X., Zheng, T., Anastassiou, D.: Inference of regulatory gene interactions from expression data using three-way mutual information. Ann. N.Y. Acad. Sci., 302–313 (2009)
24. Werhli, A.V., Husmeier, D.: Reconstructing Gene Regulatory Networks with Bayesian Networks by Combining Expression Data with Multiple Sources of Prior Knowledge. Statistical Applications in Genetics and Molecular Biology 6(1) (2007)

SRA Tool: SOFL-Based Requirements Analysis Tool

A.R. Mat[1], A.B. Masli[1], N.H. Burhan[1], and S. Liu[2]

[1] Department of Computing & Software Engineering,
Faculty of Computer Science & IT,
Universiti Malaysia Sarawak, Kota Samarahan, Sarawak, Malaysia
{marahman,bmazman,bnhazlini}@fit.unimas.my
[2] Department of Computer & Information Sciences, Hosei University,
Koganei Campus, Tokyo, Japan
sliu@hosei.ac.jp

Abstract. Capturing requirements in order to produce a complete and accurate specification for developing software systems requires a deep understanding of domain knowledge. It becomes more difficult if the system analyst is dealing with a large-scale system specification without specific guidance on how to construct a specification. At the same time the system analyst is expected to be an expert in the formal language used for the presentation of the specification. Therefore, constructing requirement specification has become a big challenge to the system analyst. In this paper, we present a tool suite for guiding a system analyst in constructing the specification. This tool manipulates the knowledge to guide a system analyst at the requirements analysis stage. The guiding process requires two steps: informal and semi-formal steps. It was indicated that our approach did facilitate guiding the process of constructing the specification during the requirements analysis.

Keywords: formal engineering method, requirements analysis, specification, knowledge-based, SOFL.

1 Introduction

Software requirements analysis is to understand the client's request and to derive a concrete specification according to the requirements gathered from the client. This process is knowledge-intensive, because the analyst needs to know how to get the essential information from the client. The analyst also needs to know how to record the information, so that not only will the client understand what has been stated, but also the developers will be able to catch everything the client has requested. Since the client cannot always be available throughout the software development process, a complete and accurate requirements specification is essential to serve as a means of agreement between the client and the system analyst. The question then is, how to effectively capture complete and accurate requirements in a specification. A possible way of achieving a complete and accurate specification is by developing an intelligent tool with knowledge support. This knowledge can be identified and categorized as a domain, dynamic, or method knowledge. This knowledge is organized as a hierarchy

P. Meesad et al. (Eds.): *IC²IT2013*, AISC 209, pp. 217–226.
DOI: 10.1007/978-3-642-37371-8_25

in which the knowledge is a root, and is classified into three categories. Each category is divided into three parts: informal, semi-formal and formal. Detailed explanation on how it works is described in [1].

The Structured Object-Oriented Formal Language (SOFL) for developing specification for software systems has been developed and applied in information systems over the last seventeen years [2][3][4]. In particular, to construct a complete specification using SOFL, three steps are required: informal, semi-formal, and formal. While the informal step is to gather all the required functions, data resources, and constraints into one specification in natural language, the semi-formal step is to group and transform the expression into SOFL language except logical expressions (i.e. type invariants, pre- and post-conditions). In the formal specification step, all informal expressions of type invariants and process specifications in the semi-formal specification are formalized properly.

In this paper, we present a tool suite for guiding the system analyst in constructing requirement specification. Aimed at supporting the requirement analysis process and manipulating knowledge, this tool will require two steps: informal and semi-formal. Each step requires several tasks for manipulating the knowledge. The tasks in the informal step include a domain identification, method selection, and constructing the informal specification. The task in the semi-formal step is by constructing a semi-formal specification.

The rest of this paper is organized as follows. Section 2 gives a short overview of the SOFL method, in particular a two-step technique for requirements analysis. This is followed in Section 3 by a description of how we manipulate knowledge to support requirements analysis. Section 4 presents the tool suite for constructing the specification, called SRA Tool, for supporting our approach. Section 5 gives a brief overview of related work. Finally, in Section 6, we conclude the paper and describe our plans for future research.

2 SOFL Overview

SOFL is a formal engineering method that offers a systematic way to develop software systems using an intuitive but formal notation. The notation results from an effective integration of VDM-SL [5], Petri nets [6], and Data Flow Diagrams [7]. To facilitate the process of achieving such a requirement specification, SOFL offers a two-step specification approach, which consists of informal specification and semi-formal specification (as in Fig. 1).

Each stage of the specification has its own goal and task. The goal for informal specification is to enable the developer to fully understand and produce complete requirements of the domain problem through communication with the client. During this step, an informal specification is written, aiming to reflect the result of an informal requirement analysis on the basis of domain knowledge and available materials. Such a specification consists of three parts: (1) a functional description, which describes the high-level operations needed for the system, (2) data resources, which indicate data items necessary for fulfilling the functions provided in (1), and (3) constraints on both operations and data resources. At this stage, the descriptions of the operations, data resources, and constraints are written in a natural language. The goal for semi-formal specification is to achieve accuracy in the description of the requirements. During this step, the entire informal specification is transformed into a semi-formal specification by (1) grouping functions, data resources, and constraints, (2) declaring data types, and (3) defining all pre- and post-conditions for each of the processes. All transformations are represented in appropriate expressions in the SOFL specification language, except that all logical expressions, such as type invariants and pre- and post- conditions for processes, are still kept informal.

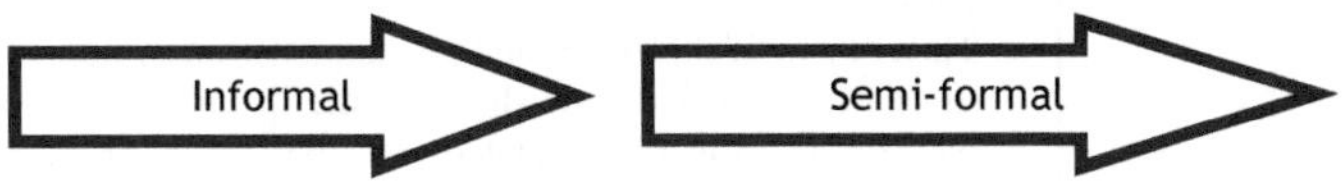

Fig. 1. SOFL-based Requirement Specification Approach

3 Knowledge-Based Approach

3.1 Knowledge

Definition 1. Let Knowledge, denoted by K be three tuples that contains Domain, Dynamic and Method Knowledge, denoted by O, Y and M, respectively. It can be represented as follows:

$$K = \{O, Y, M\} \tag{1}$$

The detailed description for the knowledge is listed in Table 1.

3.2 Guiding Process

Definition 2. The set I represent input which consists of informal input, I_{inf} and semi-formal input, I_{sf}. It can be written as $I = \{I_{inf}, I_{sf}\}$. The function Guidance, G can be static, s or non-static, t depending on the user input, u. The guidance can be expressed as:

$$G(u) = \begin{cases} s, u \in I \\ t, u \notin I \end{cases} \tag{2}$$

Based on this definition, the guidance will cover for both steps: informal and semi-formal. For each step, the guidance would be in either static or non-static form. The static guidance could be a command on what the system analyst should do, confirmation on the value keyed-in, and displaying the current state of the specification construction. However, non-static guidance will be enabled when the value keyed-in is not available and needs extra tasks in order to make it available for constructing the specification.

Table 1. Knowledge description

Knowledge	Description
Domain	• Knowledge about concept, their relationships and the descriptions of what tasks are realized and how.
Dynamic	• Contains facts about static properties of the application environment, examples being the structure of some entities of the laws that some states or performed activities must obey. • The content of this dynamic knowledge is somewhat similar to that of domain knowledge, but the dynamic knowledge has extra properties related to the keys of method knowledge. Those facts are provided by the SA each time he/she wants to construct a new system application. Other facts are derived by the tool itself as it executes its procedures. • The derived facts constitute the conclusions inferred by the tool from the system analyst's environment statements.
Method	• Contains the knowledge that describes how to acquire specifications and how to build the corresponding software model. • This method will follow the SOFL formal notation for the construction of the specification.

The process of guiding includes two steps:

Informal Step. This is a step to achieve completeness of requirements. The requirements are complete if and only if the specification (a) contains three parts: (1) a functional description which describes the high-level operations needed for the system, (2) data resources which indicate the data items necessary for fulfilling the functions described in (1), and (3) constraints on both operations and data resources; and (b) satisfies the domain knowledge [2]. As shown in Fig. 2, the system analyst is also required (1) to specify the domain of interest, (2) selecting the method to be used for constructing the specification, and, (3) constructing the specification. The specification (4) can be viewed by generating the specification.

Semi-formal Step. This is a step to achieve the accuracy of the knowledge. In this step, three activities are needed. As shown in Fig. 3, these activities include: (i) grouping the general structure into a module, (ii) defining the data type, and (iii) by describing pre- and post-conditions. For each activity, the guidance will focus: (a) on assisting step-by-step to group the functions, data resources and constraints, (b) on requesting detail of each defined data type, and within the processes, and (c) on reminding the system analyst of any contrary information given between data type declaration and in the process.

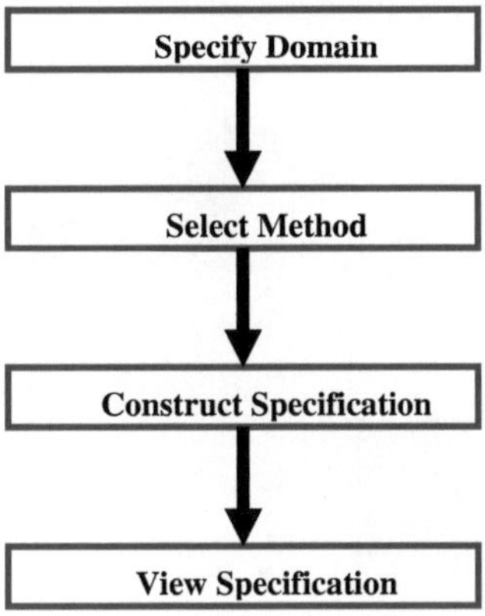

Fig. 2. Informal Step

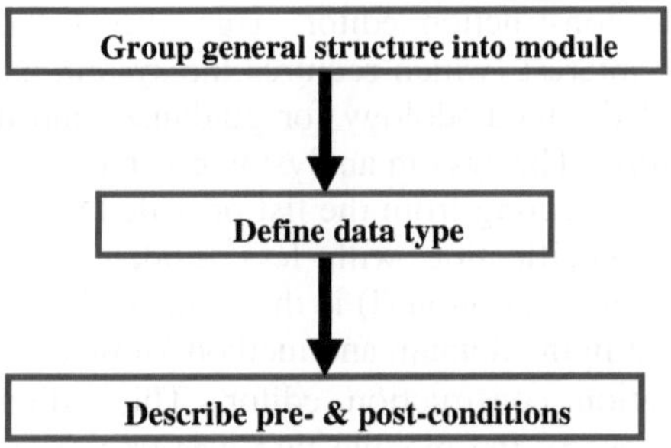

Fig. 3. Semi-formal step

4 Prototype Tool

We have built a prototype tool, called the SOFL-based Requirements Analysis Tool (SRA Tool) to support the process of requirements analysis. Fig. 4 shows a snapshot of this tool. The goal of building this tool is to guide a system analyst to capture a complete and accurate specification, which is expected to establish a fundamental for constructing an intelligent tool in the future. In this section, we first introduce the major functions of the tool, and then evaluate this tool by humanly constructing the requirement specification.

4.1 Major Functions

The tool is intended to support the requirements analysis process for constructing a specification that is required by our method, as shown in Fig. 1 in Section 2. Specifically, it provides the following potential functions.

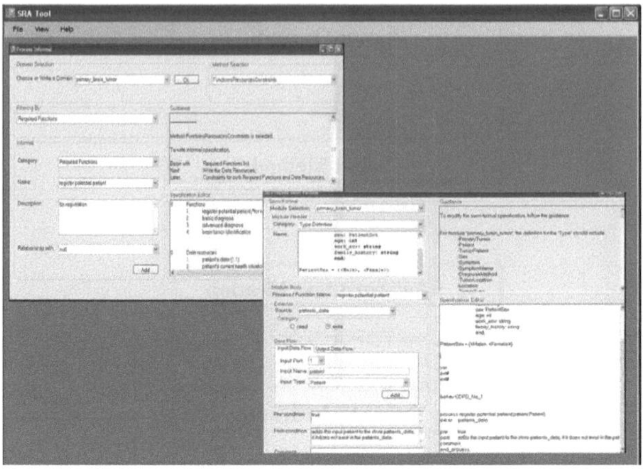

Fig. 4. A snapshot of SRA Tool

- Informal specification construction editor. This allows the user to specify the domain knowledge of interest (which requires the system analyst to use restricted language), selection of the methodology for guidance, and the editor to construct the informal specification. The system analyst is given two options for constructing the specification: (i) by selecting from the list provided, step-by-step and/or (ii) by writing the informal specification with less guidance from the tool. For an inexperienced system analyst, option (i) is the best, while option (ii) is suitable for s skillful system analyst in the domain and method knowledge.
- Semi-formal specification construction editor. This editor allows the system analyst to select and specify the module that was transformed from the informal specification editor and the semi-formal specification construction editor. For each selected module or process, the system analyst is required to declare a data type in the module header section, and define all pre- and post-conditions in the module body section. Similar to the informal specification editor, the semi-formal editor gives two options for the system analyst to construct the semi-formal specification.
- Transforming from informal to semi-formal specifications. The domain in informal step will be transformed as a module in the semi-formal step, required functions are transformed as a list of processes, function decompositions as *module_decom* and/or processes, and data resources as temporary variables. As for the constraints in the informal step, no automatic transformation is provided.
- Tool-guided specification construction support. The tool guides the system analyst for constructing the informal and semi-formal specifications process step-by-step, and recommendation on additional domain should be included and tackled by the system analyst in order to achieve a completeness and accuracy of the specifications. Each step will be provided and monitored with, at least, one guide for constructing the specifications.

- Specifications-generated support editor. In the informal step, informal specification will be generated and in the semi-formal step, semi-formal specification will be produced. Each step provides a specification editor to allow the system analyst to view and edit the specification and later, will be generated as a specification. Those specifications are kept as XML files. Fig. 5 shows a snapshot of the specifications in informal and semi-formal based on the SOFL method.

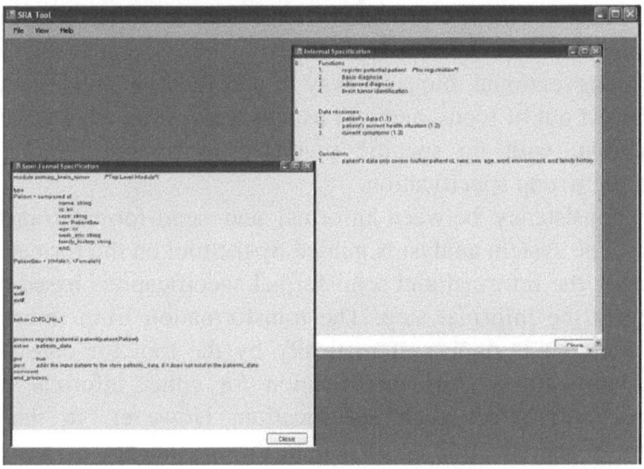

Fig. 5. Sample of informal and semi-formal specifications generated by SRA Tool

4.2 Evaluation

To explore and evaluate the distinct features of SRA Tool, we have applied this tool to construct the informal and semi-formal specifications of the primary brain tumor treatment system [8]. In the informal step, there are five major functions in which one function is decomposed into lower-level functions, six data items and each item is connected to the specific potential required function and two constraints. In the semi-formal step, the entire informal specification is transformed into a semi-formal specification by (i) grouping functions, data resources and constraints, (ii) declaring data type, and (iii) defining all pre- and post-conditions for each of the processes. Through this case study, we have found several important points.

- Firstly, this tool provides a systematic way to construct the specification from informal to semi-formal. By stating what domain of interest is to be covered and the selected methodology, the tool will guide the system analyst for both informal and semi-formal specification steps. The system analyst will be guided with the list of required functions, data resources, and constraints to be included in the informal specification. While as for the semi-formal step, the tool will remind the system analyst on the list of data types, module and/or processes that need to be defined, as well as the description of each pre- and post-conditions for each process. The system analyst does not need to know the domain knowledge and the specification language at all. In contrast, with the conventional, humanly way of constructing the specification, the system analyst needs to have knowledge of the specific domain and the specification language. This is hard for a beginner or even experienced system analyst who has no knowledge on the domain and certain language for representing the specification. It becomes more difficult if the system analyst is dealing with a large-scale system specification, without a guide from the tool.

- Secondly, this tool encourages the system analyst to be more focused, and keep thinking on a specific domain. The tool is embedded with domain, dynamic, and method knowledge. The list provided in the tool is collected from the domain knowledge and also, based on the list defined by the system analyst him/herself. In contrast, in a conventional, humanly way of specification construction, the system analyst might get out of focus, and will usually get lost on what is to be included in the specification, with no specific guidance to remind him/her if he/she is constructing the wrong specification.
- Finally, the consistency between informal and semi-formal transformation and specification. The system analyst is guided by the tool on the format to be followed for constructing the informal and semi-formal specifications based on the selected methodology at the informal step. The transformation from informal into semi-formal specification is done systematically by the tool for consistency between these two specifications. Any modification for either informal or semi-formal specifications, will affect each specification. However, in the conventional, humanly way of specification construction, the consistency between informal and semi-formal specifications is monitored seriously by the system analyst him/herself. He/she is responsible to make sure that each specification is completely covered in the specification and he/she also is required to transform the informal specification into a semiformal specification manually. This will be difficult for constructing and transforming a large-scale system specification without being supported by a tool.

5 Related Work

In this section, we compare our work with other approaches for developing knowledge-based system tools, such as Model-based Incremental Knowledge Engineering (MIKE) [9] and Requirements Analysis and Knowledge Elicitation System (RAKES) [10].

MIKE is a knowledge-based system, integrates semiformal and formal specification techniques together with prototyping into a coherent framework. This model separates different kinds of knowledge at different layers: the domain layer contains domain-specific concepts, attributes, and relationships; the inference layer contains knowledge about the functional behavior of the problem solving; and the task layer contains knowledge about the goals of a task and the control knowledge required to perform the task. The idea of separating the knowledge is similar to our approach, in which knowledge is classified into three kinds: domain, dynamic, and method knowledge. However, our work is different in terms of the target language. While the focus language for MIKE is the Knowledge Acquisition and Representation Language (KARL), our target is the SOFL language in an engineering environment. In addition, the development process of MIKE is based on prototyping, while our approach is geared towards more structured and evolutionary processes.

Another research project, to develop the Requirements Analysis and Knowledge Elicitation System (RAKES), uses the Frame and Rule Oriented Language (FRORL) to perform verification and validation of the specification and to transform the specification into a program. The data is stored in domain dependent knowledge. It seems to be easy to use for implementation, but it would be difficult to use for searching, especially if the knowledge was not organized properly.

6 Conclusion and Future Work

A derivative of guidance by manipulating the knowledge-based approach for constructing requirement specification has been introduced and prototype of tools supporting the process has been developed. Our experience has indicated that our approach could assist the system analyst to capture a requirement with providing the guidance. As for the future, more work will be done on upgrading the features and extending this tool to support system analysts in developing formal specification. We believe that the SRA Tool will not only be able to automate large parts of the data collection to mitigate the problems of manual data recording, but will also support planning and plan tracking. It will also provide various kinds of data analyses. These features will help system analysts to identify process deficiencies and create process improvement plans to remove the identified deficiencies. The tool will also provide a reliable estimate of effort and quality to allow more effective and efficient knowledge management.

This work has been carried out as the first step of a project that aims to support a complete three-step approach to developing formal specifications. However, this tool is yet to be tried on a large-scale application, and it should also be tried on various domain problems in fields such as automatic automobiles in order to test the flexibility of the domain expansion.

References

1. Mat, A.R., Liu, S.: Organizing Knowledge to Support Requirements Analysis. In: Proceedings of International Conference on Systems Engineering and Modeling (ICSEM 2011), March 11-13, pp. 65–70. IEEE Press, Shanghai (2011)
2. Liu, S.: Formal Engineering for Industrial Software Development Using the SOFL Method. Springer (March 2004)
3. Liu, S., Shibata, M., Sato, R.: Applying SOFL to Develop a University Information System. In: Proceedings of 1999 Asia-Pacific Software Engineering Conference. IEEE Computer Society Press (December 1999)
4. Mat, A.R., Liu, S.: Applying SOFL to Construct the Formal Specification of an Automatic Automobile Driving Simulation System. In: Proceedings of International Conference on Software Technology and Engineering (ICSTE 2009), July 24-26, pp. 42–48. World Scientific Publishing, Chennai (2009)
5. Dawes, J. : The VDM-SL Reference Guide. Pitman (1991)

6. Brauer, W., Rozenberg, G., Salomaa, A.: Petri Nets – An Introduction. Springer, Heidelberg (1985)
7. Yourdon, E.: Modern Structured Analysis. Prentice-Hall (1989)
8. Brain Tumor. Brain Tumor Causes, Symptoms, Diagnosis, treatment and Prognosis Information, WebMD, MedicineNet Inc., California, `http://www.medicinenet.com` (archived on December 2, 2009)
9. Angele, J., Fensel, D., Landes, D., Studer, R.: Developing Knowledge-Based Systems with MIKE. Journal of Automated Software Engineering 5, 389–418 (1998)
10. Liu, A., Tsai, J.J.P.: A Knowledge-Based Approach to Requirements Analysis. In: Proceedings of the Seventh International Conference on Tools with Artificial Intelligence (TAI 1995). IEEE Computer Society (1995)

Egyptian Vulture Optimization Algorithm – A New Nature Inspired Meta-heuristics for Knapsack Problem

Chiranjib Sur, Sanjeev Sharma, and Anupam Shukla

Soft Computing and Expert System Laboratory
ABV-Indian Institute of Information Technology & Management
Gwalior, Madhya Pradesh,
India – 474010
{chiranjibsur,sanjeev.sharma1868,dranupamshukla}@gmail.com

Abstract. In this paper we have introduced for the first time a new nature inspired meta-heuristics algorithm called Egyptian Vulture Optimization Algorithm which primarily favors combinatorial optimization problems. The algorithm is derived from the nature, behavior and key skills of the Egyptian Vultures for acquiring food for leading their livelihood. These spectacular, innovative and adaptive acts make Egyptian Vultures as one of the most intelligent of its kind among birds. The details of the bird's habit and the mathematical modeling steps of the algorithm are illustrated demonstrating how the meta-heuristics can be applied for global solutions of the combinatorial optimization problems and has been studied on the traditional 0/1 Knapsack Problem (KSP) and tested for several datasets of different dimensions. The results of application of the algorithm on KSP datasets show that the algorithm works well w.r.t optimal value and provide the scope of utilization in similar kind of problems like path planning and other combinatorial optimization problems.

Keywords: Egyptian vulture optimization algorithm, combinatorial optimization, graph based problems, knapsack problem, nature inspired meta-heuristics.

1 Introduction

The life style of the Egyptian Vulture has been a curiosity and field of study of many researchers because of its attitude, adaptive features, unique characteristics and enhanced skills and techniques for leading the lifestyle. Egyptian Vulture has been studied thoroughly and preventive measures are sought out for many decades ever since there occurred constant decrease in their numbers with some of their species became extinct with time. There has always been an effort from the government of many countries to save the species before it becomes extinct. In this work we have tried to formulate the activities of the Egyptian Vulture that can be applied for solution of real time problems and mobilized the process into some simple steps which can derive combinatorial optimization of graph based problems. These habits are discussed in details in Section 3 and are illustrated with examples so that the steps are clear to

P. Meesad et al. (Eds.): *IC²IT2013*, AISC 209, pp. 227–237.
DOI: 10.1007/978-3-642-37371-8_26

understand and easy to understand. However readily unlike any other meta-heuristics, this algorithm is best suited for the combinatorial optimization problems and graph based problems and is unique among the few algorithms (ACO [26], IWD [16], etc) in the bio-inspired computation family which can readily be utilized for the discrete problems. ACO and IWD algorithms are dependent on the local heuristic parameters created in the form of pheromone level and sand respectively and the scope of exploration decreases and the solution may saturate on the exploitation only. But in our proposed algorithm, exploitation will survive on the global best and the whole algorithm will concentrate on the exploration computation. Otherwise most meta-heuristics like PSO [4], bat algorithm [8], Honey bee swarm [5], league championship algorithm [6], cuckoo search [7], simulated annealing [9], krill herd optimization [21], Virus Optimization Algorithm [24], Magnetic Optimization Algorithms [18], etc are suitable for searches in the continuous domain where the variable can take any value in the permissible range and thus in this kind of local searches the combinations are more in number but there is no requirements for maintenance of valid sequence of nodes for the solution to be acceptable. Also algorithms like artificial immune system [11] is suitable for adaptability, pattern learning and matching and has nothing to do for optimization, for harmony search [12], genetic algorithm [3] etc valid sequence generation following the constraints and combination creation is very difficult, glowworm swarm is suitable for multi-modal continuous problems optimization. However Genetic Algorithm [3] has always involved in randomized solution formation and combination creation and has been very successful in the continuous domain optimization problems but their sister algorithms like discrete GA, discrete PSO [3] etc are not efficient in the discrete domain like problems of path finding, graph based etc. These algorithms mainly use some adaptive changes (created with combination of feedback error or solution dependent adaptive feedback and random values for controlling the variation within a certain range) in value for each dimension within range and thus produce the required combination which is totally not suitable for discrete problems. The development of the algorithm was primarily meant for the graph search problems through the process of simultaneous "development-cum-validation" through randomized selection and mixing up of the solution sets for opportunistic solution derivation amidst constraints imposed time to time by the various application systems on which it is applied. But EVOA are void of the following characteristics and these make the algorithm very special: like lack of exploitation feature (exploitation depends on the global best and depends solely on the exploration criteria which a graph based problem usually faces unlike ACO and IWD), co-agent communications, global solution influenced variations, local heuristic parameter based decision making for path selection etc.

The rest of the paper is arranged as Section 2 describing the life style of the egyptian vulture bird, Section 3 illustrating the operations of the egyptian vulture optimization algorithm, Section 4 describes algorithmic variations for the Knapsack Problem and Section 5 has computational results for a certain dimension combination of knapsack problem with conclusion in Section 6.

2 History and Life Style of Egyptian Vulture

The Egyptian Vulture, also known as White Scavenger Vulture (Scientific Name : Neophron percnopterus) [1], is one of the most ancient kinds of vulture that existed on this earth and a few species of its kind has become extinct. Like any other vulture species, the primary food habitat of the Egyptian Vulture is flesh, but the opportunistic feature for food habit which makes the species unique and thus lead to the meta-heuristic is that they eat the eggs of the other birds available. However for larger and strong (in terms of breakability) eggs they toss pebble hard on them, using the pebbles as hammer, and thus break them. Also the Egyptian Vultures have the knack of rolling things with twigs, which is another distinguish feature of the Egyptian Vulture.

Fig. 1. Egyptian Vulture at Work

Relatively the Egyptian Vulture possess a sedentary life apart from hunting for food, but their level of performance and technique have been unique among all the member of the class Aves and this makes them the best lot. However due to some unavoidable reasons of the unscrupulous activity of the human beings like poaching, cutting down of forests, global warming, etc there has been a decrease in their numbers in population. The breed has been famous among the pharaohs of Egypt and was considered as a symbol of royalty and was known as "Pharaoh's Chicken". In India, there are a lot of superstitious believes are associated with the bird. The sight of this bird is considered as unlucky and their shrilling calls believed to signify downfall, hardship or even death. How-ever there are tribal groups who can actually tame them and read the activity of the vultures to detect natural calamity or incidents. Another superstitious belief is considered in [1]. A temple at Thirukalukundram in Chengalpattu (India) was reputed for a pair of Egyptian Vulture that visited the place for centuries. These Egyptian Vultures were ceremonially fed by the temple priests and arrived before noon to feed on offerings made from rice, wheat, ghee, and sugar. Although Egyptian Vultures are normally punctual in their arrival, but in case there is failure of the Egyptian Vultures to turn up, it is attributed to the presence of "sinners" among the onlookers. Legend has it the vultures (or "eagles") represented eight sages who were punished by Lord Shiva (Hindu God), with two of them leaving in each of a series of epochs. The Egyptian Vultures possess the reputation of practicing the habit of coprophagy that is being fed on faeces of other animals.

3 Egyptian Vulture Optimization Algorithm

The Egyptian Vulture Optimization Meta-Heuristics Algorithm has been described here as steps, illustration through examples and explanations. The two main activities of the Egyptian Vulture, which are considered here or rather transferred into algorithm, are the tossing of pebbles and the ability of rolling things with twigs.

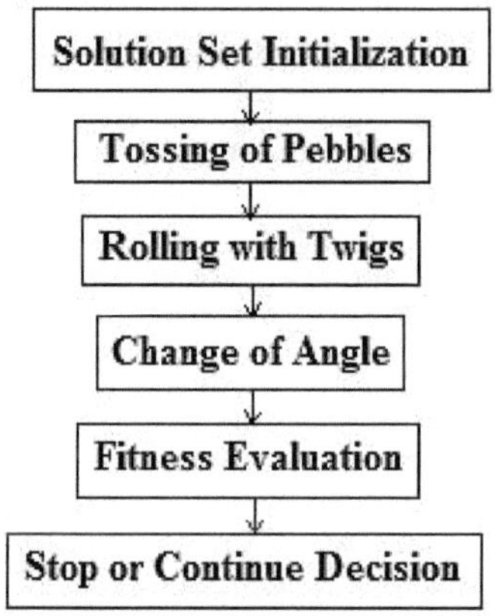

Fig. 2. Steps for Egyptian Vulture Optimization Algorithm

Pebble Tossing. The Egyptian Vulture uses the pebbles for breakage of the eggs of the other birds which produce relatively harder eggs and only after breakage they can have the food inside. Two or three Egyptian Vulture continuously toss the pebbles on the egg with force until they break and they try to find the weak points or the crack points for success. This approach is used in this meta-heuristics for introduction of new solution in the solution set randomly at certain positions and hypothetically the solution breaks into the set and may bring about four possibilities depending upon probability and the two generated parameters for execution of the operations and selection of extension of the performance. Figure 3 provides the pictorial view of the pebble tossing step of the Egyptian Vulture.

The two variables for the determination of the extent of the operation are: PS = Pebble Size (level of occupy) where $PS \geq 0$ and FT = Force of Tossing (level of removal) where $FT \geq 0$. Hence, If $PS > 0$ Then "Get In" Else "No Get In". Also If $FT > 0$ Then "Removal" Else "No Removal" where "Get In" denotes occupancy and "Removal" is for removing. Now the Level of occupy denotes how many solutions should the pebble carry and must intrude forcefully into the solution set. Level of removal implies how many solutions are removed from the solution set. Both are generated randomly within a certain limit and the pebbles carrying PS number of nodes are also generated randomly considering that the combination can produce new solution set. Now FT denotes the number of nodes that are removed from either side of the point of hitting.

Overall there are four combinations of operations are possible and are : Case 1 : Get In & No Removal, Case 2 : No Get In & No Removal, Case 3 : Get In & Removal, Case 4 : No Get In & No Removal.

The last combination is of no operation and is another way refusal of operation on the solution. Another criterion is the point of application of the step.

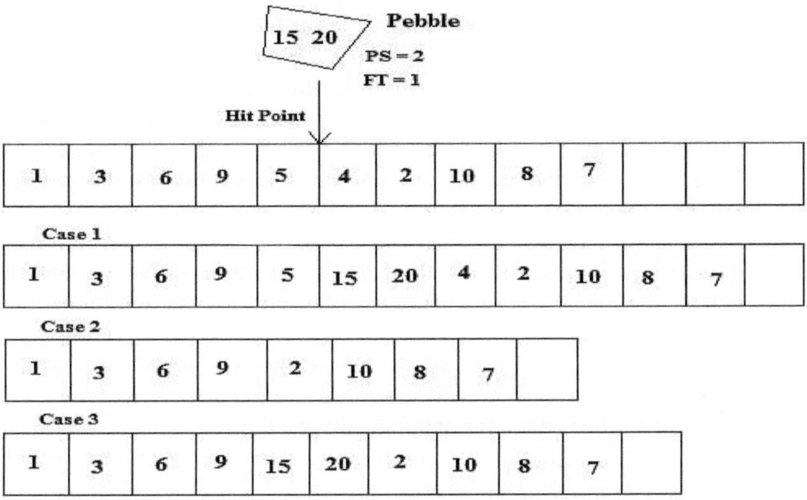

Fig. 3. Pictorial View of Pebble Tossing

It is to be pointed out that the application is sole decider of up to what extent the operation combination will take place and whether it is permissible to allow combinations where PS ≠ FT which means that the removal and occupy or both will be of unequal length and thus introduces the concept of variable string length which is perhaps very necessary for route finding problems. However for problems like TSP, QAP etc the required combination is always PS = FT to justify the constraint of constant number of cities and without duplicity. Point of hitting is another criterion which requires attention and strategy must be determined for quickening the solution convergence process. Like for TSP problem any point can be chosen for application of Tossing of Pebble step but for path finding problem the best strategy is considered if the discontinuous positions are targeted for application. In a similar way for continuous domain problems, where a certain position represents a certain variable, any point and relevant and mapped positions can be the hit point and way for removal and occupy for experimentation and combination trials.

Rolling with Twigs. The rolling with twigs is another astonishing skill of the Egyptian Vulture with which they can roll an object for the purpose of movement or may be to perform other activity like finding the position or weak points or just giving a look over the other part which is facing the floor. Rolling of the objects requires not only power but also the art of holding tight the stick with their beak and also finding the proper stick. This is perhaps the inherited skill of any bird for finding the right stick for any object they are trying to create or execute. Several birds have given testimony of making high quality nests during laying eggs. Such selection of sticks is mainly made for making the nest or positioning the right bend of the stick at the right place. Even some birds have given evidence of sewing the soft twigs with their beak.

This activity of the Egyptian Vulture is considered as rolling of the solution set for changing of the positions of the variables to change the meaning and thus may create new solutions which may produce better fitness value and also better path when it comes for multi-objective optimization. Also when the hit point is less and the numbers of options are more, it may take a long time for the finishing event to take place or in other words the appropriate matching of the random event to occur. In such case this kind of operations can be helpful. Figure 4 illustrates the effect of the "Rolling with Twigs" event for graph based problems for particular state and for certain parameter value of the probabilistic approach which is discussed in the subsequent paragraphs. Rolling can be of the whole solution string and also for the partial string which needs to be jumbled.

For the "Rolling with Twigs" to occur there is required another two more parametric variables which will direct the mathematical formulation of the event and also guide the implementation of the step. These two criteria for the determination of the extent of the operation are:

DS = Degree of Roll where DS ≥ 0 denoting number of rolls.

DR as Direction of Rolling where probabilistically we have:

DR = 0 for Right Rolling/Shift
= 1 for Left Rolling/Shift

where 0 and 1 is generated randomly and deterministically the equation can be framed as :

DR = Left Rolling/Shift for RightHalf > LeftHalf
= Right Rolling/Shift for RightHalf < LeftHalf

where RightHalf is the secondary fitness for the right half of the solution string and LeftHalf is for left half. The reason behind this is if the RightHalf is better, then this will be a provision to extent the source with the connected node portion and same is for LeftHalf, which can be connected with the destination.

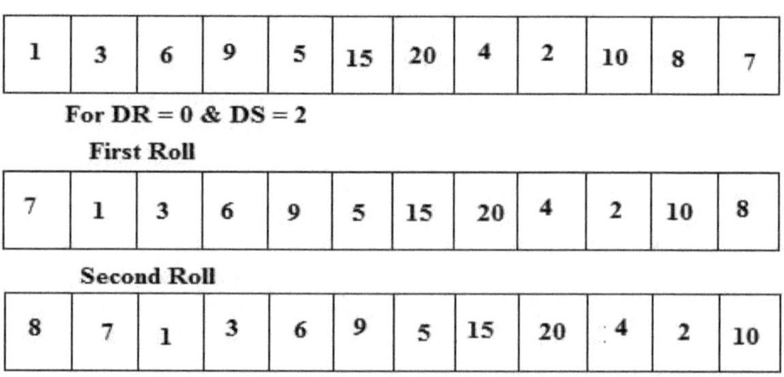

Fig. 4. Pictorial View of Rolling with Twigs for DS = 2

Another scheme that can be implemented in constraint environment without hampering the partial derived solution occurs for mainly problems like path finding etc. Here only the unstructured solution string is modified and not the partial path already found. Link and in order-ness are important factors of these kind of constraint problems, but for problems like the datasets of the TSP, where a path exist between every node and distance is the Euclidean distance between them, shifting of the partial string holds the same information as that of the whole string as each can give rise to new solution and hence the search procedure is more versatile and global solution can be attended easily.

Change of Angle. This is another operation that the Egyptian Vulture can perform which derives its analogy from the change of angle of the tossing of pebbles so as to experiment with procedure and increase the chance of breakage of the hard eggs. Now the change of the angle is represented as a mutation step where the unconnected linked node sequence are reversed for the expectation of being connected and thus complete the sequence of nodes. Figure 5 gives a demonstration of such a step. This step is not permanent and is incorporated if only the path is improved.

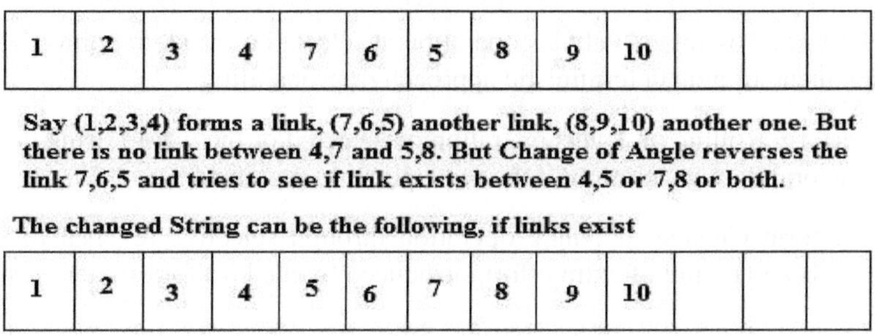

Fig. 5. Pictorial View of Change of Angle

This Change of Angle step can be multi-point step and the local search decides the points, number of nodes to be considered and depends on the number of nodes the string is holding. If the string is holding too many nodes and Pebble Tossing step cannot be performed then this step is a good option for local search and trying to figure the full path out of it.

4 EVOA for Knapsack Problem

Knapsack problem is a traditional problem which prevails as a heavy weight problem in the combinatorial optimization problem series and shares its similarity with many real life problems in some form or the other. The equation for the 0/1 KSP problem is $\mathbf{max\{\sum(v_i\, x_i)\}}$ subjected to $\mathbf{\{\sum(w_i\, x_i)\} < W_m}$ for x takes binary value {0,1} where w_i denotes the weight and v_i is the cost associated with each item i where i = 1,2,…n . So the solution string must constitute of a series of 1 and 0 and must represented x_i.

The Algorithmic stepwise details of the EVOA are as follows:

Step 1: Consider a dataset of n items and x_i is the ith bag, for m number of bags where each bag is has maximum capacity W_m.

Step 2: Generate N strings for each type of bag x_i where string for x_i for i = 1,2,...n consisting of n items and each is represent is by a number from 1 to n without repetition of any of the numbers in the string.

Step 3: Now generate a threshold between 1 and n in integer values, so that the string of integer values can be converted to string of 0s and 1s and the random value of threshold will decide how many of them should be 0s and 1s and also the positional values will create combinations and due to the EVOA the integer values for each positions will change.

Step 4: For x_i > threshold Make it 1, else make it 0. So string of 0 & 1 represent x_i set.

Step 5: Evaluate the fitness of each string, Store the value of profit if constraint satisfied else make it zero. Update the Global best if required.

Step 6: Perform Tossing of Pebbles operation at selected or random points depending upon implementation on deterministic approach or probability.

Step 7: Perform Rolling of Twigs operation on selected or the whole string depending on the pseudorandom generation of the two operation parameters.

Step 8: Perform Change of Angle operation through selective reversal of solution subset. This is some kind of extra effort introduced by the bird for efficient result.

Step 9: Evaluate the fitness of each string. Store the value of profit if constraint satisfied else make it zero. Update the Global best if required. Replace if fitness is better (here the best is the maximization of the profit subjected to the satisfaction of the capacity of the bags). The global best consists of the fitness value along with the string consisting of 1s and 0s.

Step 10: After a certain iterations, replace a certain percentage of worst fitted strings with the best string of the iteration or global best with probability.

Step 11: Stop with stopping criteria fulfilled.

Note : After each of the steps 6,7 and 8 the inserted discrete element must stay and the duplicate one must be removed and if an element is removed then its compensation must be added at the start or end (with random probability).

5 Computational Results

In this section the tabulated form of the results of simulation of the EVO algorithm on the various dimensional knapsack problem [25] are provided and its variations are shown as mean, standard deviation of best along with best and worst of the lot.

Table 1. Table for Computational Result of Knapsack Datasets

Dataset			EVO			
Name	Dim	Optimum	Mean	SD	Best	Worst
WEISH01	5,30	4554	4449.8	63.53792	4549	4327
WEISH07	5,40	5567	5255.1	52.6486	5353	5135
WEISH10	5,50	6339	5865.842	184.8216	6309	5592
WEING1	2,28	141278	140501.2	340.4875	140988	139933
FLEI	10,20	2139	2139	0	2139	2139
HP1	4,28	3418	3396	8051143	3405	3388
PB6	30,40	776	705.1	16.543	725	682
PET2	10,10	87061	87061	0	87061	87061
PET3	10,15	4015	4015	0	4015	4015
PET4	10,20	6120	6120	0	6120	6120
PET5	10,28	12400	12318.89	32.1887	12350	12270
PET6	5,39	10618	10538.1	22.83978	10570	10514
PET7	5,50	16537	16204	53.49559	16256	16078

Here the Dim column denotes the dimension of the dataset where the first element is the number of bags and the second element is the number of items. For considering the 0/1 Knapsack either the whole lot of a certain item is included or not included and there is no provision for partial inclusion of any item. The results are counted for the optimal value which is denoted by Optimal and it shows that the mean and standard deviation (SD) quite tally with the optimal value for the same number of iteration run executed for all the datasets. But the best result achievement time of the low dimension dataset is much less compared to the bigger ones.

6 Conclusion

So yet another nature inspired meta-heuristic is being introduced here for the first time in the literature which favors the graph based problems and combinatorial optimization problems and have some unique mode of operation sets which are replicating the behavior of the Egyptian vulture bird and has been successfully utilized for the knapsack problem. Results of the simulation of the various dimensional dataset reflect it potential as a good meta-heuristics in the evolutionary algorithm family and can be utilized in many constrained, unconstrained combination dependent problems and have shown it capacity for varied combination generation. However it must be mentioned here that the EVOA can also be used as swarm where several birds will operate simultaneously on a certain solution and can produce quick variations in combination generation. The uniqueness of the algorithm lays in combination creation capacity without knowledge about the different parameters and unlike ACO, IWD etc it not driven by the available paths and its parameters. Instead it is fully randomized and depends more on the path creation than on the local heuristics. However the element of local heuristics is required (in the formation of optimized link) when there are constraints like precedence factor, event dependency, etc. Local search is a superficial validation of the change in the solution string and is vital because it assures the acceptability of the solution and prevents arise of the invalid solution strings.

Acknowledgement. The first author gratefully acknowledges the motivation, guidance and support of Prof. Anupam Shukla, Professor, ABV-Indian Institute of Information Technology and Management, Gwalior, India. His time to time critics and reviews has been very helpful for the work's improvement. The work is motivated by a slide of Prof Ajith Abraham.

References

1. `http://en.wikipedia.org/wiki/Egyptian_Vulture`
2. `http://www.flickr.com/photos/spangles44/5600556141/`
3. Blum, C., Roli, A.: Metaheuristics in combinatorial optimization: Overview and conceptual comparison. ACM Comput. 35(3), 268–308 (2003)
4. Kennedy, J., Eberhart, R.: Particle swarm optimization. In: Proceedings of IEEE International Conference on Neural Networks, vol. 4, pp. 1942–1948 (November/December 1995)
5. Karaboga, D.: An idea based on honey bee swarm for numerical optimization. Technical Report TR06, Erciyes University (October 2005)
6. Kashan, H.A.: League Championship Algorithm: A New Algorithm for Numerical Function Optimization. In: Proceedings of the 2009 International Conference of Soft Computing and Pattern Recognition (SOCPAR 2009), pp. 43–48. IEEE Computer Society, Washington, DC (2009)
7. Yang, X.-S., Deb, S.: Cuckoo search via Levy flights. In: World Congress on Nature & Biologically Inspired Computing (NaBIC 2009), pp. 210–214. IEEE Publication, USA (2009)
8. Yang, X.-S.: A New Metaheuristic Bat-Inspired Algorithm. In: González, J.R., Pelta, D.A., Cruz, C., Terrazas, G., Krasnogor, N. (eds.) NICSO 2010. SCI, vol. 284, pp. 65–74. Springer, Heidelberg (2010)
9. Kirkpatrick, S., Gelatt Jr., C.D., Vecchi, M.P.: Optimization by Simulated Annealing. Science 220(4598), 671–680 (1983)
10. Storn, R., Price, K.: Differential evolution - a simple and efficient heuristic for global optimization over continuous spaces. Journal of Global Optimization 11(4), 341–359 (1997)
11. Farmer, J.D., Packard, N., Perelson, A.: The immune system, adaptation and machine learning. Physica D 22(1-3), 187–204 (1986)
12. Geem, Z.W., Kim, J.H., Loganathan, G.V.: A new heuristic optimization algorithm: harmony search. Simulation 76(2), 60–68 (2001)
13. Krishnanand, K., Ghose, D.: Glowworm swarm optimization for simultaneous capture of multiple local optima of multimodal functions. Swarm Intelligence 3(2), 87–124 (2009)
14. Haddad, O.B., Afshar, A., Mariño, M.A., et al.: Honey-bees mating optimization (HBMO) algorithm: a new heuristic approach for water resources optimization. Water Resources Management 20(5), 661–680 (2006)
15. Tamura, K., Yasuda, K.: Primary Study of Spiral Dynamics Inspired Optimization. IEEJ Transactions on Electrical and Electronic Engineering 6 (S1), S98–S100 (2011)

16. Shah-Hosseini, H.: The intelligent water drops algorithm: a nature-inspired swarm-based optimization algorithm. International Journal of Bio-Inspired Computation 1(1/2), 71–79 (2009)
17. Civicioglu, P.: Transforming geocentric cartesian coordinates to geodetic coordinates by using differential search algorithm. Computers & Geosciences 46, 229–247 (2012)
18. Tayarani-N, M.H., Akbarzadeh-T, M.R.: Magnetic Optimization Algorithms a new synthesis. In: IEEE Congress on Evolutionary Computation, CEC 2008, IEEE World Congress on Computational Intelligence, June 1-6, pp. 2659–2664 (2008)
19. Reynolds, C.W.: Flocks, herds and schools: A distributed behavioral model. Computer Graphics 21(4), 25–34 (1987)
20. Kaveh, A., Talatahari, S.: A Novel Heuristic Optimization Method: Charged System Search. Acta Mechanica 213(3-4), 267–289 (2010)
21. Gandomi, A.H., Alavi, A.H.: Krill Herd Algorithm: A New Bio-Inspired Optimization Algorithm. Communications in Nonlinear Science and Numerical Simulation (2012)
22. Tamura, K., Yasuda, K.: Spiral Dynamics Inspired Optimization. Journal of Advanced Computational Intelligence and Intelligent Informatics 15(8), 1116–1122 (2011)
23. Wolpert, D.H., Macready, W.G.: No free lunch theorems for optimization. IEEE Transactions on Evolutionary Computation 1(1), 67–82 (1997)
24. Liang, Y.-C., Josue, R.C.: Virus Optimization Algorithm for Curve Fitting Problems. In: IIE Asian Conference (2011)
25. `http://www.cs.cmu.edu/afs/cs/project/airepository/ai/areas/genetic/ga/test/sac/0.html`
26. Dorigo, M., Gambardella, L.M.: Ant Colony System: A Cooperative Learning Approach to the Traveling Salesman Problem. IEEE Transactions on Evolutionary Computation 1, 53–66 (1997)

Optical Music Recognition on Windows Phone 7

Thanachai Soontornwutikul, Nutcha Thananart, Aphinun Wantanareeyachart, Chakarida Nukoolkit, and Chonlameth Arpnikanondt

Computer Science Program, School of Information Technology,
King Mongkut's University of Technology Thonburi
thanachai@ieee.org, {n.thananart,prlnze34}@gmail.com,
{chakarida,chonlameth}@sit.kmutt.ac.th
http://sit.kmutt.ac.th/

Abstract. Optical Music Recognition (OMR) software currently in the market are not normally designed for music learning and ad hoc interpretation; they usually require scanned input of music scores to perform well. In our work, we aimed to remove this inconvenience by using photos captured by mobile phone's camera as the input. With the cloud-based architecture and the design without the assumption of perfect image orientation and lighting condition, we were able to eliminate many of the software's architectural and algorithmic problems while still maintaining an overall decent performance.

Keywords: camera, cloud-based architecture, MIDI, mobile phone, optical music recognition, sheet music.

1 Introduction

Sheet music is one of the most common ways to precisely express and share music with others. However, it can be difficult to read and mentally perceive the music inscribed in the score, even for professional musicians. This is where computers can help by providing a way of "playing" directly from the sheet music without having to play an actual instrument, which can be achieved using optical music recognition [1].

Optical Music Recognition (OMR) is a software technique that aims to provide an automated means of digitizing images of music scores, which is useful for archival and recreational purposes. Nevertheless, in implementing a practical OMR solution, many concerns have to be addressed. In particular, the software design and the algorithms to be incorporated have to be vigorously evaluated. The resolution and lighting limitations of the images obtained using phone cameras also add to the overall complexity. While, several commercial software packages with OMR capability are available, they are not designed for ad hoc uses. That is, they are based on the assumption that the input images of the music scores are scanned from a scanner, where the obtained images usually exhibit good clarity, contrast, and orientation. In *Optical Music Recognition Application on Android* [2], those problems are described and a mobile-based solution that

P. Meesad et al. (Eds.): *IC²IT2013*, AISC 209, pp. 239–248.
DOI: 10.1007/978-3-642-37371-8_27

tries to improve the practicality and portability of the application is proposed. However, its basic implementation is not without problems. Some of the major problems are as follows:

- The entire processing, which is computationally expensive, is done within the mobile phone. This can quickly drain the phone's battery, making it impractical for ordinary usage.
- The processing time can be perceived as high compared to the users' general expectation of on-phone processing time.
- The design is not modular and exhibits strong dependency among different modules, resulting in a low level of extensibility of the software: inclusion of additional or alternative algorithms may require a major rework.
- The architecture implemented does not allow easy extension of the recognition capabilities without loss of accuracy.
- Symbol recognition algorithm is static. That is, the rules and artificial neural networks used in the recognition process are stored statically and locally on the phone.
- Automatic self-correction after the recognition process is not included.
- The design does not support systematic evaluation and timing of the algorithms in each of processing phases. Performance evaluation has to be done manually and can only cover the final outputs of the software, disregarding intermediate steps, making it more difficult to find out the root cause of poor performance.

Implementation-wise, we arbitrary picked Windows Phone 7.5 for mobile phone platform and the C# language for both phone and service-related code. The service is hosted on Windows Communication Foundation (WCF) and employs the REST architecture for its HTTP interface, which would facilitate the creation of additional client-side software as needed.

2 Related Work

Apart from *Optical Music Recognition Application on Android* [2] on which this project is based, there have also been several related research and projects including, but not limited to:

- *Guido: A Musical Score Recognition System* [3,4], which presents an OMR system called Guido, which makes use of the GUIDO Music Notation Format [5], and provides detailed information on how it works.
- *OpenOMR* [6], a Java-based open-source OMR software, which supports recognition of printed music scores and playback through computer speakers.
- *Audiveris Music Scanner* [7], a Java-based open-source OMR software, which supports recognition of printed music scores with multiple staffs and voices. It also supports playback and export to either MIDI or MusicXML format.

Different from those listed above, our solutions aims for ad-hoc uses and expects inputs from mobile phone cameras instead of scanned inputs. Some accuracy is traded off in favor of portability and processing speed, while at the same time solving the problems that exist in *Optical Music Recognition Application on Android.*

3 Methodology

3.1 Overall Process

Assuming that the user has one or more pages of the music score of a song ready, and that the photos of the pages are input to the system, the outline of the main process flow will be as follows.

1. The images are spatially down-sampled as necessary. The target width is empirically set to 800 pixels, with the corresponding proportional height.
2. The images are converted into gray-scale images and then into binary images. The intermediate step is required by the binarization algorithm.
3. The binary images are merged vertically into one image.
4. The staff lines in the merged binary image are detected and filtered.
5. The detected staff lines are grouped into their corresponding staffs, each containing five lines.
6. The orientation of the image is corrected using the detected staffs as guides, resulting in a binary image segment for each staff. Each binary image segment is called a *staff-based binary image.*
7. The staffs are re-detected using the same procedure.
8. The staff lines are removed from the staff-based binary images while remembering the original locations. This isolates the symbols from the staff lines, simplifying the segmentation process.
9. Symbol segments are extracted, each assigned a music symbol class. If the confidence in classifying a particular segment is low, the segment is either enlarged, shrunk, combined, or split, so that a higher confidence level is obtained. If this still fails, the segment is discarded.
10. The symbols are grouped and corrected from potential classification errors.
11. The song is interpreted using the detected symbols.

3.2 Algorithms and Improvements

In this section, we discuss in more detail about the algorithms used in our implementation, particularly those that contain improvements from the original.

Conventions. For conciseness, the following definitions are used:

- The value of foreground pixels is 1 and the value of background pixels is 0.
- If i is an image[1], then $i[x, y]$ is the value of the pixel at the location (x, y) of the image with the origin $(0, 0)$ at the top-left corner of the image.
- If i is an image, then w_i is the width of i in pixels.

[1] A binary image is also considered as an image.

Score Image Binarization. The algorithm we based on for image binarization is the BLIST (Binarization based in LIne Spacing and Thickness) algorithm [8,9], which is designed specifically for optical music recognition purpose. That is, in addition to extracting foreground pixels from background ones, another objective of the algorithm is to maximize the visibility of the staff lines, disregarding the area in which no staff line exists. Two performance optimizations to the algorithm that decrease the time required to binarize an image while still maintaining the binarization accuracy have been devised and applied in the software, as follows.

Rough Estimation of Run-Length Encoding Runs. Originally, when estimating the staff line thickness and spacing in BLIST, all columns within a particular image window is used. However, since most of columns will contain parts of the staff lines (except along the left and the right margins of the page), not all columns need to be taken into account in the estimation. Let a be an image window from the input gray-scale image and b be another image with the relations:

$$w_b = \lfloor \frac{w_a}{f} \rfloor \ . \tag{1}$$

$$b[x, y] = a[xf, y] \ . \tag{2}$$

where $f \in \mathbb{N}^+$ is the optimization factor. b can then be used instead of a in estimating the staff line thickness and spacing of a.

Rough Sampling of Binarization Thresholds. In BLIST, for each sampling window, the histogram of the runs obtained from run-length encoding (RLE) is accumulated from all possible binary images constructed by varying thresholds from the lowest possible value to the highest possible value. However, it turns out that there is no need to sample from all possible threshold values. This is justified by the fact that the contrast (that is, the difference in gray-scale intensity of the foreground pixels and the background pixels of any particular area of a score image) is usually fairly high; otherwise, it would be difficult even to human eyes to read the score. Let Δt denote the amount of threshold to skip, $hist_i[L]$ be the value of the class L of the run histogram[2] of an image i, and $run(L, i, t)$ be a function that yields the number of times the runs of length L occur when globally binarizing an image i with the threshold t, we use this equation:

$$hist_i[x] = \sum_{k=1}^{\lfloor \frac{254}{\Delta t} \rfloor} run(x, i, k\Delta t) \ . \tag{3}$$

In our case, given the intensity range of $(0, 255)$ (from darkest to brightest), the feasible threshold range would be $(1, 254)$, and according to Fig. 4, a good range of Δt is $[10, 20]$, where using the Δt in this range yields a good time optimization while not sacrificing the binarization accuracy.

[2] The value of the class L of a run histogram is the number of times the (line thickness + spacing) runs of length L occur.

Staff Detection. The Stable Path staff line detection algorithm [10] is used as the basis for staff detection. Two enhancements have been made to the algorithm, lowering the processing time as well as improving line detection performance.

Horizontal Compression of Score Images. Generally, score images can be compressed horizontally to a certain degree without affecting the visibility of the staff lines and the performance of the staff line detection algorithm, as can be seen in Fig. 1 and 2. As a side effect of the operation, disconnected portions of the staff lines can also be connected together, potentially increasing the chance of detecting the lines.

Fig. 1. This is an uncompressed portion of a simple music score

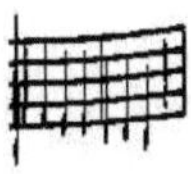

Fig. 2. This is the same portion of the music score as in Fig. 1, but compressed with $f = 8$. Notice that even though the symbols are a lot less legible, the staff lines can still be clearly seen. Note that the position of the ledger line can be estimated from the nearby staff lines and thus need not be separately detected.

Depending on the amount of compression, the algorithm can run significantly faster than without any compression. The compression method works as follows.

$$w_b = \lfloor \frac{w_a}{f} \rfloor . \tag{4}$$

$$b[x,y] = 1 \text{ if } \sum_{k=xf}^{xf+f-1} a[k,y] \geq 1, \text{ otherwise } 0 . \tag{5}$$

where a is the original binary image, b is the scaled binary image, and $f \in \mathbb{N}^+$ is the compression factor.

For each detected path from the scaled image, we can derive a path with the original width by interpolating for the missing pixels of the path. Note that the purpose of the compression is only for detecting the staff lines in the image; the original image is still used in the subsequent processing steps.

Identifying False Staff Lines. In [10], lines that consist of less than a fixed threshold (80%) of foreground pixels are filtered out. In addition, lines that have the average y-distance from the line with the median number of foreground pixels (after removing the means) above a set threshold are also eliminated. For the second rule, we instead use the absolute difference between the average

y-distance and the median y-distance between the two lines. This allows inclusion of curved lines (provided that all lines in the image curve in the similar way) and straight lines with different slopes (which can occur when the score paper is bent when the photo is taken).

3.3 Modular Architecture

The top part of the diagram in Fig. 3 represents the client-side software, which in this case is run on a Windows Phone 7-based phone. The bottom part represents the service side and is responsible for handling most of the OMR-related processes. The OMR-related module interfaces are designed such that new or alternative algorithms can be swapped in with little effort. This also allows isolating an algorithm for the purpose of evaluating the performance of the particular algorithm alone, in contrast to evaluating the performance of the entire process, which may not be able to identify the poor-performing parts among all of the algorithms.

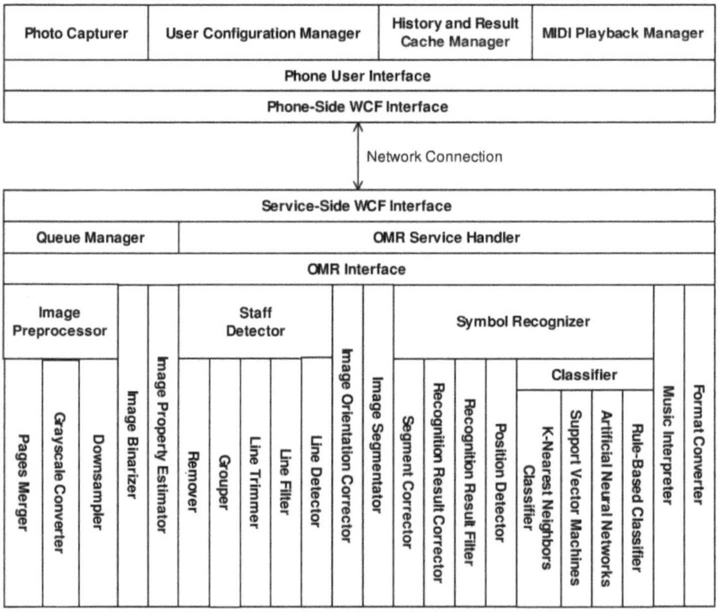

Fig. 3. Overall module architecture

4 Evaluation

In this section, the evaluation methods and results of the algorithm enhancement techniques discussed in the previously are discussed. To facilitate automatic evaluation, the *ground truths* (also known as *gold standards*) are created and used to compare with the actual results obtained from using the algorithms.

4.1 Evaluating Binarization Performance

In evaluating the performance of a binarization method, the measures recommended in [8,9] are used, which include:

- Misclassification Error (ME)
- Missed Object Pixels (MOPx)
- False Object Pixels (FOPx)

As shown in Fig. 4, the error rates become unstable as the threshold step Δt gets higher, potentially due to different threshold values needed for different images, resulting in a good threshold set being selected for one set of images but not others. In contrast, from Fig. 5 there seems to be no clear correlation between the optimization factor f and any of the error rates, though using the optimization factor higher than 16 yields no significant improvement in processing time. Note that only the pixels in the areas containing staff lines are considered in both evaluations.

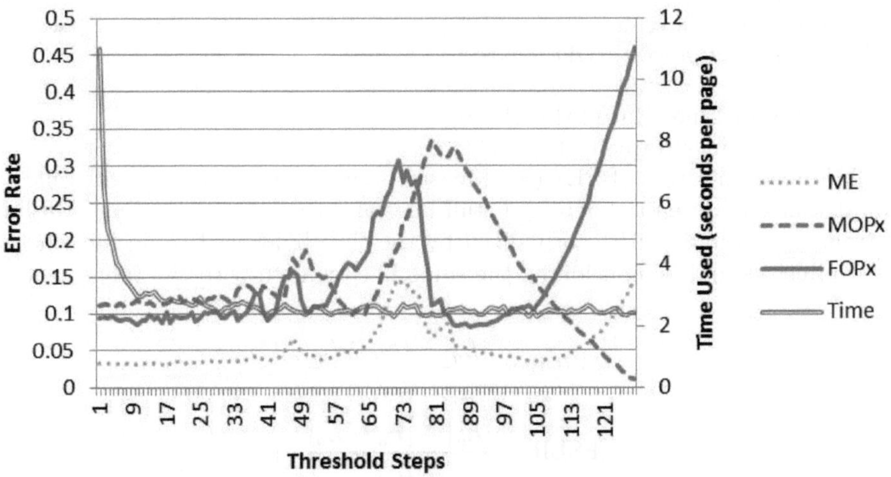

Fig. 4. A plot showing the relationship between the binarization threshold step Δt and both the three binarization error measures (left y-axis) and the average time per page used in the binarization process (right y-axis)

4.2 Evaluating Staff Line Detection Performance

Staff line detection aims to detect as many staff lines in music score as possible, where the location of the detected lines can be slightly different from the actual lines as long as they can still accurately represent the original lines.

When evaluating, two staff lines are considered the same line if the sum of squared difference in y-distance is less than a threshold, defined as the x-width of the ground truth line multiplied by a constant μ. If the two lines do not have the same length, the distance calculation only takes place on the x-positions that exist in both lines. The evaluation results are given in Fig. 6.

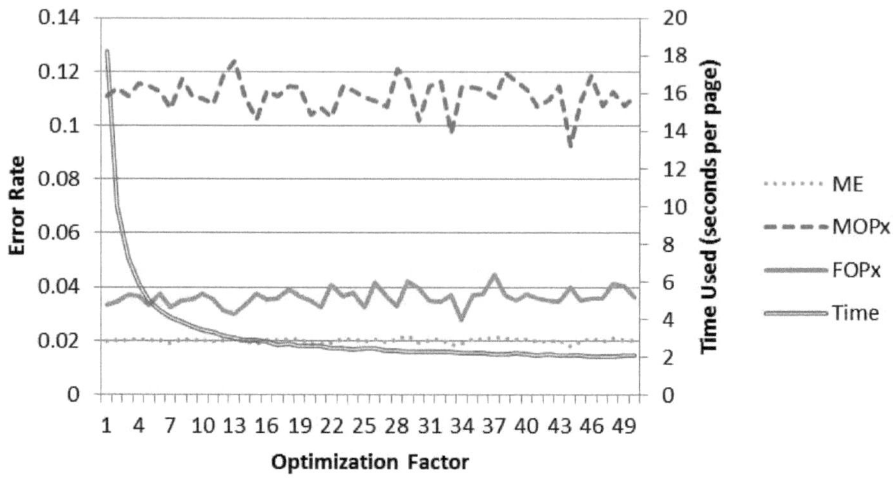

Fig. 5. A plot showing the relationship between the optimization factor f and both the three binarization error measures (left y-axis) and the average time per page used in the binarization process (right y-axis)

- Missed Staff Lines Rate (MSLR) (6)
- False Staff Lines Rate (FSLR) (7)
- Line Position Error (LPE) (8): Measures the sum of squared difference in y-distance between the detected staff lines and their corresponding staff lines from the ground truth (normalized by the threshold used to decide whether two lines should be considered the same line or not), with the range of $[0, 1]$.

$$\text{MSLR} = \frac{|L_g| - |L_a \cap L_g|}{|L_g|} \ . \tag{6}$$

$$\text{FSLR} = \frac{|L_a| - |L_a \cap L_g|}{|L_a|} \ . \tag{7}$$

$$\text{LPE} = \frac{\sum_{l_g \in L_g} \delta(l_g)}{|L_g|} \ . \tag{8}$$

$$\delta(l_g) = \min(1, \min_{l_a \in L_a} \frac{\sum_x (y(l_g, x) - y(l_a, x))^2}{\mu w(l_g)}) \ . \tag{9}$$

$$a \in L \Leftrightarrow \exists b \in L : \sum_x (y(a, x) - y(b, x))^2 \leq \mu \min(w(a), w(b)) \ . \tag{10}$$

where L_g is the set of lines in the ground truth, L_a is the set of actually detected lines, $y(l, x)$ is the y-position of the line l at the specified x-position[3], $w(l)$ is

[3] The Stable Path algorithm never yields a line that has more than one y-position for any particular x-position, nor do the ground truths will ever contain lines with this characteristic. Thus, this function is evaluable in all cases.

the horizontal width of the line l, and μ is the line approximation sensitivity (empirically set to 5).

From the graph in Fig. 6, it seems that the optimization factor between 6 and 8 are good candidates, as they result in a relatively lower error rates than other values while also taking less time to process. Note that not performing any optimization ($f = 1$) turns out to be actually worse in terms of both the error rates and the processing time, and that false staff lines can usually be detected and ignored and therefore false staff lines do not always affect the overall performance.

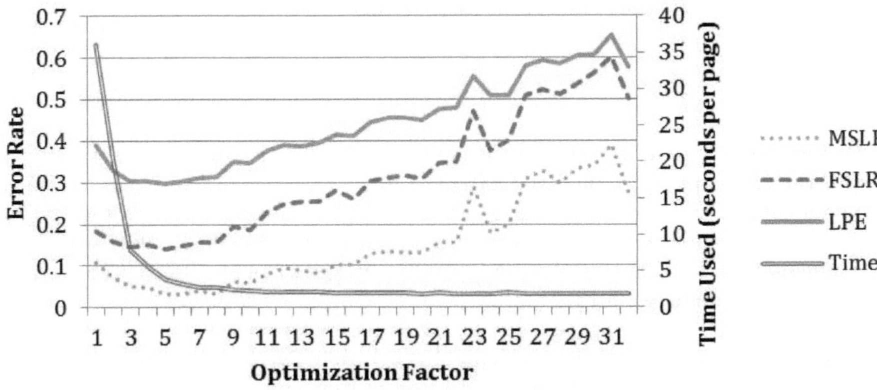

Fig. 6. A plot showing the relationship between the optimization factor f and both the three measures of staff line detection error (left y-axis) and the average time per page used in detecting the staff lines (right y-axis)

5 Conclusion

While still taking advantage of the portability of mobile phone, by exploiting the ubiquity of the mobile Internet connection, the entire processing does not have to be done on the phone. Since the speed and the cost of mobile network connection is usually of concern, input preprocessing is applied on-phone. By binarizing the score images first, an average of less than 30 kilobytes per page (800 × 1066 pixels) are needed to be transferred over the network, making it fast and cost-effective. Coupled with speed improvements of the algorithms, the entire processing could be done within 10 seconds, which should be acceptable for spontaneous uses.

Regarding the improvements, much more can still be done to improve certain aspects of the OMR process. In particular, the service could be modified such that it allows autonomous and continuous learning in the classifiers, improving the recognition performance over time as more users use the service. In addition, if the processing time could be made fast enough, new applications such as a real-time playback aid for music learners and performers might be possible. Finally, when evaluating the binarization performance, the measures used in evaluating staff line detection performance could also be used, as one of the objectives of the binarization process is to maximize the detectability of the staff lines.

Acknowledgments. We would like to express our appreciation to Microsoft Thailand for the Windows Phone 7 device used for software implementation testing, and to Nawapon Luangnapa and Thongchai Silpavarangkura, who sacrificed their precious time in providing us valuable insights about the optical music recognition process they used in their project.

References

1. Baingridge, D., Bell, T.: The Challenge of Optical Music Recognition. Computers and the Humanities 35, 95–121 (2001)
2. Luangnapa, N., Silpavarangkura, T., Nukoolkit, C., Mongkolnam, P.: Optical Music Recognition on Android Platform. In: Papasratorn, B., Charoenkitkarn, N., Lavangnananda, K., Chutimaskul, W., Vanijja, V. (eds.) IAIT 2012. CCIS, vol. 344, Springer, Heidelberg (2012)
3. Guido Engine Library, `http://guidolib.sourceforge.net/`
4. Szwoch, M.: Guido: A Musical Score Recognition System. In: 9th International Conference on Document Analysis and Recognition, pp. 809–813 (2007)
5. Hoos, H., Hamel, K.: The GUIDO Music Notation Format, `http://guidolib.sourceforge.net/doc/GUIDO-Music%20Notation %20Format.html`
6. Desaedeleer, A.: OpenOMR, `http://sourceforge.net/projects/openomr/`
7. Audiveris Music Scanner, `http://audiveris.kenai.com/`
8. Pinto, T., Rebelo, A., Giraldi, G., Cardoso, J.S.: Music score binarization based on domain knowledge. In: 5th Iberian Conference on Pattern Recognition and Image Analysis, pp. 700–708 (2011)
9. Pinto, T.T.B.: Music Score Binarization Based On Content Knowledge (2010)
10. dos Santos Cardoso, J., Capela, A., Rebelo, A., Guedes, C., da Costa, J.P.: Staff Detection with Stable Paths. IEEE Transactions on Pattern Analysis and Machine Intelligence 31(6), 1134–1139 (2009)

Applying Smart Meter and Data Mining Techniques to Predict Refrigeration System Performance

Jui-Sheng Chou and Anh-Duc Pham

Department of Construction Engineering,
National Taiwan University of Science and Technology, Taiwan
jschou@mail.ntust.edu.tw, paduc@dut.udn.vn

Abstract. This study presents six data mining techniques for prediction of coefficient of performance (COP) for refrigeration equipment. These techniques include artificial neural networks (ANNs), support vector machines (SVMs), classification and regression tree (CART), multiple regression (MR), generalized linear regression (GLR), and chi-squared automatic interaction detector (CHAID). Based on COP values, abnormal situation of equipment can be evaluated for refrigerant leakage. Experimental results from cross-fold validation are compared to determine the best models. The study shows that data mining techniques can be used for accurately and efficiently predicting COP. In the liquid leakage phase, ANNs provide the best performance. In the vapor leakage phase, the best model is the GLR model. The models built in this study are effective for evaluating refrigeration equipment performance.

Keywords: Refrigeration management, smart meter, monitoring experiment, data mining, performance diagnosis.

1 Introduction

Data mining (DM) is expected to be one of the most important technologies in marketing, healthcare, civil engineering and many others [1]. However, DM is rarely applied in the energy field, particularly to support energy efficiency. Moreover, in many countries, current policies for reducing emissions coupled with growing public awareness of increased utilities price have increased the use of smart meters as monitoring tools. Therefore improving smart meter data mining is an important research issue.

Taiwan, which imports 99.4% of its energy needs, has already begun replacing conventional meters with smart meters. Taiwan Power Company plans to install 1 million smart meters for its customers before 2015 [2]. The aim of this policy is to improve energy efficiency and reduce carbon emissions in Taiwan. The performance of these systems has not been fully investigated [3-5]. Thus, developing an appropriate methodology for predicting refrigeration system performance based on refrigerant conditions is essential.

In recent years, various electricity load forecasting methods have become highly advanced. Current forecasting tools include: regression based models [7], time series

P. Meesad et al. (Eds.): *IC²IT2013*, AISC 209, pp. 249–257.
DOI: 10.1007/978-3-642-37371-8_28

models [8], artificial intelligence techniques [9], fuzzy logic method [10], and nonlinear approach [11]. The DM techniques were used as analytical tools to predict the coefficient of performance (COP) under different refrigerant amounts. The DM techniques compared in this study included artificial neural networks (ANNs), support vector machines (SVMs), classification and regression tree (CART), multiple regression (MR), generalized linear regression (GLR), and chi-squared automatic interaction detector (CHAID) techniques. To avoid bias from data, cross-fold validation was also executed [6].

In another attempt to predict thermodynamic properties of refrigerant by using data mining, Şencan *et al.* (2011) used the ANNs method to determine thermodynamic properties of five alternative refrigerants R143A, R417A, R422A, R422D and R423A.The ANNs accurately estimated the properties such as heat conduction coefficient, dynamic viscosity, kinematic viscosity, thermal diffusivity, density, specific heat capacity of new mixed refrigerants [12]. In the refrigeration industry, system performance is expressed in terms of COP, which is defined as the ratio of change in heat at the "output" to the supplied work. Therefore, the literature has increasingly focused on predicting and utilizing COP. Ozgoren *et al.* (2012) utilized $COP_{cooling}$ as a performance prediction evaluation of solar absorption refrigeration (SAR) system. This studies confirm COP as a refrigeration system performance indicator [3]. These analysis shows that a review of refrigeration system models can greatly aid users in evaluating refrigeration equipment condition. To increase the energy efficiency and predict performance of refrigeration systems, this study therefore used six data mining methods to predict system performance.

The rest of this study is organized as follows. Section 2 shows the data mining methodology used in this study. Section 3 describes the experimental design and monitoring system. Section 4 discusses the model implementation and analytical outcomes, and the last section gives concluding remarks.

2 Research Methodology

The primary objective of data mining is achieved by combining technological methods in many fields, including computer science, statistics, online analytical processing, information retrieval, machine learning, and expert systems. Data mining technology is now used to predict behavior in many areas [1]. In the models for the six data mining techniques selected for comparison in this study, the default values were set in the numeric predictor node using IBM SPSS modeler. Researchers often use k-fold cross-validation algorithm to minimize bias associated with the random sampling of the training and holdout data samples. Orenstein *et al.* (1995) showed that ten folds are optimal [13].

ANNs consist of information-processing units that function similarly to neurons in the human brain except that a neural network consists of artificial neurons. This study applied a quick and simple training method. In SVMs Model, the SVM classifies data with different class labels by determining a set of support vectors that are members of a training input set that represents a hyper plane in a feature space. CART is a decision tree method for constructing a classification or regression tree according to its dependent variable type, which may be categorical or numeric. Decision tree methods are far superior to other modeling techniques in terms of logic rules. An extension of simple regression model is the MR, which is built from two or more predictor variables. This method is popular due to its simplicity. The GLR model which is more flexible than LR, assumes data points have an arbitrary of distribution pattern, and the relationship between X and Y is constructed by a link function according to its distribution pattern. CHAID is a decision tree technique for performing data set classification. This model uses a rule set to classify outcomes for an input dataset.

To evaluate the model performance, Mean Absolute Percentage Error (MAPE) is adopted as given below.

$$\mathrm{MAPE} = \frac{1}{n}\sum_{i=1}^{n}\left|\frac{y_i - y_i'}{y_i}\right| \tag{1}$$

where y = actual value; y'= predicted value; and n = number of data samples.

3 Experimental Design

The experiment was set up in the Employment and Vocational Training in Bureau of the northern Taiwan city of Taoyuan. The experimental data were measured and recorded by a smart meter equipped with distributed sensors for automatically retrieving temperature and pressure values. Detailed processes of the experiment are described below.

3.1 Monitoring System

In this study, system performance was monitored with a smart meter, which is an electrical meter that records electrical energy consumption at intervals of an hour or less and sends the information back to the utility centre for monitoring and billing purposes. The system enables monitoring of all electricity usage information, including real-time and historic data and daily and monthly power demand, which can be communicated through RS-485 or Ethernet or ZigBee.

The experiment simultaneously monitored 16 temperature and 4 pressure values controlled digitally and mechanically. The sensors are located at the compressor inlets and outlets and at the oil separator, condenser, accumulator, filter dryer, injector and evaporator (Fig. 1). The function of the compressor is a mechanical device used to compress the vapor. The oil separator separates oil and water to prevent the discharge of oil from the compressor together with the refrigerant circulating in the refrigeration system. The condenser converts vapor to liquid by condensation. An accumulator acts a pressure storage reservoir in which a non-compressible hydraulic fluid is continuously pressurized by an external source. The function of the filter dryer is to remove debris from the refrigeration system. The evaporator is to compress the cooling chemical substances, which convert it from liquid to vapor form, and to absorb heat.

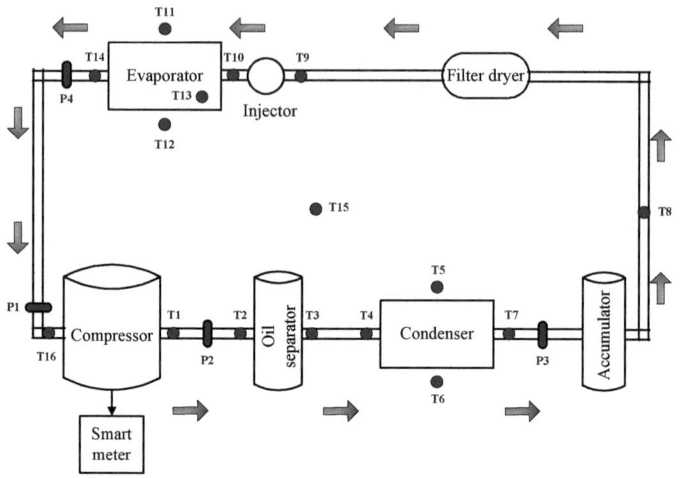

Fig. 1. Monitored points (T: temperature; P: pressure)

The collected data can be classified as electricity data, temperature data and pressure values. The liquid leakage phase obtained 528 values after processing the monitoring points. The vapor leakage phase obtained 217 values. The vapor leakage phase obtained 217 values. Table 1 displays the experiment settings. The selected refrigerant in this experiment is R404A, which is a compound used in a heat cycle that reversibly undergoes a phase change from vapor to a liquid. The R404A refrigerant has an ozone depletion index of 0 and a global warming potential value of 3300, which indicates that it is an environmentally safe non-azeotropic refrigerant with minimal effects on the ozone layer. Therefore, this study refilled the refrigerant in liquid state and analyzed the effect of different amounts of refrigerant without considering the composition of the remaining refrigerant.

Table 1. Experimental information

Experimental settings	
Location	A municipal bureau of employment and vocational training
Purpose	Effect of COP on refrigeration under varying R404A refrigerant.
Variables	Amount of refrigerant, pressure sensor, temperature sensor
Conditions	Duration of evaporator temperatures ranging from +20°C to -20°C and the temperature (T1~T16) and pressure (P1~P4) sensor readings for all equipment
Equipment	Vacuum pump, electronic platform scales, pressure gage, refrigerant recycle bottle
Monitored values	Temperature, pressure
Refrigerant type	R404A
R404A	6kg; 5kg; 4kg; 3kg; 2kg; 1kg

3.2 COP Forecasting

This study estimated COP values based on the above temperature and cooling capacity values. A high COP value was interpreted as high equipment efficiency. Six data mining techniques were used to predict COP under varying amounts of refrigerant. The COP formula is given below.

$$COP = \frac{\Delta H_{evap}(kJ/kg)}{\Delta H_{compvap}(kJ/kg)} \tag{2}$$

where, COP= coefficient of performance. ΔH_{evap}= heat removed by the refrigeration. $\Delta H_{compvap}$= heat consumed by the refrigeration.

4 Data Preprocessing and Analytical Results

4.1 Data Preprocessing

The experiments in this study measured both liquid leakage and vapor leakage. The experimental data retrieved by the smart meter were converted to general reading patterns *via* spreadsheet. Pearson correlation coefficient was used to measure the correlation between input and output variables. The Pearson formula is given below.

$$|r| = \frac{\left|\sum_{i=1}^{n}(X_i - \bar{X})(Y_i - \bar{Y})\right|}{\left|(n-1)S_x S_y\right|} \tag{3}$$

where n is the sample number, X_i is the value of input variable i; Y_i is the value of the output variable; $\bar{X}$ is the mean of the input variable; $\bar{Y}$ is the mean of the output variable; S_x is the standard deviation of X; and S_y is the standard deviation of Y. The $|r|$ describes the degree of the relationship between two variables. A high $|r|$ indicates a strong correlation between two variables. Original data are normalized to improve the precision of prediction values.

4.2 Results and Discussions

Figure 2 compares the MAPEs for liquid phase. The comparison shows that 21 models have values less than 10%, 21 models have values between 10% and 20%, 17 models have values between 20% and 50%, and only 1 has a value larger than 50%. The comparison results indicate that 98.3% models yield satisfactory prediction accuracy. Furthermore, under various attribute settings, six of the best models were ANNs models, three of the best models were CHAID models, and one of the best models was a CART model.

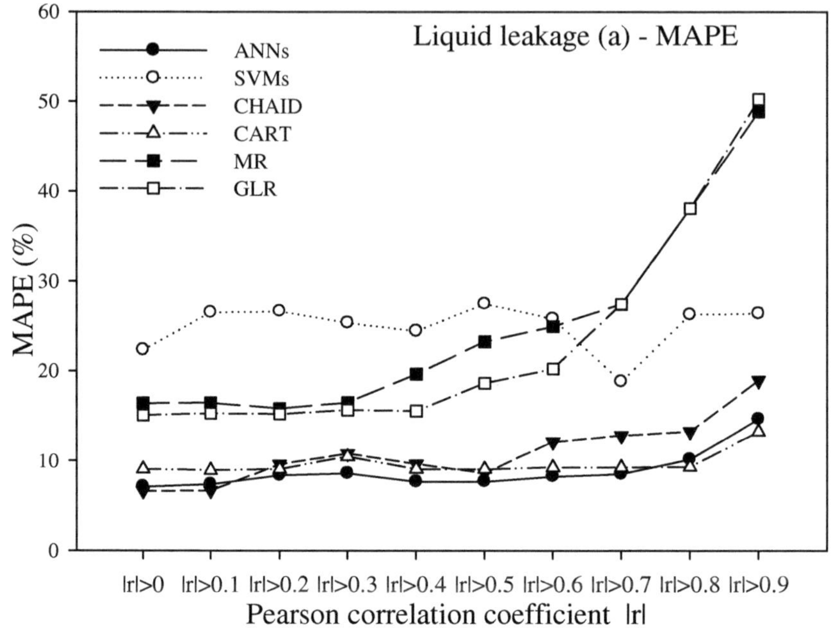

Fig. 2. The MAPE analysis in liquid leakage phase

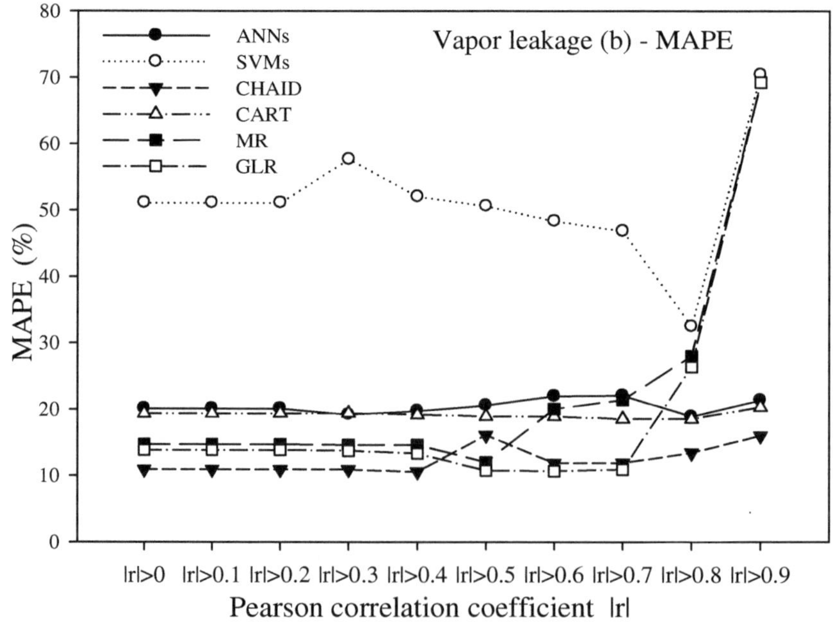

Fig. 3. The MAPE analysis in vapor leakage phase

In Fig. 3, the MAPE values for vapor phase data show that no model has values less than 10%, 36 models have values between 10% and 20%, 15 models have values between 20% and 50%, and 9 models have value higher than 50%. These results indicate that 85.0% of models had medium prediction accuracy, but none had high prediction accuracy. Moreover, under various attribute settings, seven of the best models were CHAID models, and three of the best models were GLR models.

To evaluate equipment performance by combining energy management with data mining techniques, this study further analyzed the relationship between cooling capacity (kJ/kg) and COP at varying refrigerant settings. Figure 4 describing the 19th ANNs model in liquid phase shows that the cooling capacity of R404A in 2 kg appears sufficient, but the time needed to lower the temperature is long. Cooling capacity stabilizes at 177kJ/kg. The figure shows that the system performs best when the amount of R404A is 3kg.

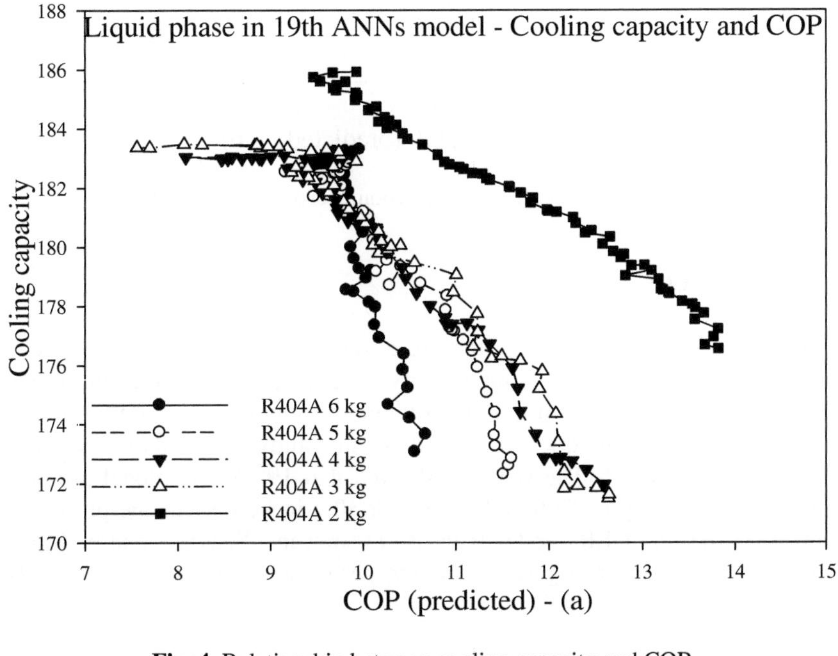

Fig. 4. Relationship between cooling capacity and COP

Especially, the analytical results show that cooling behavior in vapor phase differs from that in liquid phase. For example, Fig. 5 shows that no cooling capacity is available in vapor phase when the amount of refrigerant is 2kg. When the liquid refrigerant converts to vapor phase, the refrigeration state is unstable. Hence, vapor leakage can substantially affect equipment performance.

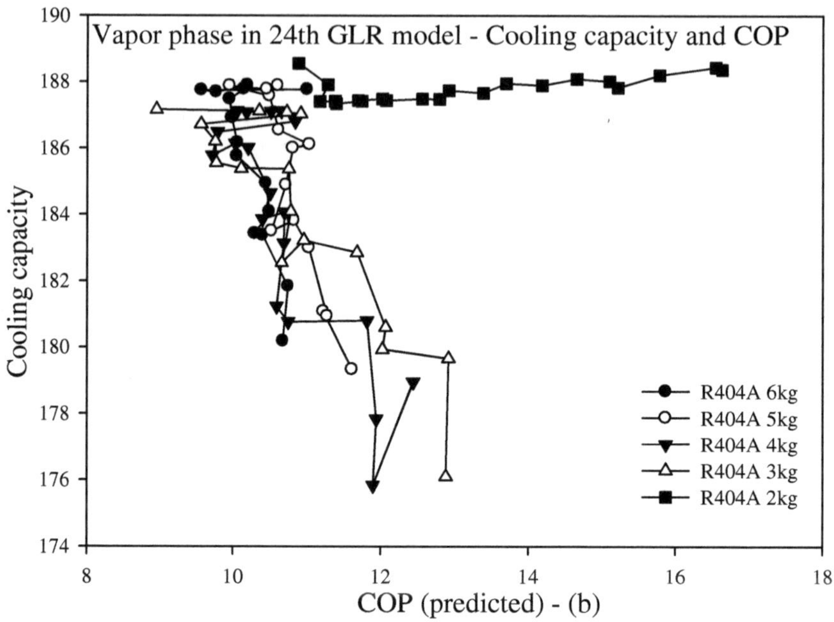

Fig. 5. Relationship between cooling capacity and COP

5 Conclusions

The aim of this study is to evaluate the use of data mining techniques for predicting COP of refrigeration equipment under varying amounts of refrigerant. Predicting coefficient of performance is useful in equipment monitoring in utility companies. The six data mining techniques compared in this study were ANNs, SVMs, MR, CHAID, CART and GLR. The findings showed that 19th ANNs model provides the best prediction accuracy of COP values with $|r|>0.7$ in liquid leakage phase and the best model is the 24th GLR model with $|r|>0.6$ in vapor leakage phase. Moreover, the best of the proposed refrigeration system performs when the amount of R404A is 3kg.

References

1. Liao, S.-H., Chu, P.-H., Hsiao, P.-Y.: Data mining techniques applications. A decade review from 2000 to 2011. Expert Systems with Applications 39(12), 11303–11311 (2012)
2. Lee, H.I.: Energy report. lower the price of electricity. Bureau of Energy Ministry of Economics Affairs (2011)
3. Ozgoren, M., Bilgili, M., Babayigit, O.: Hourly performance prediction of ammoniaewater solar absorption refrigeration. Applied Thermal Engineering 40, 80–90 (2012)
4. Şahin, A.T.: Performance analysis of single-stage refrigeration system with internal heat exchanger using neural network and neuro-fuzzy. Renewable Energy 36(10), 2747–2752 (2011)

5. Ahmed, A., Korres, N.E., Ploennigs, J., Elhadi, H., Menzel, K.: Mining building performance data for energy-efficient operation. Advanced Engineering Informatics 25(2), 341–354 (2011)
6. Chou, J., Chiu, C., Farfoura, M., Al-Taharwa, I.: Optimizing the Prediction Accuracy of Concrete Compressive Strength Based on a Comparison of Data-Mining Techniques. Journal of Computing in Civil Engineering 25(3), 242–253 (2011)
7. Taylor, J.W., Buizza, R.: Using weather ensemble predictions in electricity demand forecasting. International Journal of Forecasting 19(1), 57–70 (2003)
8. Saab, S., Badr, E., Nasr, G.: Univariate modeling and forecasting of energy consumption. The case of electricity in Lebanon. Energy 26(1), 1–14 (2001)
9. Chen, C.S., Tzeng, Y.M., Hwang, J.C.: The application of artificial neural networks to substation load forecasting. Electric Power Systems Research 38(2), 153–160 (1996)
10. Ye, B., Guo, C., Cao, Y.: Short-term load forecasting using a new fuzzy modeling strategy, Hangzhou, P.R. China, pp. 5045–5049 (2004)
11. Pao, H.T.: Comparing linear and nonlinear forecasts for Taiwan's electricity consumption. Energy 31(12), 1793–1805 (2006)
12. Şencan, A., Köse, S.L., Selbaş, R.: Prediction of thermophysical properties of mixed refrigerants using artificial neural network. Energy Conversion and Management 52(2), 958–974 (2011)
13. Orenstein, T., Kohavi, Z., Pomeranz, I.: An optimal algorithm for cycle breaking in directed graphs. Journal of Electronic Testing 7(1-2), 71–81 (1995)

Extended Knowledge Management Capability Model for Software Process Improvement

Kamolchai Asvachaiporn and Nakornthip Prompoon

Department of Computer Engineering, Faculty of Engineering,
Chulalongkorn University, Thailand
Kamolchai.A@student.chula.ac.th,
Nakornthip.S@chula.ac.th

Abstract. Software Process Improvement (SPI) is always included as a necessary activity in IT organizations that give importance to the process, as it has been proven from its application in many organizations that good process will likely bring good product and project performance. While performing under SPI, knowledge usually occurs from different sources, in different formats and from different activities. Therefore, managing knowledge is a challenging issue for software organizations. This paper presents an Extended Knowledge Management Capability Model (EKMCM) for SPI, focusing on the Supplier Agreement Management (SAM) Process Area of CMMI. The proposed EKMCM is a set of defined processes classified into five levels integrating three aspects: the knowledge management process, human resource development and the use of IT to support the process. The proposed EKMCM model can be applied in other CMMI processes. Organizations may choose to perform at any knowledge management level according to their readiness.

Keywords: Extended Knowledge Management Capability Model, EKMCM, Software Process Improvement, CMMI, Supplier Agreement Management.

1 Introduction

Software Process Improvement (SPI) is important to software development organizations, as if it is applied to organizations which have appropriate strategy and resource management, it will provide many benefits: improved productivity and project performance, which may lead to a reduction in total cost and time spent in software development, as well as more opportunity to deliver projects on time, leading to customer satisfaction. There are many SPI models available for software organizations to adopt. One of the well-known models is CMMI [1], which is presented by the Software Engineering Institute, Carnegie Mellon University. Generally, organizations may gain new knowledge during SPI process implementation activities. Knowledge is one of the key resources of the project/organization that can be later used in a project that has similar characteristics. Therefore, Knowledge Management (KM) in SPI is a challenging issue for software organization, for utilization and reuse [2] [3] of the knowledge to provide decision making support.

P. Meesad et al. (Eds.): *IC²IT2013*, AISC 209, pp. 259–268.
DOI: 10.1007/978-3-642-37371-8_29

Nowadays, many organizations are implementing the acquisition of IT services or products using an outsourcing strategy, due to the lack of skill team and difficulty of maintaining IT staff. Thus, this increases the need to effectively manage suppliers. Therefore, this research presents an Extended Knowledge Management Capability Model (EKMCM) for SPI, focusing on the Supplier Agreement Management (SAM) process area of CMMI. The presented EKMCM is a model of KM that provides a framework for organizations to create, capture, analyze and present knowledge that meets specified needs. The model consists of five levels in three main aspects of integration: the knowledge management process, human resource development, and the selection of technology to the support KM Process. EKMCM represents the identification of processes and examples of applying models, which can be used as guidance for the organizations to select the suitable knowledge management capability level according to their readiness, for continuous process improvement. Furthermore, EKMCM can be applied in other processes defined in the CMMI.

The paper is structured as follows. Section 2 presents underlying concepts. Section 3 mentions some related works. Section 4 proposes our approaches. Finally, the paper conclusions and future works are presented in section 5.

2 Underlying Concepts

2.1 Knowledge Management

Knowledge is classified into 2 categories [4]: Tacit Knowledge, which is considered in the form of individual ideas, and Explicit Knowledge, which is stored in the form of articles, texts, publications and databases. Transforming Tacit Knowledge into Explicit Knowledge is quite difficult; therefore, it requires appropriate Knowledge Management (KM), composed of at least four elements[5] [6]: Knowledge Creation and Capture, Knowledge Sharing and Enrichment, Information Storage and Retrieval, and Knowledge Dissemination. To maximize the benefit of KM, technological, organization and individual factors should also be considered [7], as well as motivational factors, derived from studies on various factors that affect KM [8] [9] [10] [11] [12] [13], in order to motivate personnel to participate in using KM.

2.2 Software Process Improvement

Software Process Improvement (SPI) is a set of activities which is concerned with improving the process quality to ensure that an organization or project can fulfill its goals effectively. The application of SPI may also help improve the quality and reliability of software, personnel, and customer satisfaction, resulting in a significant return from investment [14]. The major outcome of SPI that each organization must deliver is set of defined processes that specify roles and responsibility. Organizations implementing SPI should summarize organizational requirements in order to develop the process in the organization, and allocate human resources to define the process. The result of defining the process are guidelines issued within the organization, which can be used to execute a pilot, so that personnel can understand and operate according to the defined processes, before officially being implemented in the organization.

2.3 CMMI Models

CMMI Models [1] are models providing approaches for effective process improvement, by recommending best practices to accomplish the goals for each process. This paper focuses on the Supplier Agreement Management (SAM), which presently many organizations require other organizations to develop or provide software for them. In other words, this process is related to supplier product acquisition management. It is a part of CMMI for Development.

2.4 Knowledge Management Maturity Model

The Knowledge Management Maturity Model was invented in order to classify the level of KM for each organization, so that the organization would realize the risk of KM which cannot be implemented according to process attributes, for preparing to handle the risk that may happen. Moreover, it is also used as a model for continuous KM, but context of model has not explicitly described how to implement KM. The Knowledge Management Maturity Model is adopted from Venkatraman, 1994, dividing the level of KM into 5 levels. Beginning from the lowest level, level 1, to the highest level, level 5, each level focuses on three principal skills: KM Process skills, Human Resources development skills and KM Technology skills, for measuring how effective the KM of each organization is.

3 Related Works

P. Chongsringam [7] presented a KM Framework in order to support SAM Process area in CMMI, to solve the problem of scattered knowledge, which is difficult to be utilized in an organization. The proposed system can be used as a guideline for software development organizations using the CMMI standard, namely the SAM Process area in CMMI, with the development of a framework of the KM system in three parts: 1) KM Process Definition 2) KM System Development and 3) Knowledge Repository Management. In all three parts, factors for the success of KM have been considered, consisting of technology issues, organizational issues and individual issues. However, the research does not specify, if the organization uses the proposed framework into action, what level of KM they should manage, and whether it is efficient and effective enough. N. Mamaghani and et al.'s research [8] accumulates information from related research on KM from 1997 to 2009 and applies the Delphi Technique for summarizing the crucial success factors in KM. The resulting factors are classified into 4 groups that have a highly significant impact on the level of KM: Knowledge Strategy, Management Support, Motivational Encouragement and Technical Infrastructure. K. Ramamurthyet al. [16] presented A Model of Data Warehousing Process Maturity as a guide in defining the Warehousing Process for high quality, for those interested in this field. This model is defined in five levels: Initial, Repeatable, Defined, Managed and Optimizing respectively. The description of each level is presented according to Key Process Area (KPAs) for Development and KPAs for Operations/Services. Each level mentions the process areas required in order to attain that level. Therefore, achieving each of the levels requires undergoing all of the process areas in each level first, before following the Maturity Model approach which does not support the Capability Model. For this reason, it is not appropriate for the organization which only focuses on some of the process areas and requires implementation under the model.

4 EKMCM for SPI in SAM Process Area of CMMI

This paper presents a classification of KM levels for SPI. Figure 1 shows the overview of this research, which has been categorized into 2 steps.

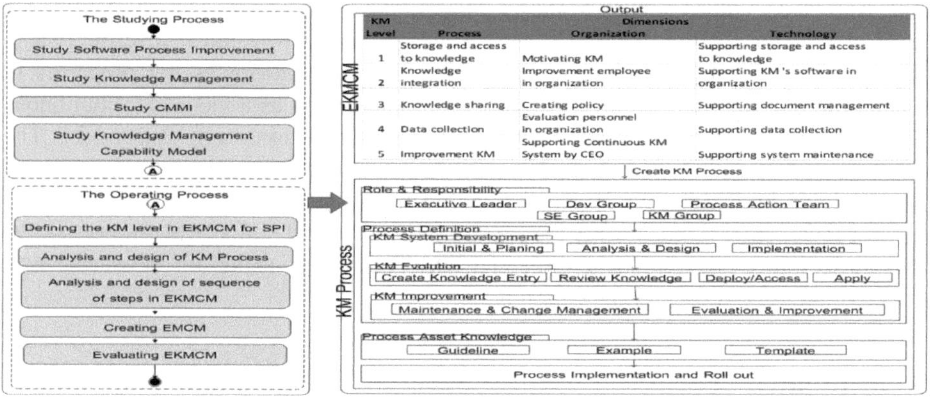

Fig. 1. Overview of our approach

- The studying process, which will study SPI, KM, CMMI, and Knowledge Management Capability Model, as input to be used in the operating process.
- The operating process, with details of each step explained as follows: 1) Defining the KM level in EKMCM for SPI - We used the research [15] to define a model classifying KM into 5 levels, where each level integrates three aspects of skills: the KM process, human resource development and usage of IT to support the process. 2) Analysis and design of KM process - We analyzed and designed a KM process to support SAM [7] and used the context [15] as input for the next step. 3) Analysis and design of sequence of steps in EKMCM - We used output of 1 and 2 as the input of this step to create a sequence of steps for KM. 4) Creating EKMCM - We integrated the output from 1, 2 and 3 to input for making EKMCM. 5) Evaluating EKMCM - We evaluated the qualities of the model by using a heuristic checklist. The results and the proposed implementation guideline may help apply EKMCM for any organization in an effective way because of its format of process attributes and examples of implementation, and defining five levels of KM, similar to CMMI. Furthermore, EKMCM has a strong fundamental for KM because it has integrated three aspects, all of which are factors that lead to the success of KM [8]. Many organizations that have already utilized these factors have found that they are very important to the success of KM. This paper uses EKMCM which creates KM Process, focusing on SAM to be used in the organization.

4.1 Context of EKMCM

This paper has presented the context of EKMCM in the form of process attribute. The results are shown in Table 1, which shows process attributes and examples of how to achieve each level of the proposed KM for SAM process area.

Table 1. Context of EKMCM for SPI focusing on SAM process area of CMMI

KM Level/ Purpose / Process Attributes (Aspects: P – Process , H – Human Resource, T – Technology)	Examples
1. Localized Exploitation: The purpose is to achieve the KM, by enabling the storage and access of knowledge obtained from SPI, as well as creating a positive attitude towards KM in order to serve as a motivation for implementing KM. Moreover, the technology supporting KM can be used for the storage and access of knowledge at the individual level. 1.1.1 The storage and access of knowledge in media, e.g., in printed or electronic form, must be specified. (P) 1.2.1 The self-motivation plan must be specified. (H) 1.3.1 The knowledge storage and access technology must be specified. (T)	1. Storage and access to the knowledge about the specific goals that must be implemented for the SAM in the form of electronic documents. (1.1.1) 2. Setting a target to storing at least 3 pieces of knowledge on CMMI of SAM per day. (1.2.1) 3. Storage and access to the knowledge in format of PDF file. (1.3.1)
2. Internal Integration: The purpose is to achieve the integration of knowledge in KM as well as to develop human resources and organizations, following the process of KM, and to enable the use of technology supporting KM within the organization. 2.1.1 The knowledge integration must be specified. (P) 2.2.1 The development of self-motivation and organization for implementing knowledge must be specified. (H) 2.3.1 The organizational capability in using technology for KM must be specified. (T)	1. Integration from SAM's knowledge into lessons learnt for others to reuse. (2.1.1) 2. Support for KM within organization by CEO. (2.2.1) 3. Encourage usage of software to support the SAM process within the organization. (2.3.1)
3. Re-Engineering: The purpose is to achieve KM with knowledge sharing among the people within the organization and by creating policies to help achieve the KM goals effectively, and to	1. Create a workflow to review knowledge entering into system, before publishing them to have access to knowledge. (3.1.1)

Table 1. *(continued)*

KM Level/ Purpose / Process Attributes (Aspects: P – Process , H – Human Resource, T – Technology)	Examples
enable the use of KM in knowledge storage and retrieval as well as sharing and reusing knowledge within the organization. 3.1.1 The sharing of personal knowledge must be specified. (P) 3.1.2 The sequence of KM activities in the organization must be specified. (P) 3.2.1 The policies or rules for people in the organization must be specified. (H) 3.3.1 The Document Management must be specified. (T) 3.3.2 The index for searching content in context must be specified. (T) 3.3.3 The storage of information in the form of Knowledge Packages for internal user access must be specified in the KM system. (T)	2. Publish knowledge that has been reviewed for everyone to access. (3.1.2) 3. Create policy for people to add knowledge according to the KM process definition. (3.2.1) 4. Create version control for SAM knowledge in the KM System. (3.3.1) 5. Create index for searching content in the Knowledge Package. (3.3.2) 6. Specific the knowledge package for usage in SPI, e.g. CMMI knowledge package for SAM process. (3.3.3)
4. KM Measurement: The purpose is to evaluate the KM of the people in the organization, to create procedures for the collection of KM information within the organization for use in evaluation and to improve KM process effectively, and to enable the technology supporting KM for gathering information to be used in the KM process. 4.1.1 The requirements for supporting the KM process to comply with the needs of the organization must be specified. (P) 4.1.2 The purposes of the measurement process according to 4.1.1 must be specified. (P) 4.1.3 The tracking of results from the implementation of KM process must be specified. (P) 4.2.1The KPI for people in the organization must be specified. (H) 4.3.2 The storage of data for the improvement of KM process must be specified. (T)	1. Collect the requirements for improving KM process for SAM. (4.1.1) 2. Set the goals of the organization and the plan for evaluating KM for SAM. (4.1.2) 3. Make a report summarizing the results of the process according to KM process in order to inform the CEO. (4.1.3) 4. Set the principles of the evaluation KPI for personnel who participate in KM. (4.2.1) 5. Create a web board to gather opinions about SAM knowledge for integrating knowledge to create the best practice. (4.3.1) 6. Categorize the information for use in evaluating KM.(4.3.2)
5. Continuous Improvement: The purpose is to continuously improve KM process, by using the information gathered from KM evaluation to improve the efficiency of the KM process, to enable the technology for supporting access to	1. Create the implementation plan based on analysis of the feedback data from relevant persons. (5.1.1) 2. Implement plan from 1 to

Table 1. (*continued*)

KM Level/ Purpose / Process Attributes (Aspects: P – Process , H – Human Resource, T – Technology)	Examples
knowledge from anywhere to be used to motivate people in the organization to participate in KM, and to use the technology for maintaining the system to support sustainable KM. 5.1.1 The plan for KM improvement based on users' feedback must be specified. (P) 5.1.2 The improvement of the KM process must be specified. (P) 5.2.1 The support for resource and budget in the system for continuously improving the KM process must be specified. (H) 5.3.1 The technology supporting access to knowledge in the system from anywhere must be specified. (T) 5.3.2 The screening channel for knowledge that has not been accessed in the internal system for a long time must be specified. (T)	improve the KM process. (5.1.2) 3. Support organization, manpower and budget for continued improvement from CEO. (5.2.1) 4. Support thorough access to knowledge, for example, enabling to access to knowledge of SAM disseminated within the organization via mobile devices and tablets. (5.3.1) 5. Inspecting the knowledge access information by screening less accessed knowledge out of the system. (5.3.2)

4.2 KM Process

The proposed EKMCM are composed of the defined roles and responsibility, KM process definition, process asset knowledge and guideline for process implementation. The details of each one are described as follows.

Roles and Responsibility. Implementing EKMCM in definition of the process initially requires defining persons responsible for the process. In this research, we classify 5 groups as follows: 1) Executive Leader Group: Initiate the management system. 2) Development Group: Support the KM technology. 3) KM Group: Verify the knowledge introduced into the system before disseminating to personnel in the organization. 4) Expert Software Engineering Process Group: Provide knowledge advice in terms of CMMI. 5) Process Action Team: Bring the knowledge into the system and utilize the stored knowledge. The interrelationship among each role was presented in KM process definition.

KM Process Definition. The next step after defining related people is to define the required processes by referencing EKMCM, in which at each level. It is specified what is required to define the process. The results are shown as Activity Diagrams in Figures 2, 3 and 4, defining the process in 3 main parts, with details as follows: 1) KM System Development: this step covers the establishment of the KM system, and selecting the strategy for bringing the KM system to support KM. 2) Knowledge Evolution: this step covers bringing knowledge into the system, which requires a workflow for checking the knowledge before disseminating to personnel in the

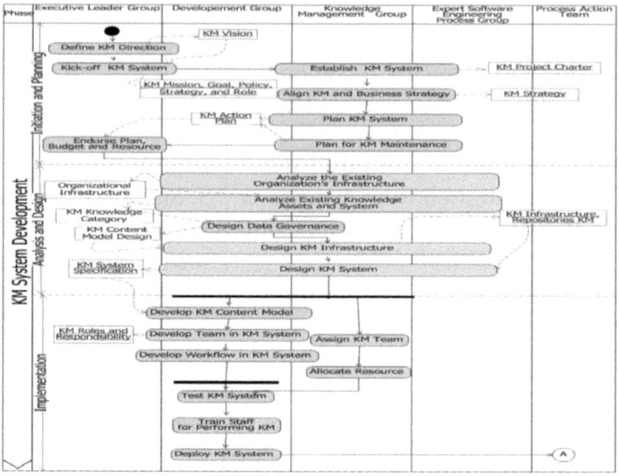

Fig. 2. KM System Development Phase

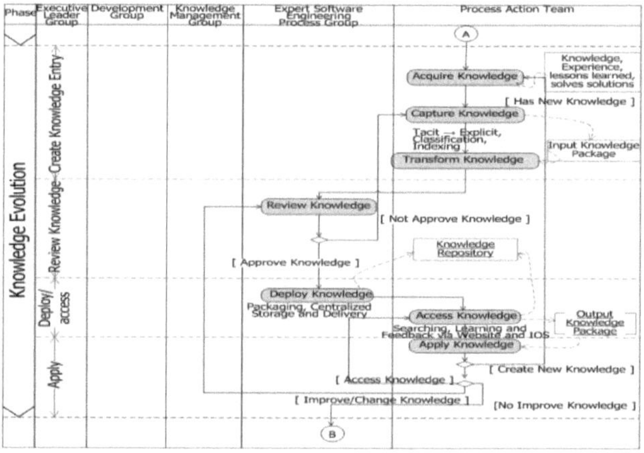

Fig. 3. Knowledge Evolution Phase

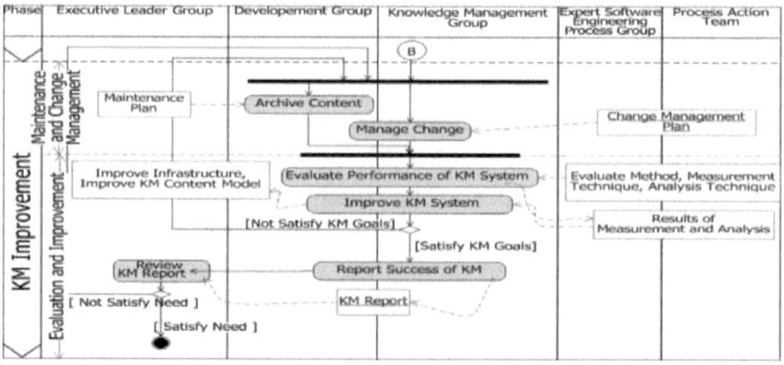

Fig. 4. KM Improvement Phase

organization. 3) KM Improvement: this step covers the maintenance of the KM system, by separating the outdated or long-unused knowledge to enable faster access, as well as evaluating KM in order to improve the KM process continuously, which is the core of SPI.

Process Asset Knowledge. Process Asset Knowledge [17] refers to anything that is useful for implementing the process according to the KM process definition, such as templates, examples and guidelines. The activity diagrams in Figures 2, 3, and 4, resulting from using EKMCM to defining KM process for SPI for use in SAM, are divided into three phases, and this research presents examples of assets resulting from definition of the process in each phase as follows: 1) KM System Development: KM Policy for organization, KM System Infrastructure Template for organization. 2) Knowledge Evolution: Practical Guideline for SAM process management, KM Repository. 3) KM Improvement: Example methods to evaluate KM usage in organization, Evaluation report on KM in the organization.

Process Implementation and Roll Out. This part will discuss the implementation of the result from the EKMCM. This research presents a plan for implementing the KM process as defined by EKMCM in the organization, with consideration to the culture of software organizations for which other organizations develop or provide software. Details of the plan are as follows: 1) Establish requirements for selecting KM Level to be implemented according to organizational readiness, as well as specifying organizational goals for KM. 2) Plan and Apply the KM process definition to correspond with the organizational goals for KM. 3) Train persons involved with KM process definition, including training in knowledge about KM process and his or her responsibility and the use of tools supporting KM. The purpose of training is to help to understand about the activities, input and output from defining processes in the EKMCM. 4) Test the process defined by EKMCM by using it in a pilot project, to familiarize the personnel with activities within the process. 5) Collect data during the pilot project implementation, for subsequent analysis of strong and weak points. 6) Analyze and summarize data from the pilot project, to present to the project sponsor or other relevant persons for planning the strategy to improve the process for the future adoption of the practice within the organization.7) Store the results from the summary of the data as lessons learnt in database for future use. 8) Improve KM process from the pilot project upon approval for use in the organization by the project sponsor. The improvement has to initiate from the review process, re-development of the KM process based on the desired level, then evaluation of the update process, before repeating from step 1, which will improve the process continuously, which is at the main issue of SPI.

5 Conclusions and Future Works

This paper presents EKMCM, extended from the model presented in [15]. The proposed model focused on SPI for SAM area of CMMI for KM. Our research purpose is to provide a model composed of 5 levels ranging from localized exploitation, internal integration, re-engineering, KM measurement, continuous improvement, and 3 aspects: process, organization and technology. These may help enable organizations interested in KM and conducting SPI based on the CMMI standard within the organization, to select the proposed KM level according to the readiness of the organization. Our future research will focus on the integration of measurement framework in order for the efficiency of EKMCM implementation. In addition, a software tool is also going to develop that help EKMCM implementation and EKMCM process attributes assessment for KM level rating.

References

1. CMMI Product Team.: CMMI for Development, Version 1.3: Improving Processed for Better Products and Services. Carnegie Mellon: Software Engineering Institute (2010)
2. Dingsoyr, T., Bjornson, F., Shull, F.: What do we know about Knowledge Management? Practical Implications for Software Engineering. IEEE Software Magazine, 100–103 (2009)
3. Ko, D., Dennis, A.: Profiting from Knowledge Management: The Impact of Time and Experience. Information Systems Research 22, 134–152 (2011)
4. Ioana, R., Lindvall, M.: Knowledge Management in Software Engineering. IEEE Software, 26–38 (May/June 2002)
5. Uriarte, F.: Introlduction to Knowledge Management. ASEAN Foundation, pp. 45–65 (2008)
6. Dayan, R., Evans, S.: KM Your Way To CMMI. Journal of Knowledge Management 10, 69–80 (2006)
7. Chongsringam, P., Prompoon, N.: A Knowledge Management System for supporting CMMI Organization Knowledge. In: 11th NCSEC, Bangkok, pp. 499–506 (2007)
8. Abd, B.A., Ezz, I., Papazafeiropoulou, A., Paul, R., Stergioulas, L.: Investigation the Critical Success factors and Infrastructure of Knowledge Management for Open Innovation Adoption: The Case of GlaxoSmithKline (GSK) in Egypt. In: 45th International Conference, Hawaii, pp. 4022–4031 (2012)
9. Mamaghani, N.D., Saghafi, F., Shahkooh, K.A., Sadeghi, M.: Extracting Success Factors for Knowledge Management Organizational Readiness Assessment. In: 4th NISS, Gyeongju, pp. 170–175 (2010)
10. OuYang, Y., Yeh, J., Lee, T.: The Critical Success Factors for Knowledge Management Adoption – A Review Study. In: 3rd KAM, Shuai Zhang, pp. 445–448 (2010)
11. Ajila, S.A., Sun, Z.: A four-factor model on the success of knowledge management. In: IEEE International Conference, pp. 320–325 (2004)
12. Changzheng, Z.: Impact of Knowledge Delivery Factors on Software Product Development Efficiency. In: 2nd ICNDS, Jiahu, pp. 349–352 (2010)
13. Aggestam, L., Persson, A.: Increasing the Quality in IT-Supported KnowledgeRepositories: Critical Success Factors for Identifying. In: 43rd HICSS, Hawaii, pp. 1–9 (2010)
14. Mathiassen, L., Ngwenyama, O., Aaen, I.: Managing Change in Software Process Improvement. IEEE Software Magazine, 84–91 (2005)
15. Pierre, J.: Towards a maturity model of knowledge management competences as an organisational capability. In: ICEE, Shanghai, pp. 1–5 (2011)
16. Ramamurthy, K., Sinha, A.P.: A Model of Data Warehousing Process Maturity. IEEE Transactions on Software Engineer, 336–353 (March/April 2012)
17. Leyman, B.: Implementing an Organizational Software Process Improvement Program. IEEE Software Engineering, 279–288 (2005)

Record Searching Using Dynamic Blocking for Entity Resolution Systems

Aye Chan Mon and Mie Mie Su Thwin

University of Technology (Yatanarpon Cyber City)
Pyin Oo Lwin, Mandalay Region, Myanmar
achanmon@gmail.com,
drmiemiesuthwin@mmcert.org.mm

Abstract. Entity Resolution also known as data matching or record linkage, is the task of identifying records from several databases that refer to the same entities. The efficiency of a blocking method is hindered by large blocks since the resulting number of record pairs is dominated by the sizes of these large blocks. So, the researchers are still doing researches on handling the problems of large blocks. Same blocking methods can yield bipolar results against different datasets, selecting a suitable blocking method for the given record linkage algorithm and dataset requires significant domain knowledge. Many researches in entity resolution has concentrated on either improving the matching quality, making entity resolution scalable to very large databases, or reducing the manual efforts required throughout the resolution process. In this paper, we propose an efficient record searching using dynamic blocking in entity resolution systems.

Keywords: entity resolution, data integration, data reduction, indexing, pre-processing.

1 Introduction

Entity resolution (ER), the problem of extracting, matching and resolving entity mentions in structured and unstructured data, is a long-standing challenge in database management, information retrieval, machine learning, natural language processing and statistics. Accurate and fast ER has huge practical implications in a wide variety of commercial, scientific and security domains [1]. The problem of finding similar entities not only applies to records that refer to persons. In bioinformatics, record linkage can help find genome sequences in large data collections that are similar to a new, unknown sequence at hand. Increasingly important is the removal of duplicates in the results returned by Web search engines and automatic text indexing systems, where copies of documents have to be identified and filtered out before being presented to the user.

If unique entity identifiers (or keys) are available in all the databases to be linked, then the problem of linking at the entity level becomes trivial: a simple database join

P. Meesad et al. (Eds.): *IC²IT2013*, AISC 209, pp. 269–279.
DOI: 10.1007/978-3-642-37371-8_30

is all that is required. However, in most cases no unique keys are shared by all databases, and more sophisticated linkage techniques need to be applied. These techniques can be broadly classified into deterministic, probabilistic, and modern, machine learning based approaches [8].

The main contribution of this paper is the development of partitioning in dynamic block to reduce the number of record pairs search space with minimal accuracy loss. A key innovation is that partitioning in dynamic block can reduce search space and linking processes to cover all the duplicate records.

The rest of the paper is organized as follows. Section 2 is about literature review. Section 3 presents our proposed system using dynamic block partitioning for addressing limitations of current systems. Section 4 is the analysis with previous methods. Conclusions are made in Section 5.

2 Literature Review

Entity Resolution is a very active research topic. H.Zhao and S.Ram [2] applied a recently-developed constrained cascade generalization method in entity matching and reported on empirical evaluation using real world data. L.Getoor and A. Machanavajjhala [3] discussed both the practical aspects and theoretical underpinnings of ER and also described existing solutions, current challenges and open research problems for entity resolution. L.Jin, C.Li and etal.[4] described an efficient approach to record linkage. Their proposed system can also be generalized to other similarity functions between strings.

M.A. Hern'andez and S.J.Stolfo [6] implemented that performs a generic Merge/Purge process that includes a declarative rule language for specifying an equational theory making it easier to experiment and modify the criteria for equivalence. M.Bilenko, Beena Kamath, et al. [7] introduced an adaptive framework for automatically learning blocking functions that are efficient and accurate. They described two predicate-based formulations of learnable blocking functions and provide learning algorithms for training them.

L.Klob and A.Thor , et al. [5] proposed that how Block based Techniques can be combine with Map Reduce tasks. The approach is based on blocking techniques and MapReduce jobs. The disadvantage is that we have to do redundant MapReduce task because all the duplicates records are not cover with one step.

In contrast to our proposed system, blocks are partitioned into dynamic size so that linking blocks between matching process can be reduced. We can reduce the search space for entity matching and can apply a complex match strategy for each pair of entities within a block.

3 Proposed System for Dynamic Block Record Searching System

Entity Matching is an important and difficult step for integrating data. To reduce the typically large space for doing entity matching is time consuming. Blocking entails a logical partitioning of the input entities such that all matching entities should be assigned to the same output partition called block. By restricting the match comparisons to the entities of the same block the match overhead can often drastically be reduced.

A general schematic outline of the record linkage process is given in Figure 1. As most real-world data collections contain noisy, incomplete and incorrectly formatted information, data cleaning and standardisation are important pre-processing steps for successful linkage. Since potentially every record in one dataset has to be compared with every record in a second dataset, blocking or searching techniques are often used to reduce the number of comparisons. The data sets are partitioned into smaller blocks (clusters) using blocking variables. Only records within the same blocks are then compared in details using the defined comparison variables. The comparison vectors generated by such detailed comparison functions are then passed to the decision model to determine the final status of the record pairs. The results of the record linkage can be assessed by the evaluation model [4].

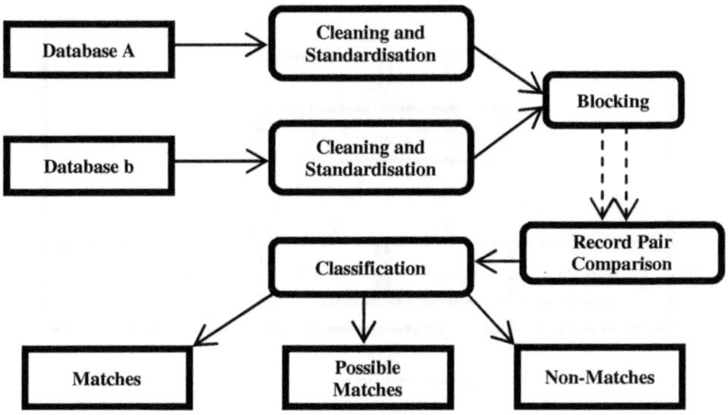

Fig. 1. General Record Linkage Process. The output of the blocking step is candidate record pairs, while the comparison step produces vectors with numerical similarity weights.

A key distinction between prior works and our approach is that previously described methods focus on improving blocking efficiency while assuming that an accurate blocking function is known and its parameters have been tuned manually. They focused on fixed size blocking for matching process and it consumes so much time of matching processes. In contrast, our approach attempts to construct an optimal blocking function automatically and adaptive blocking to speed up matching process and searching process. Because blocking functions can be learned using any combination of similarity predicates on different record fields, and no assumptions are made about the number of record fields or their type, our approach can be used for adapting the blocking function in any domain. In this paper, will focus on the blocking steps in the process.

In this paper, we propose a dynamic block based searching method which consists of following strategies:

- Standardization
- Key Creation
- Sorting
- Dynamic Block Creation for Data Partitioning
- Record Matching

The input entity matching consists of a set of entities. We focus on the common case where all entities to be matched reside already in a single dataset. The proposed system is shown in figure 2.

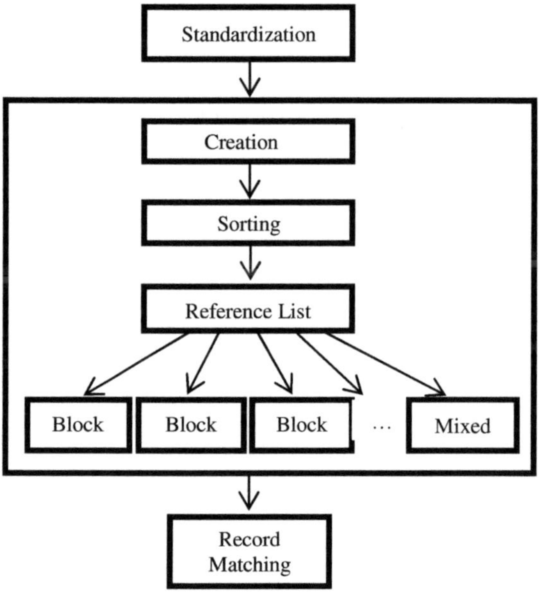

Fig. 2. Proposed System

3.1 Standardization

Data standardization is important preprocessing steps for successful entity resolution and before such data can be loaded into data warehouses or used for further analysis. This standardization process is employed before performing record linkage in order to increase the probability of finding matches. Without standardization, many true matches could be wrongly designated as non-matches because the common identifying attributes do not have sufficient similarity.

The main task of data standardization is the conversion of the raw input data into well-defined consistent forms. In name standardization, all letters are converted into upper case and remove certain characters (like punctuations) as shown in Table 2.

Table 1. Record Structure of Client Database for Bookstore

ID	1	2	3	4	5
Name	Bela Williams	Kennith Showers	Shane Garcia	Bela Moore	Bela Willam
Mid name	M	S	J	M	
SSN	934545761	989351027	982930405	981234935	915211198
DOB	10/01/57	09/10/46	03/5/48	08/2/89	10/1/57
Gender	Male	Male	Male	Male	Male
Race	Caucasian	African	Asian	African	Caucasian
NickName	Bela	Kennith		Bela	Bela
City	Evererr	Everett	Aberdeen	Sedrowool	Everett
Date_txt3	10/01/1957	09/14/1946	03/15/1948	05/12/1969	11/11/1994
Date_txt4	1957/10/01	1946/09/14	1948/03/15	1969/05/12	1994/11/11
...	...	...	...	...	..

Table 2. Name attribute standardization

ID	**Name**	**Replacement**
1	Bela Williams	bela williams
2	Kennith Showers	kennith showers
3	Shane Garcia	shane garcia
4	Bela Moore	bela moore
5	Bela Willam	bela willam
...	...	...

In gender attribute, it is replaced by one of the standard format such as Male is replaced by 'm' and Female is replaced by 'f' as shown in Table 3.

Table 3. Gender attributes standardization

Original	**Replacement**
Male	m
Female	f

For date of birth it is converted into standardize format (day-month-year) as shown in Table 4.

Table 4. Date attributes standardization

Original	**Replacement**
10/01/57	10.1.1957
01.01.57	1.1.1957

After standardization the records from various database, it is easy to search for the duplicate records.

3.2 Key Creation

The idea is to choose a high quality key in terms of accuracy, completeness and consistency. The choice of a key with a low completeness value; after a sorting on the basis of such a key, the potential matching records can be not close, due to null values. The choice of the key is a fundamental step of the matching process, as the results of the matching are directly influenced by it. A key is defined to be a sequence of a subset of attributes or substrings within the attributes, chosen from the record. Keys must provide sufficient discriminating power. The name, date of birth, sex, and all concern data are stored in database. It is created by the first three consonants of a last name are concatenated with the first letter of the first name field, followed by a gender field and day, month, last three numbers of the year of birth . If the name is only one word then it will take consonants of three words of the name. The basic key creation for this system is shown in Table 5.

Table 5. Key Creation Example

ID	Name	DOB	Sex		Key
1	Bela Williams	10.1.1957	m		wllbm101957
2	Kennith Showers	9.10.1946	m		shwkm910946
3	Shane Garcia	3.5.1948	m	→	grcsm35948
4	Bela Moore	8.2.1989	m		mrbm821989
5	Bela Willam	10.1.1957	m		wllbm101957
..	...	...	...		...

3.3 Sorting

The keys are now used for sorting the entire dataset with the intention that all equivalent or matching data will appear close to each other in the final sorted list. The keys are now used for sorting the entire dataset. Sort the records in the data list using the created list. All records are sorted according to the alphabetic manner of created key. Therefore, all equivalent or matching records will appear close to each other. If the data was not sorted, a record may be near the beginning of the array of records and a duplicate record may be near the end of the array of records.

Table 6. Key Sorting Example

ID	Key
1	grcsm35948
2	mrbm821989
3	shwkm910946
4	wllbm101957
5	wllbm101957
...	....

3.4 Reference List

In **Reference List**, we keep record of which blocks maintain which ranges. The Reference List is the central access point of the infrastructure for managing the execution of spilt blocks and maintains the block data. The system use flexible "**Adaptive Block**" to obtain high performance, to reduce search space and to cover the entire same entities block size. All records are divided into blocks according to Lexicographical Order. Match tasks are done within the blocks which are Partitioning Dynamically. Reference List will perform the preprocessing and post-processing of partitioning data as well as matching the final results. It is also responsible for assigning the input records into corresponding blocks and retrieving the data from corresponding block.

```
Input  : Created Key
Output : Corresponding Block

Begin
Search the Key initial in Block List
If Key is found in Block List
        Then go to the corresponding block
Else
        Go to the Mixed block
End
```

Algorithm 1. Reference List Process

3.5 Mixed Block

In Mixed Block, incomplete information will be kept in that block. Information may be omitted because user is unwilling or unable to supply it. Due to missing data values or other data quality issues in real-world data, it may not always be possible to assign entities to a unique block. We therefore assign such entities to a dedicated *mixed* block. Entities of this block have to be matched against the entities of all blocks.

3.6 Dynamic Block Creation for Data Partitioning

Data Partitioning can be of great help in facilitating the efficient and effective management of big data. But data partitioning could be a complex process which has several factors that can affect partitioning strategies and design, implementation, and management considerations in an entity resolution. In this paper, the system first partition the dynamic blocks of the input records and record searching in that dynamic blocks. Data partitioning also results in faster data addressing and efficient data retrieval.

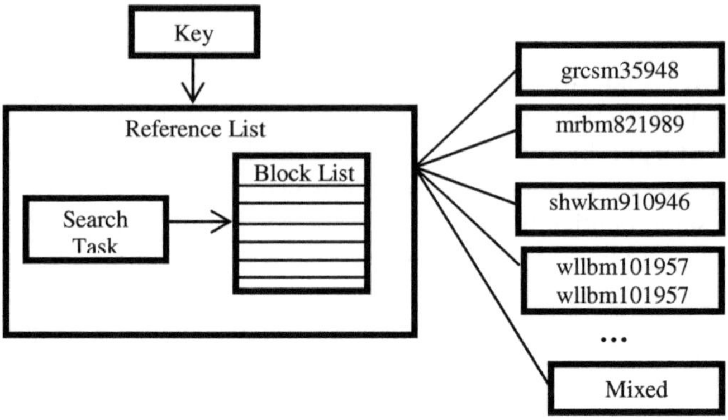

Fig. 3. Dynamic block creation

After sorting the all records in the database, it was partitioned the collection of records into set of blocks in the alphabetic manner. Firstly, the created key is search in the reference list to know which blocks it should be kept. If the initial of the created key is the same as the reference list's key then it will save in the corresponding block. Otherwise, the record must be added to the mixed block. This is done until no record remains in the data source. The same records are in the same group as shown in figure 4.

```
Input:
-Number of records n
Output:
-Block []

Begin
 n=Number of records
 Initialize Block[]={}
  For i=1..n:
    Create Key
    Sort records on key
    Look in the Reference List to add into the
    corresponding block
     If created key initial key==Reference List
     initial key
        Then add to the corresponding Block[]
    Else
          Add to the Mixed Block
    Until no records
    Return dynamic block
End
```

Algorithm 2. Dynamic Block Creation

Blocking entails a logical partitioning of the input entities such that all matching entities should be assigned to the same output partition, also called block. By restricting the match comparisons to the entities of the same block the match overhead can drastically be reduced. Algorithm 1 describes the creation of the dynamic block.

3.7 Record Matching Process

The basic idea of string comparison is to be able to compare pairs of strings such as 'Mon, Min' that contain minor typographical error. String Comparison in record linkage can be difficult because lexicographically "near by" record look like "matches" when they are in fact not. Possible errors range from small typographical slips to complete name and address changes. The equational theory does not detect the second pair as duplicates. So, a *distance function* applied to the fields of the records. A string comparator function returns a value depending on the degree of the match of the two strings. Because pairs of strings often exhibit typographical variation, the record linkage needs effective string comparison functions that deal with typographical variations.

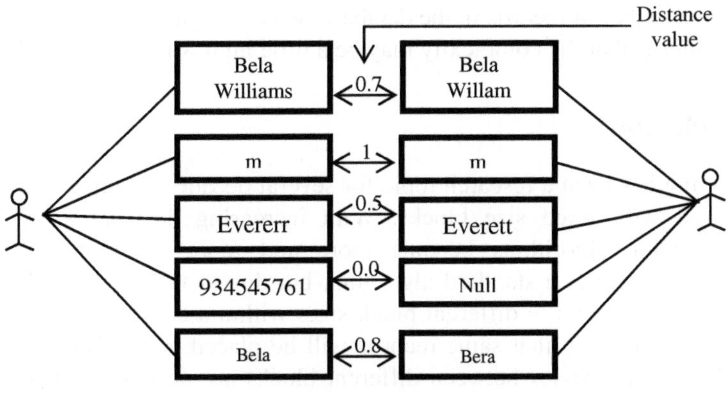

Fig. 4. Record Matching process

Each selected field will have a quality of match value. The individual field quality of matches will be combined into a composite score for each record. This value will be used to determine the overall probability of record matching. There are several methods available for determination of record matching from the quality of match of the individual fields. The individual field quality of matches will be combined into a composite score for each record. This value will be used to determine the overall probability of record matching. Firstly, we have to compare selected field comparison. If the result is greater than a fixed threshold, the two values are considered equal. After that we have to compare overall fields' comparison. The number of equal pairs of values in the two records is greater than a threshold, and then the two records are considered as duplicate. Both thresholds are set to fixed value. Example of record matching is shown in figure 4.

According to the above Proposed Step, records can be retrieved efficiently. The output blocks may largely differ in their size depending on the entity value. The Proposed System is efficient due to less search space and avoiding unnecessary links to other blocks. The proposed system will reduce the search space for the entity matching and linkage process between different blocks. The complexity of the preprocessing steps will increase because it needs to check before splitting the entities into different blocks.

4 Evaluation of the System

In traditional method, assuming two data sets with n records each are to be linked, the blocking key results in b blocks, and each block contains n/b records, the resulting number of record pairs is O($\frac{n^2}{b}$).This is of course the ideal case, hardly ever achievable with real data. In general, the number of record pairs is O($\sum_{i=1}^{b} n_i^2$) where n_i is the number of records in block i.

In the above proposed system, create key phase is an O (n) operation, the sorting phase is O (n log n). The resulting number of record pair comparisons is O(w n), where n is the number of records in the database, w is the dynamic block size. But one thing to consider is that the complexity may be a little high when dividing blocks.

5 Conclusion

Entity Matching has been a research topic for several decades. Many researches have been proposed with fixed size blocks. With increasing data set sizes, efficient duplicate detection algorithms become more and more important. The Sorted Neighborhood method is a standard algorithm, but due to the fixed window size, it cannot efficiently respond to different block sizes within a data set. With fixed sized blocks, sizes are fixed so that same records will be placed into different blocks and additional linkage processes between different blocks are needed. In this paper, we have proposed the partition the block into dynamic block to reduce search space and reduce linkage process when comparing records pairs. Dynamic Blocking is needed to efficiently eliminate linkage process between different blocks. Our partitioning strategy is independent on the applied area chosen. The proposed system enables matching entities assigned to the same block. It uses Dynamic Blocking to obtain high performance, to reduce search space and to cover the entire same entity within block. Future directions of this work include tree like structure as an additional step to do efficient record searching in entity resolution systems.

References

1. Elmagarmid, A.K., Ipeirotis, P.G., Verykios, V.S.: Duplicate Record Detection: A survey. IEEE Transactions on Knowledge and Data Engineering 19(1), 1–26 (2007)
2. Zhao, H., Ram, S.: Entity matching across heterogeneous data sources: An approach based on constrained cascade generalization. Data & Knowledge Engineering 66, 368–381 (2008)

3. Getoor, L., Machanavajjhala, A.: Entity Resolution: Theory, Practice and Open Challenges. In: The 38th International Conference on Very Large Data Bases, Istanbul, Turkey, August 27-31. Proceedings of the VLDB Endowment, vol. 5(12) (2012)
4. Jin, L., Li, C., Mehrotra, S.: Efficient Record Linkage in Large Data Sets. In: Proceedings Eighth International Conference on Database Systems for Adavanced Applications (DASFAA 2003) (2003)
5. Kolb, L., Thor, A., Rahm, E.: Block-based Load Balancing for Entity Resolution with Map Reduce. In: CIKM 2011, October 24-28 (2011)
6. Hernández, M.A., Stolfo, S.J.: The merge/purge problem for large database. In: Proceedings of SIGMOD 1995 (1995)
7. Bilenko, M., Kamath, B., Mooney, R.J.: Adaptive Blocking: Learning to scale Up Record Linkage. In: Proceedings of the 6th IEEE International Conference on Data Mining, Hong Kong (December 2006)
8. Christen, P.: Towards Parameter-free Blocking for Scalable Record Linkae. TR-CS-07-03, Technical Reports
9. Christen, P.: A Survey of Indexing Techniques for Scalable Record Linkage and Deduplication. IEEE Transactions on Knowledge and Data Engineering (2011)

A New Fast and Robust Stereo Matching Algorithm for Robotic Systems

Masoud Samadi and Mohd Fauzi Othman

Center for Artificial Intelligence and Robotics
Universiti Teknologi Malaysia, Kuala Lumpur, Malaysia
Solariseir@IEEE.org, fauzi@fke.utm.my

Abstract. In this paper, we propose a new area-based stereo matching method by improving the classical Census transform. It is a difficult task to match corresponding points in two images taken by stereo cameras, mostly under variant illumination and non-ideal conditions. The classic Census nonparametric transform did some improvements in the accuracy of disparity map in these conditions but it also has some disadvantages. The results were not robust under different illumination, and because of the complexity the performance was not suitable for real-time robotic systems. In order to solve these problems, this paper presents an improved Census transform using Maximum intensity differences of the pixel placed in the center of a defined window and the pixels in the neighborhood to reduce complexity and obtain better performance and needs a smaller window size to obtain best accuracy compared to the Census transform. Experimental results show that the proposed method, achieves better efficiency in term of speed and robustness against illumination changes.

Keywords: Census, Disparity Map, Robots, Robotic Systems, Stereo Vision, Stereo Matching.

1 Introduction

Most mobile robots are designed to operate under different circumstances. In order to perform their task, robots should have a sufficient realization of their environment to work safely within their workhouse [1]. 3D information of all objects in the field of view of the robot is essential for a reliable operation of autonomous vehicles. Classic sensors which are based on ultrasonic, laser, or time-of-flight, may generate 3D data, but they suffer from a low resolution and are high at cost. Nowadays, the fast development of robotics systems, using perception modules in different fields of autonomous robot operations like navigation [5], [6], visual servoing [7], or grasping [8] are became more important. Mobile robot systems, especially those which require fast and robust navigation methods are spread around the world. The main requirements for such a visual system are reliability and the rapid ability to convert images of a scene to 3D data to be used in the immediate and future reactions of the robot [4].

Stereo vision is a technology that uses two cameras which are horizontally aligned and mounted on a common baseline to estimate the depth of a field of view. Some of

P. Meesad et al. (Eds.): *IC²IT2013*, AISC 209, pp. 281–290.
DOI: 10.1007/978-3-642-37371-8_31

the most important advantages of this technique are high resolution and low cost. In addition, images taken by this method can be used for other applications too. As a result of this passive technology, it does not affect the operation area of the robot, thus it is an acceptable choice for home applications. However, due to the similarity, matching the corresponding points which had been collected by a pair of cameras is difficult to solve. Also, the environment conditions are not fixed and the illumination is always changing in the field of robot views, which leads to mismatching for many stereo matching algorithms [9]. In here we propose a method to overcome the mismatching problem in different lighting condition without losing the performance of the stereo matching algorithm in terms of speed and accuracy.

2 Related Works

In recent years, a lot of work has been done in the field of stereo matching techniques with researchers approaching many new stereo matching algorithms. These algorithms are evaluated and compared in the work of Hirschmuller in [12], [5]. Many matching algorithms just use the intensity value of a pixel to find the similarity between a pair of images. Since these methods depend directly on environment illumination, changes in the luminance will affect their results. The sum of squared differences (SSD) and the sum of absolute differences (SAD) [7] are two famous algorithms which work with a pixel intensity value. The correspondence feature in the second image should have the same value with the reference image. This assumption can be true only under ideal lightning conditions, which is hard to find in robotic application.

In [8], Hannah proposed a slightly different method named normalized cross-correlation (NCC). This algorithm reduced the effect of environment illumination changes on the result in the calculation of the images normalized intensity value. However NCC did not works directly with pixel values and it still had weakness regarding illumination [10]. The most different stereo matching algorithms are that presented by Zabih and Wodfile [11], which was a non-parametric local transform. These methods were very well-known Census and Rank transform, which look for a certain relations between image pixels in a defined window. These kind of non-parametric transforms have a good ability to reduce the effect of noises on images and achieve good results in various test conditions.

Census transform obtains its result with a bit-wise calculation and compared to previous methods which only use one pixel value to perform the operation, has more efficient and accurate results and a great capability to be used in robotic systems because of its robustness to illumination changes. This particular feature makes Census efficient in different kinds of environments. Equations (1) and (2) explain the Census transform formula that uses intensity relation between the pixel in the center and neighbor pixels in a defined window size:

$$\xi(p_1, p_2) = \begin{cases} 0 & p_1 < p_2 \\ 1 & p_1 > p_2 \end{cases} \tag{1}$$

p_1 presents the pixel located in the center of the window and adjacent pixels are demonstrated by p_2 . If the intensity value of p_1 is more than its neighbor pixel p_2 the value of Census for that specific pixel will be 1 and if it is less than the neighbor pixel it will be 0 . The equation in (2) describes how the Census value for the center pixel of defined window is calculated:

$$I_{census}(u,v) = \otimes_{i=n} \otimes_{j=m} \left(\xi\big(I(u,v), I(u+i, v+j)\big)\right) \tag{2}$$

Operator $\otimes$ denotes a bit-wise concatenation, where $n * m$ is Census windows size and u, v are pixel coordinates.As mentioned before, the classic Census transformation has the ability to reduce noise and increase the performance of stereo matching algorithms in non-ideal lighting condition and different environments. One of the most important drawbacks in this method is the complexity of the algorithm which consumes a lot of system resources for computation. Then Zabih proposed the Sparse Census transform to reduce the calculation time. This method avoids the twice calculation for one pixel by defining a certain space to reduce the number of comparisons, although the formula is the same as normal Census. One of the best ways to decrease the computation time and solve the speed problem in robotic systems is changing the Census transform equation. In the next sections, our solution is explained and the test results are compared with existing methods.

3 An Improved Census Transform Stereo Matching Algorithm

In order to increase the performance of Census method in terms of speed, without losing accuracy of the disparity map, we propose a novel method to achieve this goal. The workflow of the proposed method is demonstrated in Figure 1. As shown in the chart, at the very first stage of the procedure, the stereo images are captured from a pair of cameras which are vertically aligned. These cameras are calibrated before taking photos and the calibration values are saved in a matrix to use for future use. In the next step, images are going through undistortion function to correct the lens distortion and rectify them by using the calibration matrix that was mentioned in the previous stage. More explanation about image rectification and camera calibration can be found in [13]. Next, the pair of images is ready for further stereo matching process. In previous methods, the rectified images are transformed to an 8 bit images by means of the Census method before computing the initial three dimensional image.

3.1 Differential Transform

To reduce the computation cost and increase the speed performance of the Census transform, we changed the Census equation and detract the number of comparisons, so the complexity of the code is reduced and the program is executed with more speed than the old Census transform. Thus, in order to increase performance, we change the Census equation (1) to (3)

$$\zeta(p_1, p_2) = |p_1 - p_2| \quad (3)$$

The result of ζ is the absolute differences between the center pixel of the Census window and the neighborhood pixels. The equation (2) is reformulated as shown in (4)

$$I_{diff}(u,v) = \max_{i=-\frac{n}{2}} \max_{j=-\frac{m}{2}} (\zeta(I(u,v), I(u+i, v+j))) \quad (4)$$

The I_{diff} determines the maximum differences between the pixel located in the center of the window and the neighbor pixels in a window with n columns and m rows. As the maximum intensity difference between two pixels in an image could be 255, thus the transformed matrix could be saved as an 8 bit image and does not depend on the window size. As a result of these changes, the execution speed of the code is increased and the resource consumed by the program is reduced, all these modification lead to computation the disparity map in less time compared to previous methods.

3.2 Stereo Matching

The results of the last step, so called Differential transform, are needed to produce the three-dimensional matrix. This matrix, which is called disparity space image (DSI), is generated with the size of $number\ of\ disparities * image\ size$. DSI is created by computing the hamming distance between the transformed images as shown in the formula:

$$DSI_L(u,v) = Hamming(L_{diff}(u,v), R_{diff}(u-d, v)) \quad (5)$$

DSI is computed when the right image is shifting horizontally from right to left (Figure 2), and the left image is used as the reference. The shifting distance is defined by d in equation (5) and this value is called the disparity value. In fact, this is the furthest range which the stereo vision system can calculate. To achieve a better quality, we generate two different DSI. In the first, the left image is used as the reference, and in the second, the left image is used as the shifting and the right image is fixed. After computing the two DSI, the cost of aggregation is done on both DSI from the previous step. The cost aggregation is a method that sums the intensity values of pixels in a certain $[M, N]$ window (6)

$$DSI_{L,aggr}(u,v) = \sum_{n=-\frac{N}{2}}^{n=\frac{N}{2}} \sum_{m=-\frac{M}{2}}^{m=\frac{M}{2}} DSI_L(u+n, v+m) \quad (6)$$

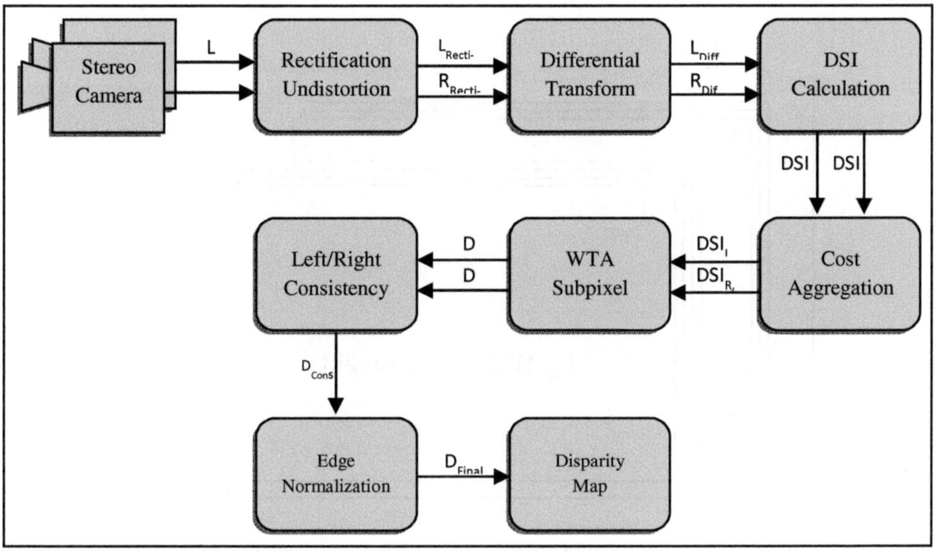

Fig. 1. The workflow of the proposed algorithm

We then compare the aggregated value of different DSI layers to find the lowest value. When the lowest value is detected, it is seemed to be the best match for disparity in that specific region. This procedure will continue until all pixels in the image are checked. Thus, the initial disparity map is computed by using the winner take all method on the sum of the intensity value in a specific region in each DSI level (7).

$$D_L(u,v) = \min_{d=d_{min}} D_{L,aggr}(u,v,d) \tag{7}$$

After computing two initial disparity maps, with one left image acts as the reference and one right image as another reference, it is time to do a consistency check between D_R and D_L . This method will help to remove uncertain areas and occluded matches. In the final step noises are removed by a median filter and edges are normalized by dilate and blur steps.

4 Experiment Result

The proposed method, Census, Rank and Census Sparse have been tested on the Middlebury stereo vision dataset (Figure 3) and the results show that this method gained better execution speed compared to other methods. As a result of our experiments, it is clear that the proposed method needs a smaller window size to achieve the best accuracy in comparison to other algorithms. Thus, the executing speed of the algorithm increases and the time to calculate the disparity map reduces.

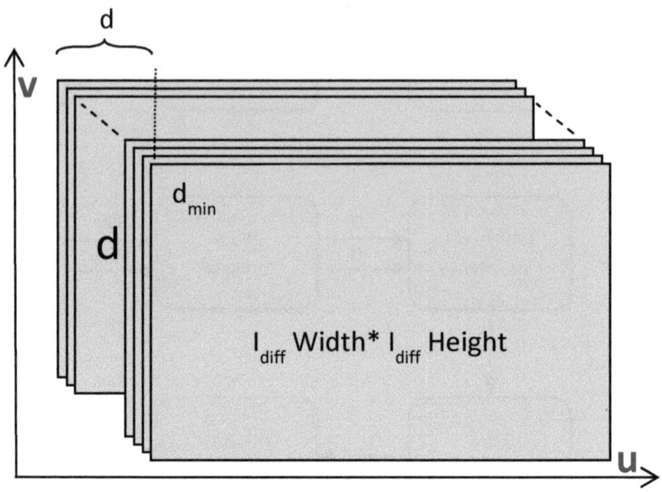

Fig. 2. The three-dimensional data called Disparity space image (DSI) in the size of *disparities * image size*

By our experiment and as mentioned in [2], [3] the best result of Census transforms are achieved with $16 * 16$ window size, while the proposed algorithm can achieve the best accuracy with $5 * 5$ window. This particular feature can reduce the large window size overload on the processor unit and decrease the calculation time. The time consumed by each method and the error rate of them are demonstrated in Table 1.

Table 1. Speed and Accuracy Comparison on the Middlebury Dataset

Metrics	*Dataset*	*Methods*				
		Proposed	**Census**	**Census Sparse**	**Rank**	**Rank Sparse**
Time (s)	Tsukuba	*0.00542*	*0.12891*	*0.02683*	*0.07890*	*0.02846*
	Cones	*0.00803*	*0.12891*	*0.07734*	*0.17412*	*0.06475*
	Teddy	*0.00790*	*0.31196*	*0.08282*	*0.16543*	*0.05402*
	Venus	*0.00732*	*0.28534*	*0.08445*	*0.08445*	*0.06348*
	Average	*0.00716*	*0.21378*	*0.06786*	*0.12572*	*0.05267*
Error (%)	Tsukuba	*10.67*	*11.39*	*10.37*	*11.99*	*12.19*
	Cones	*15.18*	*15.68*	*15.24*	*16.22*	*15.32*
	Teddy	*16.55*	*16.46*	*17.32*	*16.74*	*17.02*
	Venus	*5.78*	*5.07*	*4.85*	*6.22*	*5.69*
	Average	*12.045*	*12.15*	*11.945*	*12.7925*	*12.555*

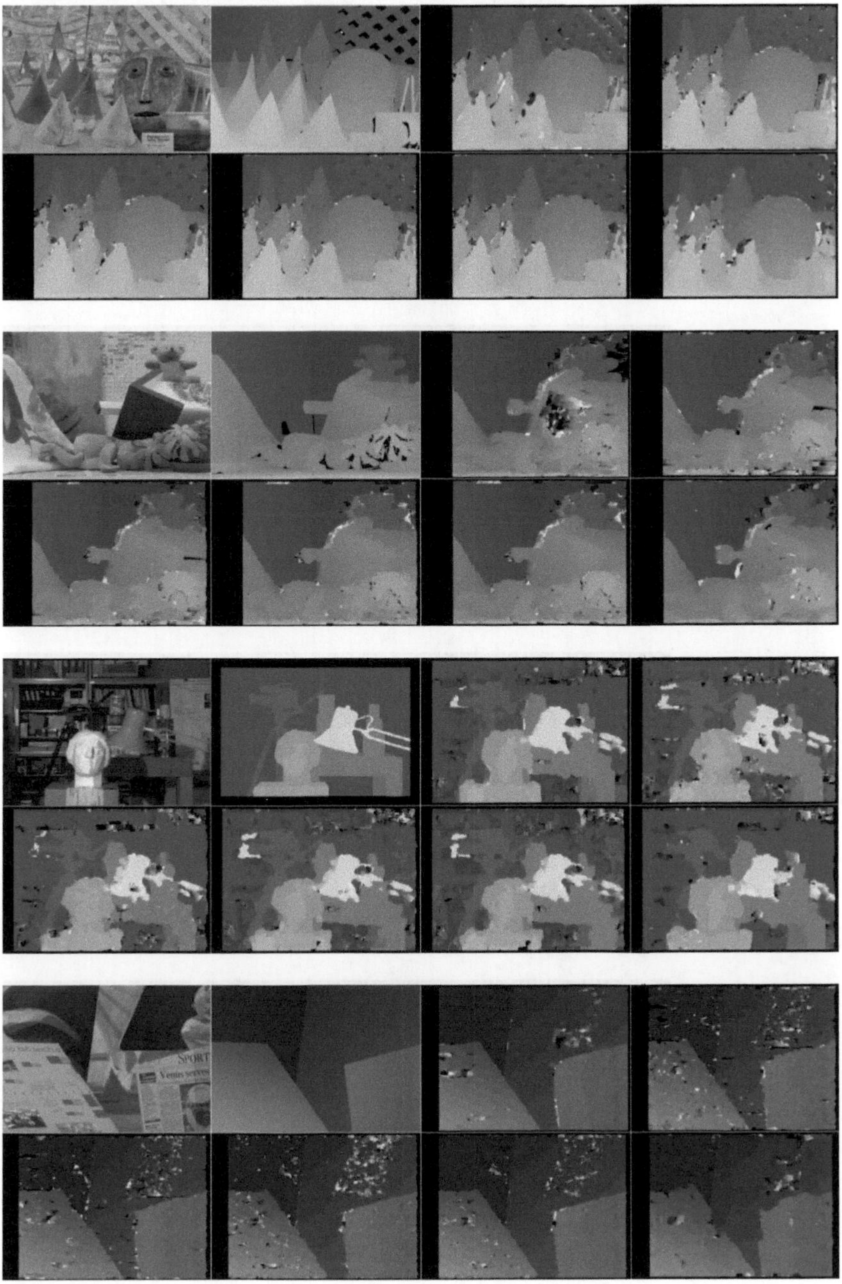

Fig. 3. Comparison of the disparity map for the Middlebury dataset (Cones, Teddy, Tsukuba, Venus). From left to right (First row): Left Stereo image and its local ground truth disparity map, SAD, Census, (Second row): Census Sparse, Rank, Rank Sparse, Proposed method.

Fig. 4. Comparison of disparity map in different brightness and exposure for Middlebury dataset (Cones, Teddy, Tsukuba, Venus). From left to right SAD, Census Sparse, Rank, Proposed method.

The proposed method also inherits some particular features of Census transform such as being robust to different luminance conditions, and has the ability to reduce effects of camera gain and bias, as shown in Figure 4.

5 Conclusion and Future Works

In this work, we deal with the stereo matching speed problem. Our research lies attention on non-parametric image transform methods to gain robustness of the stereo matching algorithm in different lighting conditions without losing the execution speed of the program. To achieve this goal we reformulate the old Census transform method and gained better performance in computing the disparity map compares to previous works. This method can be implementing in real-time stereo vision which is using in robotic applications. In future work, the proposed method will be implemented on a stereo vision-based robot which uses Intel x86 CPU architecture, the code already developed in C++ language under Visual Studio 2010 Environment therefore the program can be executed on the mentioned platform to analyze the robot behavior in a real world experiment.

Acknowledgements. We would like to thank the Center for Artificial Intelligence and Robotics (CAIRO), Universiti Teknologi Malaysia (UTM) and Ministry of Higher Education (MOHE) for providing the research facilities. This research work has been supported the grant No. VOT 00G20.

References

1. Calderon, J., Obando, A., Jaimes, D.: Road Detection Algorithm for an Autonomous UGV Based on Monocular Vision. In: The Electronics, Robotics and Automotive Mechanics Conference, CERMA 2007, pp. 25–28 (September 2007)
2. Zinner, C., Humenberger, M., Ambrosch, K., Kubinger, W.: An Optimized Software-Based Implementation of a Census-Based Stereo Matching Algorithm. In: Bebis, G., Boyle, R., Parvin, B., Koracin, D., Remagnino, P., Porikli, F., Peters, J., Klosowski, J., Arns, L., Chun, Y.K., Rhyne, T.-M., Monroe, L. (eds.) ISVC 2008, Part I. LNCS, vol. 5358, pp. 216–227. Springer, Heidelberg (2008)
3. Weber, M., Humenberger, M., Kubinger, W.: A Very Fast Census-Based Stereo Matching Implementation on a Graphics Processing Unit. In: 2009 IEEE 12th International Conference on the Computer Vision Workshops (ICCV Workshops), September 27-October 4 (2009)
4. Grigorescu, S.M., Macesanu, G., Cocias, T.T., Dan, P., Moldoveanu, F.: Robust Camera Pose and Scene Structure Analysis for Service Robotics. Robotics and Autonomous Systems 59(11), 899–909 (2011)
5. Brown, M.Z., Burschka, D., Hager, G.D.: Advances in Computational Stereo. IEEE Transactions on Pattern Analysis and Machine Intelligence 25(8), 993–1008 (2003)
6. Scharstein, D., Szeliski, R.: A Taxonomy and Evaluation of Dense Two-Frame Stereo Correspondence Algorithms. Int. J. Comput. Vision 47(1-3) (2002)

7. Kanade, T., Kano, H., Kimura, S., Yoshida, A., Oda, K.: Development of a Video-Rate Stereo Machine. Paper Presented at the Intelligent Robots and Systems 1995, Proceedings. 1995 IEEE/RSJ International Conference on Human Robot Interaction and Cooperative Robots, August 5-9 (1995)
8. Hannah, M.J.: Computer Matching of Areas in Stereo Images. Stanford University (1974)
9. Xin, L., Zhou, H., Yu, F., Li, X., Xue, B., Song, D.: A Robust Local Census-Based Stereo Matching Insensitive to Illumination Changes. In: 2012 International Conference on the Information and Automation, ICIA, June 6-8 (2012)
10. Murray, D., Little, J.J.: Using Real-Time Stereo Vision for Mobile Robot Navigation. Autonomous Robots 8(2) (2000)
11. Zabih, R., Woodfille, J.: Non-Parametric Local Transforms for Computing Visual Correspondence. In: Eklundh, J.-O. (ed.) ECCV 1994. LNCS, vol. 801, pp. 151–158. Springer, Heidelberg (1994)
12. Hirschmuller, H., Scharstein, D.: Evaluation of Stereo Matching Costs on Images with Radiometric Diferences. IEEE Transactions on Pattern Analysis and Machine Intelligence 31(9), 1582–1599 (2009)
13. Fusiello, A., Emanuele, T., Alessandro, V.: A Compact Algorithm for Rectication of Stereo Pairs. Mach. Vision Appl. 12(1), 16–22 (2000)

Ten-LoPP: Tensor Locality Preserving Projections Approach for Moving Object Detection and Tracking

M.T. Gopala Krishna[1], M. Ravishankar[1], and D.R. Rameshbabu[2]

[1] Department of Information Science and Engineering,
[2] Department of computer Science and Engineering,
Dayananda Sagar College of Engineering, Bangalore, India
{gopalmtm,ravishankarmcn,bobramysore}@gmail.com

Abstract. In recent years, automatic moving object detection and tracking is a challenging task for many computer vision applications such as video surveillance, traffic monitoring and activity analysis. In this regard, many methods have been proposed based on different approaches. Despite of its importance, moving object detection and tracking in complex environments is still far from being completely solved for low resolution videos, foggy videos, and also Infrared video sequences. A novel scheme for Moving Object detection based on Tensor Locality Preserving Projections (Ten-LoPP) approach is proposed. Consequently, a Moving Object is tracked based on the centroid and area of a detected object. Numbers of experiments are conducted for indoor and outdoor video sequences of standard PETS, OTCBVS, Videoweb Activities datasets and also our own collected video sequences comprising partial night vision video sequences. Results obtained are satisfactory and competent. Comparative study is performed with existing well known traditional subspace learning methods.

Keywords: Moving Object Detection, Moving Object Tracking, Video Surveillance, Tensor Locality Preserving Projections (Ten-LoPP).

1 Introduction

The increasing need for an intelligent automated video analysis has generated a great deal of interest in detection & tracking of dim moving objects at low contrast and foggy image sequences are challenging research topics in computer vision & machine learning community. The major potential applications include intelligent automated video surveillance, motion-based recognition, video indexing etc [1]. Several object detection and tracking methods have been reported based on point detectors, segmentation, background modeling and supervised classifiers in the literature.

A comparative evaluation of interesting point detectors is studied in [2]. The dim and foggy image sequences are obtained by weather variations in the environment and dynamic range of the camera lens. with this, the obtained image sequences does not provide a clear visual information of objects. However, in contrast to visual image sequences, the dim and foggy images sequences have extremely low visibility,

P. Meesad et al. (Eds.): *IC²IT2013*, AISC 209, pp. 291–300.
DOI: 10.1007/978-3-642-37371-8_32

resulting limited information for detection & tracking tasks. To overcome the shortcoming nature of the dim and foggy video sequences, different approaches impose different constraints to provide solutions. As per the literature, various techniques have been reported for automatic moving object detection and tracking in video sequences. The techniques include: Subspace Learning Models, Support Vector Models and Gaussian Models [3]. Methods pertain to Gaussian Models are [4, 5, 6], wavelet transformation [7, 8]. There are still many other algorithms have described in [9-12]. Most of the earlier works on statistical image analysis represent an image by a vector in high-dimensional space. However, In many real-world applications an image is intrinsically a matrix or a second order tensor. Initially, these tensors have to be unfolded into one-dimensional vectors then the embedding methods can be applied. Because of this reason, some useful information in the original data may not be captured well. Moreover, high-dimensional vectorized representations also suffer from the curse of dimensionality and the high computational demand. Recently, a number of algorithms for dimensionality reduction with tensor representation have been proposed in [13-16]. Tensor-based algorithms directly treat the data as a tensor, and thus effectively avoid the problems derived from treating data as vectors. In order to preserve the local manifold structure, He et al. extended their proposed locality preserving projections (LPP) algorithm to handle second order data tensors [17, 18, 19]. This technique was further extended to tensor LPP (TLPP)[20]. As an alternative to the PCA and the LPP, in this paper, Ten-LoPP is applied to detect moving object in foggy and infrared video sequences. For an image of size m × n, it is represented as the second-order tensor in the tensor space $R^{m \times n}$. The intrinsic local geometrical structure of images is retained as same as original in tensor space. Main concern is developing a efficient algorithm to detect fast/small moving objects in foggy and infrared video sequences and to achieve better performance. In this work an idea of Ten-LoPP is explored to detect and track moving objects.

2 Proposed Method

In past few years, some embedding methods have proposed for feature extraction and dimensionality reduction in various machine learning and pattern classification tasks. In such applications, LPP has used for moving object detection [21]. The LPP is linear projective maps that optimally preserve the neighborhood structure of the data of a image sequence. LPP are obtained by finding the optimal linear approximations to the Eigen functions of the Laplace Beltrami operator on the manifold. Though LPP are linear projections and similar to nonlinear techniques such as Laplacian Eigenmaps, it only processes the vectorized data, suffering from the curse of dimensionality and the high computation. To overcome the drawback of LPP, in this work, a new multilinear approach is proposed for moving object detection based on Tensor Locality Preserving Projections (Ten-LoPP). Ten-LoPP is a natural extension of LPP in multilinear case. It is particularly useful in applications when the data samples are naturally represented as matrices or higher-order tensors.

So, in order to perform a most challenging tasks in video surveillance in complex, low contrast, infrared and foggy video sequences including partial night vision video sequences, the present system is proposed with a novel tensor embedded technique for efficient moving object detection and tracking. Ten-LoPP mainly considers data as tensors of arbitrary order as input. This tensor embedded technique allows the relationships between dimensions of a tensor representation to be efficiently characterized. The algorithm of the proposed work is outlined with the help of a block diagram shown in Fig.1.

In the first step, Ten-LoPP is applied on video sequences to locate the potential moving object regions, by exploiting the intrinsic local geometric and topological properties of the manifold. In the second step, post processing is performed on the responses of Ten-LoPP to eliminate the noise and unwanted small objects present in the input video sequences. Further, a moving object is detected by applying median filter followed by morphological dilation operation in the third step. Finally, a moving object is tracked by using the area and centroid features of a detected moving object.

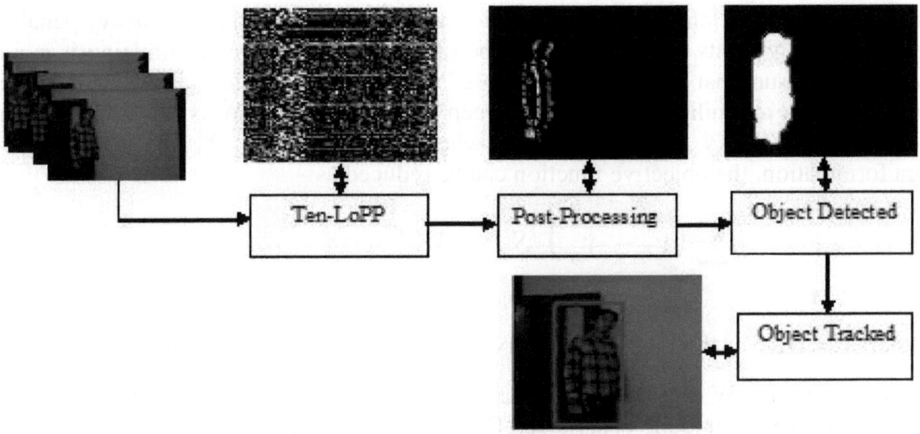

Fig. 1. Block Diagram of the Proposed System

2.1 Tensor Locality Preserving Projections (Ten-LoPP)

The section explains about the notation and basic definitions of multilinear algebra along with the mathematical structure of the tensor. A tensor is known as m-way array or multidimensional matrix. Tensors are multilinear mappings over a set of vector spaces [22]. An Nth tensor A is denoted as $A_{i1...i2,..in}$, where $1 \le i_n \le I_n$, and is written as $A_n \in R^{I1\times...\times I2...\times In}$. In tensor terminology, column vectors are referred as mode-1 vectors and row vectors as mode-2 vectors. The mode-n vectors are the column vectors of matrix $A_n \in R^{I1...I2...In}$ that result from flattening or unfolding the tensor A. It is easy to see that the mode-n vectors of an Nth order tensor A are the I_n-dimensional vectors obtained from A by varying index while keeping the other indices fixed [22]. The scalar product <A, B> of two tensors A, B $\in R^{I1\times...\times I2...\times In}$ is defined as:

$$\langle A, B \rangle = \sum_{i_1} \sum_{i_1} \cdots \sum_{i_1} a_{i_1 i_2 i_3 \ldots i_n} b_{i_1 i_2 i_3 \ldots i_n} \tag{1}$$

Tensor Locality Preserving Projection Formulation: The formulation of Ten-LoPP is essentially based on Locality preserving projection (LPP) [23, 24]. LPP seeks to preserve the intrinsic geometry of the data. Given a data set $\{X_1,X_2, \ldots\ldots X_n\}$ in higher dimensional Euclidean space we construct a weighted graph G=(V,E) with edges connecting nearby points to each other. we choose the heat kernel S_{ij} as the weights of the edges. Consider the problem of mapping the weighted graph G to a line so that connected points stay as close together as possible. Let $y = (y_1, y_2, \ldots y_n)$. A reasonable criterion for choosing a "good" map is to minimize the following objective function

$$\min_y = \sum_{i,j} (y_i - y_j)^2 S_{i,j} \tag{2}$$

The objective function with the symmetric weights S_{ij} ($S_{ij} = S_{ji}$) incurs a heavy penalty if neighboring points x_i and x_j are mapped far apart. Therefore, minimizing it is an attempt to ensure that if x_i and x_j are "close" then y_i and y_j are close as well. S_{ij} can be thought of as a similarity measure between objects. Suppose W is a transformation vector and set $y = (y_1, y_2, \ldots y_n) = WTX$, where $X = \{X_1,X_2, \ldots\ldots X_n\}$ By simple algebra formulation, the objective function can be reduced as:

$$\sum_{i,j} (y_i - y_j)^2 S_{ij} = W^T XLX^T W \tag{3}$$

Where L = D - S is the Laplacian matrix, $D_{ii} = \sum_j S_{ji}$ is a diagonal matrix. The transformation vector W that minimizes the objective function is given by the minimum Eigen value solution to the generalized Eigen value problem [25].

$$XLX^T W = \lambda XDX^T W \tag{4}$$

LPP considers an object as a vector. So firstly, the image matrices must be vectorized. The resulting image vectors usually lead to a high-dimensional vector space, which is suffer from the curse of dimensionality and the high computational demand. This problem is more apparent in small-sample-size cases such as image recognition. Such a matrix-to-vector transform may cause the loss of some structural information residing in original images [26]. However, it is realized that most objects in pattern recognition are more naturally represented as second or higher-order tensors and the filtered Gabor image is a third-order tensor [27]. We now discuss about the conduct of subspace analysis in general case, in which objects are represented as tensors of second or higher order.

Given n data points $A_1; \ldots;A_n$ from an unknown manifold M embedded in a tensor space $R^{I_1 \times \ldots \times I_n}$, Ten-LoPP finds K optimal projections $U_i \in R^{Ii \times Ii}$; ($I_i < I_i$, i = 1; ....;K) such that the local topological structure of M is preserved and the intrinsic geometric property is effectively captured. In order to construct a neighborhood graph Ġ to capture the intrinsic geometric structure of M and to apply the heat kernel to define the affinity matrix $S = (S_{ij})_{n \times n}$, it can be defined as

$$S_{ij} = \begin{cases} exp(-\|A_i - A_j\|_f^2 / t), & if\ A_j \in O(k, A_i)\ or\ A_i \in O(k, A_j) \\ 0, & Otherwise \end{cases} \quad (5)$$

Where $O(k, A_i)$ denotes the k nearest neighbor of A_i and t is a positive constant. Both k and t can be determined empirically. For the above said generic approach, the moving object detection problem can be summarized as follows:

Learning & classification Phases

- Acquire two training sample images (current and previous) as a original input images.
- Perform the subtraction process by subtracting previous image from current images, then compute average image X.
- Follow the steps described in section 2, from constructing adjacency grap W.
- Apply Ten-LoPP on obtained X and W of each iteration.
- Finally, Ten-LoPP gives the embedded result Y.

Fig. 2. Resulted image of Ten-LoPP

2.2 Post Processing

In this section, post processing stage is described. The resulted Ten-LoPP image is absolute and is normalized by multiplying with an empirical value in order to eliminate much of the noise present. Resulted image of Post Processing is shown in Fig. 3.

Fig. 3. Resulted Post processing Image

2.3 Moving Object Detection

The section describes about the successful moving object detection process. As the resulted post processing image still has few noise and small unwanted components, further these will be eliminated by applying median filter followed by morphological dilation operation. The successful detection of moving object is shown in Fig 4. Finally, an object is tracked by using the area and centroid of the moving objects.

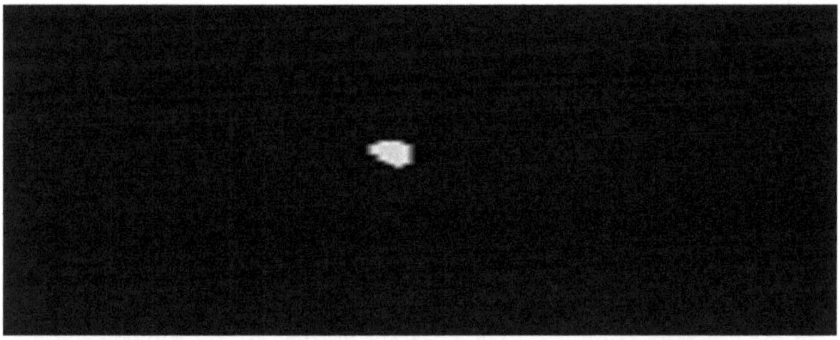

Fig. 4. Successful detection of moving object

2.4 Moving Object Tracking

In this section, tracking of the moving object is briefed. The Area and Centroid of the detected moving object is calculated by using Eqns. (6) and (7). The obtained Area is compared with an empirical threshold ‘T’, if Area is less than threshold ‘T’ then the corresponding region is considered as non moving object otherwise it is treated as a region of moving object. A centroid of the corresponding Area is used to track the moving object. The successful tracking of the moving objects is shown in Fig. 5.

$$Area = \sum_{(x,y)\varepsilon R} \sum I(x,y) \tag{6}$$

where I(x,y) is a detected moving object.

$$Centroid \quad : \overline{X} = \frac{1}{N} \sum_{(x,y) \varepsilon R} x, \ \overline{Y} = \frac{1}{N} \sum_{(x,y) \varepsilon R} y \tag{7}$$

where N is the total number of pixels in the moving object.

$$I(x,y) = \begin{cases} \text{Moving Region,} & \text{if Area} > T \\ \text{Non Moving Region,} & \text{Otherwise} \end{cases} \tag{8}$$

Fig. 5. Moving Object Tracking

3 Experimental Results and Comparative Study

The system is experimented on standard PETS, OTCBVS, Videoweb Activities datasets and also on our own collected video sequences. The system is able to detect & track moving objects in indoor & outdoor environments efficiently. A comparative study is performed with the existing well known approaches [21] and the corresponding results are shown in Figure 6. Columns 1 & 2 of Figure 6 show the results obtained for PCA & LPP [21] approaches. Column 3 of Figure 6 shows the results obtained for the proposed method with successive detection of moving objects and their tracking in video. The performance evaluation of the proposed method shows the better detection rate of the moving objects than the existing approaches and is shown in Table 1. It is also noticed that the false alarm rate is considerably reduced compared to the existing well known approaches.

Table 1. Percentage of Detection and False Alarm Rate

Input Sequences	Approaches	Detection Rate	False Alarm Rate
Fig 6 (d ,e ,f) Arial (Videoweb Activities Dataset)	PCA	60.00	45.45
	LPP	60.00	40.00
	Proposed	**96.88**	**28.74**
Fig 6 (m, n ,o) (OTCBVS dataset)	PCA	40.00	71.43
	LPP	60.00	57.14
	Proposed	**80.77**	**41.67**

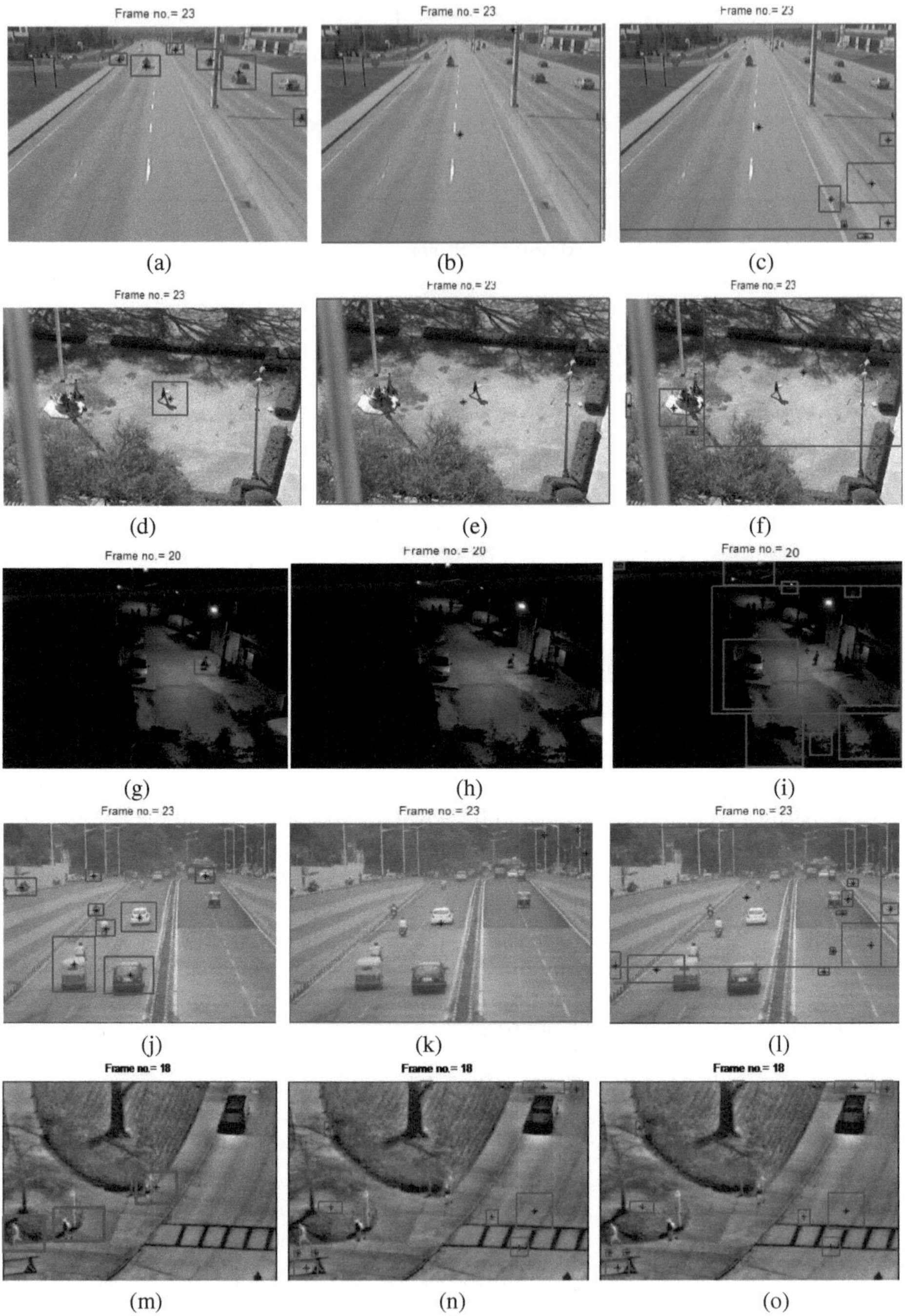

Fig. 6. Comparison results of proposed approach with other existing approaches on standard PETS, Videoweb Activites & our own collected datasets: Column-1(a,d,g,j,m): Moving Object Detection and Tracking using PCA. Column-2(b,e,h,k,n): Moving Object Detection and Tracking using LPP. Column-3(c,f,i,l,o): proposed approach(Ten-LoPP).

4 Conclusion

The proposed system is a framework for moving object detection and tracking in low resolution, foggy and infrared video sequences. The system also detects and tracks moving objects in partial night vision video sequences effectively. Experimental results on standard PETS, OTCVBS, Videoweb Activities dataset and our own collected video sequences show efficient moving object detection and tracking. The present system performance outreaches the existing well known traditional subspace learning methods. The proposed algorithm is adaptable to detect and track moving objects in various environmental conditions especially in the highway traffics during foggy weather and also to detect far visible small objects in an Arial view. A fully night vision video sequences can be considered for further work.

References

1. Yilmaz, A., Javed, O., Shah, M.: Object Tracking. A Survey. ACM Computing Surveys 38(4), 1–45 (2006)
2. Mikolajczyk, K., Schmid, C.: A performance evaluation of local descriptors. In: IEEE Conference on Computer Vision and Pattern Recognition (CVPR), pp.1615–1630 (2003)
3. Bouwmans, T.: Subspace learning for background modeling, A survey. Recent Patents on Computer Science 2(3), 223–234 (2009)
4. Stauffer, C., Grimson, W.: Adaptive background mixture models for real-time tracking. In: IEEE Conf. on CVPR, pp. 246–252 (1999)
5. Allili, M., Bouguila, N., Ziou, D.: A Robust video foreground segmentation by using generalized Gaussian mixture modeling. In: Fourth Canadian Conf. on Computer and Robot Vision (CRV), pp. 503–509 (2007)
6. Elgammal, A., Harwood, D., Davis, L.: Non-parametric model for background subtraction. In: Vernon, D. (ed.) ECCV 2000. LNCS, vol. 1843, pp. 751–767. Springer, Heidelberg (2000)
7. Hewer, G., Kenney, C., Hanson, G., Kuo, W., Peterson, L.: Detection of small objects using adaptive wavelet based template matching. In: SPIE Conference on Signal and Data Processing of Small Targets 1999, Denver, Colorado, vol. 3809, pp. 95–106. SPIE (1999)
8. Li, L.Q., Tang, Y.Y.: Wavelet-Hough transform and its applications to edge and target detections. International Journal of Wavelets, Multi-Resolution and Information Processing 4, 567–587 (2006)
9. Yilmaz, A., Shafique, K., Shah, M.: Target tracking in airborne forward looking infrared imagery. Image and Vision Computing 21, 623–635 (2003)
10. Zhang, F., Li, C., Shi, L.: Detecting and tracking dim moving point target in IR image sequence. Infrared Physics & Technology 46, 323–328 (2005)
11. Zhang, T., Li, M., Zuo, Z., Yang, W., Sun, X.: Moving dim point target detection with three-dimensional wide-to-exact search directional filtering. Pattern Recognition Letters 28, 246–253 (2007)
12. Chen, Y., Liu, X., Huang, Q.: Real-time detection of rapid moving infrared target on variation background. Infrared Physics & Technologym 51, 146–151 (2008)
13. Yamazaki, M., Xu, G., Chen, Y.-W.: Detection of moving objects by independent component analysis. In: Narayanan, P.J., Nayar, S.K., Shum, H.-Y. (eds.) ACCV 2006. LNCS, vol. 3852, pp. 467–478. Springer, Heidelberg (2006)

14. Bucak, S., Gunsel, B., Gursoy, O.: Incremental non-negative matrix factorization for dynamic background modeling. In: International Workshop on Pattern Recognition in Information System (2007)
15. Li, X., Hu, W., Zhang, Z., Zhang, X.: Robust foreground segmentation based on two effective background models. In: MIR, pp. 223–228 (2008)
16. Vasilescu, M.A.O., Terzopoulos, D.: Multilinear subspace analysis of image ensembles. In: IEEE Conference on Computer Vision and Pattern Recognition, vol. 2, pp. 93–99 (2003)
17. He, X., Cai, D., Niyogi, P.: Tensor subspace analysis. In: Advances in Neural Information Processing Systems, vol. 18 (2005)
18. Yan, S., Xu, D., Yang, Q., Zhang, L., Tang, X., Zhang, H.-J.: Multilinear discriminate analysis for face recognition. IEEE Transactions on Image Processing 16, 212–220 (2007)
19. He, X., Yan, S., Hu, Y., Niyogi, P., Zhang, H.-J.: Face recognition using Laplacian faces. IEEE Transactions on Pattern Analysis Machine Intelligence 27, 328–340 (2005)
20. Dai, G., Yeung, D.Y.: Tensor embedding methods. In: Proceedings of the Twenty First National Conference on Artificial Intelligence (AAAI), Boston, Massachusetts, USA, pp. 330–335 (2006)
21. Gopala Krishna, M.T., Manjunath, V.N., Ravishankar, M., Ramesh, D.R.: Locality Preserving Projections for Moving Object Detection (C3IT 2012). Procedia Technology 4, 624–628 (2012)
22. Lathauwer, D.L., Moor, D.B., Vandewalle, J.: A multilinear singular value decomposition. SIAM Journal on Matrix Analysis and Applications 21(4), 1253–1278 (2000)
23. Belkin, M., Niyogi, P.: Laplacian eigenmaps and spectral techniques for embedding and clustering. In: Advances in Neural Information Processing Systems, Vancouver, British Columbia, Canada, vol. 14 (2001)
24. Belkin, M., Niyogi, P.: Laplacian eigenmaps for dimensionality reduction and data representation. Neural Computation 13, 1373–1397 (2003)
25. He, X., Niyogi, P.: Locality preserving projections. In: Advances in Neural Information Processing Systems, vol. 16 (2003)
26. Hu, D., Feng, G., Zhou, Z.: Two-dimensional locality preserving projections (2DLPP) with its application to palmprint recognition. Pattern Recognition 40, 339–342 (2007)
27. Yan, S., Xu, D., Yang, Q., Zhang, L., Tang, X., Zhang, H.-J.: Multilinear discriminate analysis for face recognition. IEEE Transactions on Image Processing 16, 212–220 (2007)

CAMSHIFT-Based Algorithm for Multiple Object Tracking

Sorn Sooksatra and Toshiaki Kondo

School of Information, Computer, and Communication Technology
Sirindhorn International Institute of Technolgy, Thammasat University,
Bangkok, Thailand
ohm.sorn@gmail.com, tkondo@siit.tu.ac.th

Abstract. This paper presents a technique for object tracking by using CAMSHIFT algorithm that tracks an object based on color. We aim to improve the CAMSHIFT algorithm by adding a multiple targets tracking function [1].When one object is selected as a template, then it will search objects that have the same hue value and shape by shape recognition. So,the inputs of the algorithm are hue values and shape of the object. When all objects are absent in the frame, the algorithm will search whole frame to find most similar-looking objects and track them. The important task of the object tracking is to separate a target from background or frame that in some cases where the noise is present in the tracking frame. Then object identification method was added to the algorithm for filtering the noise and counting numbers of objects to make decide how many targets track.

Keywords: CAMSHIFT, Multiple target, Object tracking, Shape recognition.

1 Introduction

Nowadays, the technology is growing fast. The application of the technology will be diverse and then interfere of machine is not only keyboard or mouse any more. There is more advance technology that supports more applications at present and in future. Object tracking is one of them. Therefore, object tracking is one of important research topics in computer vision applications.

There are a lot of applications related to object tracking. For example Meanshift tracking for Tracking of Moving Images on an Autonomous Vehicle Testbed Platform in [2] and Armed robot for selecting object based on color and shape of object in [3]. There are other application such as auto-mobile driver assistance, vehicle navigation, robotics, human-computer interaction, video surveillance, biometric, video games, industrial automation and security. There are several techniques of object tracking. A certain technique can be usable and suitable for each application. The most popular one is CAMSHIFT that has been used for this research paper.

The CAMSHIFT (Continuously Adaptive Mean Shift) that is detailed in [4] is derived from the Mean Shift algorithm and it is one of popular techniques

P. Meesad et al. (Eds.): *IC²IT2013*, AISC 209, pp. 301–310.
DOI: 10.1007/978-3-642-37371-8_33

that are used for object tracking. This algorithm is based on object hue value from HSV image. It has been applied as basis and core component in many applications.

This technique is a robust nonparametric technique for finding the peak probability distribution. Then, the target color will have high probability and other will be almost zero. We use this for separate target from background. Therefore, CAMSHIFT will be most efficient when target's color is totally different to background's color.

There are five main drawbacks of traditional CAMSHIFT. First, its tracking failure when color of background and target are similar. Second, object can change their appearance anytime (rubric example in [5]) that makes it difficult to track a target by using only one histogram. Third, traditional CAMSHIFT cannot track more than one target. We can use multiple tracker for tracking multiple object same in [1]. Fourth, traditional CAMSHIFT can track similar color object that overlap real target instead. Lastly,when saturation is low. Hue will becomes quite noisy, because of small range in Hue, the small number of discrete hue pixels cannot adequately represent slight changes in RGB [4].

The main aim of this paper is to modified the CAMSHIFT to be able to track more than one target, identify the shape of object and make CAMSHIFT work every situation even if the color of a target is similar to that of background and background's color being for quite same color.

2 CAMSHIFT Algorithm

In the CAMSHIFT process as shown in Fig. 1, the process will collect the data or parameters of each step for update data in each frame since video consists of many frames. The important parameter is location of search window and image order moment. The process is following

2.1 Select Region of Interesting (ROI)

This step will find the initial location of target located in the frame and obtain data related to hue value in the target for making color histogram in the next step. The initial location will be used for calculation in tracking process that starts in this region to avoid other object in the frame.

2.2 Create Color Histogram

Histograms of each color will be created. The height of each column represents number of pixels in a frame region having that "hue." Hue is one of the three values describing a pixel's color in the HSV (Hue, Saturation, and Value) color model as shown in Fig. 2(b).

This will take number of pixels with the Hue value within ROI. These data will be used for generating probability map process.

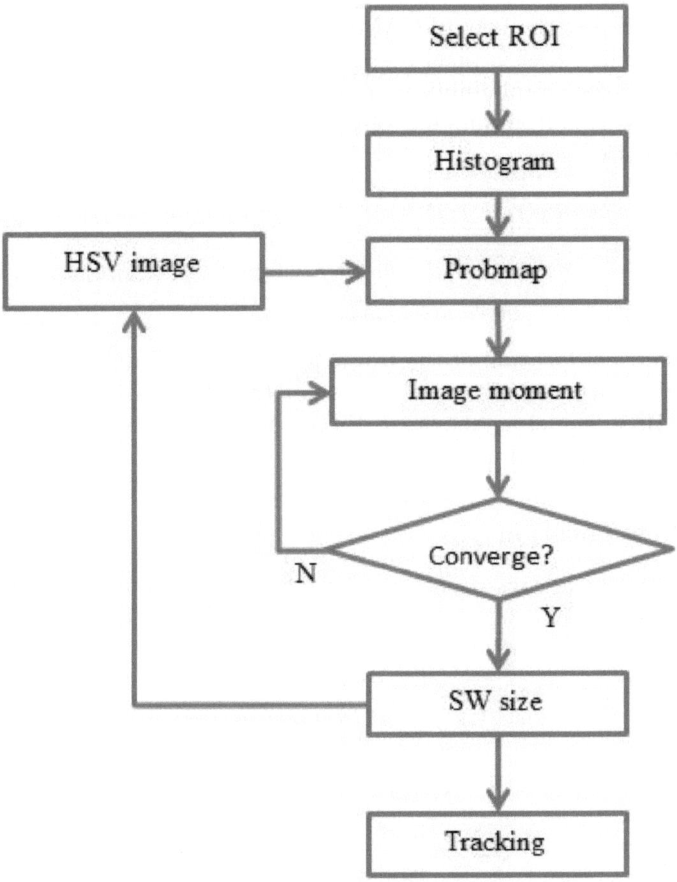

Fig. 1. CAMSHIFT Flowchart

Fig. 2. (a) input image and region of interest (ROI) and (b) Histogram of hue within the ROI [6]

2.3 Generating Probability Map (probmap)

This is the first step in the loop of CAMSHIFT technique. The probability map shows the probability of each pixel in each frame used for separating target and background. Therefore, probability map is the heart of this technique.

For probability, it is from number of colors that obtained from previous step by following equation.

$$probmap(row, col) = numberof(hue(row, col)) \tag{1}$$

$$probmap = probmap/max(probmap) \tag{2}$$

$$probmap = probmap \times 255 \tag{3}$$

where row is vertical location of the frame and col is horizontal location in the frame. Since this algorithm is based on target that has colourful color or high saturation in HSV model as shown in Fig. 3. So, it ignore pixel that has low saturation for calculating in probability map (i.e. make it to be zero).

2.4 Shift New Location

Since the target can always move, we must find the new centroid within the search window (SW) by computing 0th (M_{00}), 1st (M_{10}) and 2nd (M_{01}) order image moment within the ROI by following equation

$$M_{00} = \sum_{row}\sum_{col} probmap(row, col) \tag{4}$$

$$M_{10} = \sum_{row}\sum_{col} col \times probmap(row, col) \tag{5}$$

$$M_{01} = \sum_{row}\sum_{col} row \times probmap(row, col) \tag{6}$$

$$X_C = \frac{M_{01}}{M_{00}}; Y_C = \frac{M_{10}}{M_{00}} \tag{7}$$

where X_C and Y_C is location of the new centroid of SW. This step will be repeated until it converges.

2.5 Calculating the Size of Tracking Window

The size of tracking window depends on the size of the target or zeroth order image moment (M_{00}) calculated from previous step. Then size of SW is equal to function of M_{00} that in following equation.

$$H = \sqrt{\frac{M_{00}}{255}}; W = 1.2 \times H \tag{8}$$

where H and W are the height and width of tracking window respectively. This equation comes from Ref. [4] that are adjusted for our implementation.

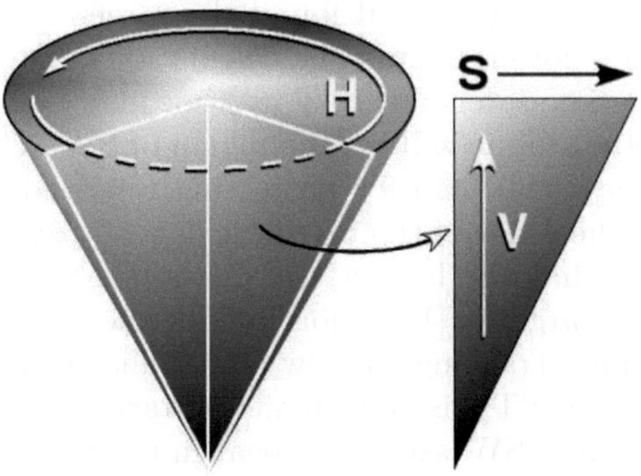

Fig. 3. HSV color model

2.6 Tracking Process

For the tracking window, it will locate the tracking window by the location that gets from previous step. When the target is disappear on the frame. It will track whole frame instead of SW again same as in [5] until target enter in the frame again. After this step, the process will get back to step C again.

3 CAMSHIFT Improvement

In this paper, we tried several ways to improve CAMSHIFT to support more applications. We tried to solve the noise problem in the probability map of CAMSHIFT and add more feature with multiple target tracking and object identification. The detail of improvement is following.

3.1 Multiple Tracking Technique

The main purpose of this paper is to add multiple tracking ability to CAMSHIFT technique. According to today's applications. The object tracking does not track single or only one target anymore. There are application that use multiple tracking such as counting the object, track many human faces in the camera and other.

The technique that we use is making all parameters used for tracking into array. As mentioned before, object tracking needs data about related to location and size of the SW. That information comes from minimum row (rmin), maximum row (rmax), minimum column (cmin), maximum column (cmax) and 3 image order moment .The format of array is following:

$$P[i] \in P[1], P[2], P[3], ..., P[n] \tag{9}$$

$$T[i] \in T[1], T[2], T[3], ..., T[n] \tag{10}$$

where P is one of parameters , T is one of target and n is number of target in the frame.The initial search area of T[1] is in the ROI and other T[i] is the whole frame and it will be reduced if found the target.The process is same as traditional CAMSHIFT for each T[i]. The calculation of P[i] of T[i] is processed in order of array (10) (i.e. order T[1] to T[n]) not at the same time. When T[i] is finished to obtain the P[i], it will set the region of SW of T[i] in the probmap to be zero. This process will help for protecting repeated or same location of T[i+1] and so on.For the order of the target, it depends on size or density (M_{00}) of the target. If target has the largest size in the probmap, it will be tracked first. However, the first target (T[1]) is normally be the object that in the ROI on the initial frame of video.Tracking window appearance is depending on the size (M_{00}) of each T[i] . If size of T[i] is very small (i.e. target is far away from camera or disappear in the frame), SW will not appear in the frame and T[i] will search whole frame again.There are some rule for this multiple tracking technique as following.

1. First target that enter to the frame or in ROI region will be MT, Second will be ST-1, third will be ST-2 and so on. If two or more ST has already in the frame, it will choose order for ST randomly.
2. When MT disappears in the frame, it will choose one of ST to be MT.
3. When center of each tracking window are close enough, it will merge 2 or more target to be one target.

3.2 Object Identification Method

This method is used as noise filter in probability map. This method is concern about the size of object both noise and target. In most situations, size of noise is less than target size. We can eliminate noise concerning size of object by using [7] in this process as follows

1. The algorithm will search from top-to-bottom scan order in probability map.
2. When it found connected or only one pixel that has intensity more than zero. They will be counted as one object.
3. The size of each object is equal to number of pixel of that object.
4. The algorithm will filter out the object that has size less than threshold (noise object).

This method can be used as object counting by counting the object after filter out.We must adjust size of object for this method to make most efficacy for object tracking.

3.3 Shape Recognition

For the shape recognition, this process begins after tracking process. It help identify the shape of the target (T[i]) by using compactness theorem in [8] and [9]. Eq.(11) show the equation of compactness calculation.

$$C = \frac{P^2}{4\pi A} \tag{11}$$

where C is compactness, P is perimeter and A is area of the target. We use region of SW of each target in probmap as image input because probmap shows the target clearly form the background. Then, we use this to check whether shape of T[i] is same as template that we select in ROI or not. If difference of compactness between those two is too much, it will identify that this target is not target that we are interest due to the shape.

4 Experiment Results

At this point, this paper implement algorithm that can track more than one target at that time. Since we add more feature to this algorithm, then the flow chart is changed as in Fig 5. Fig. 4 shows the result about tracking window and probability map of traditional CAMSHIFT and improved CAMSHIFT. It shows that improved CAMSHIFT has less noise but less density of the target when compare with traditional CAMSHIFT and It can track more than one target. However, it has the problem when two target overlap to each other, then the order of the two targets are switched.

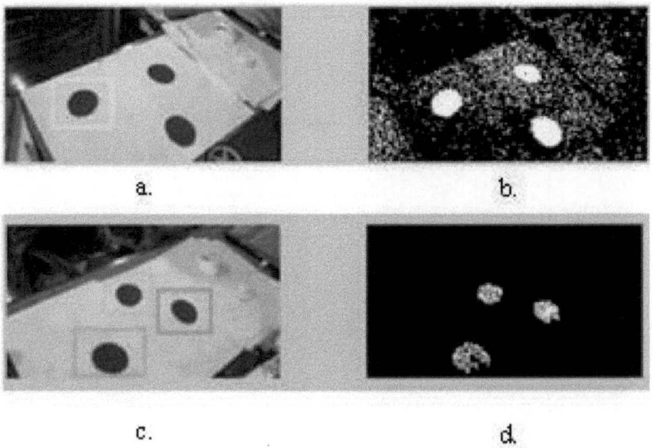

Fig. 4. (a) Input frame,(b) probmap of traditional,(c) input frame, and (d) probmap of improved

Fig. 6 shows the comparison between this algorithm without shape recognition (Fig. 6.a)and with shape recognition (Fig. 6.b). It shows that when a target is found, it checks the shape of the target if the target matches to template or not (Template in this case is circle shape). Therefore, square and triangle shaped objects are not be tracked. However, this tracker is quite sensitive when object's shape is changed due to environment. For example, saturation changed by the light reflection or object is shown only some part that causes the shape of object in probmap be changed.

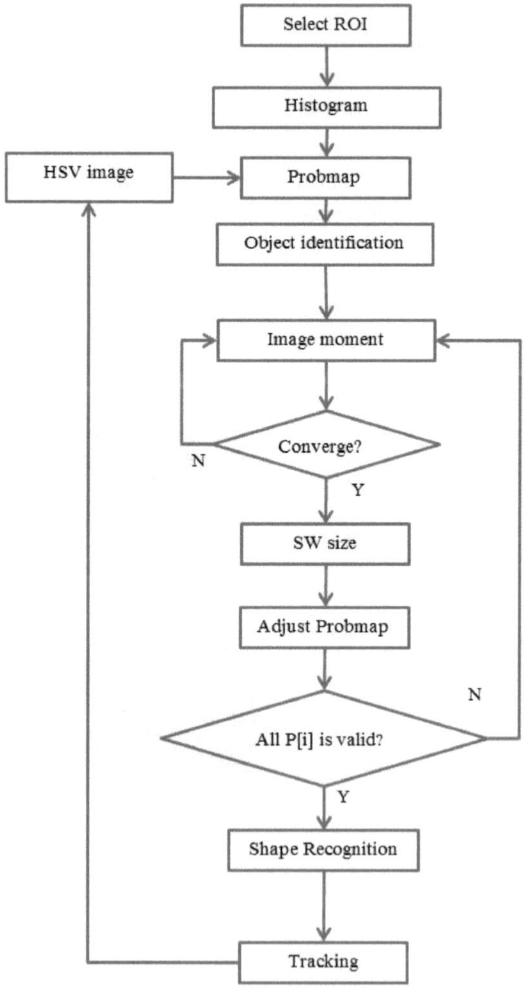

Fig. 5. Improved CAMSHIFT flowchart

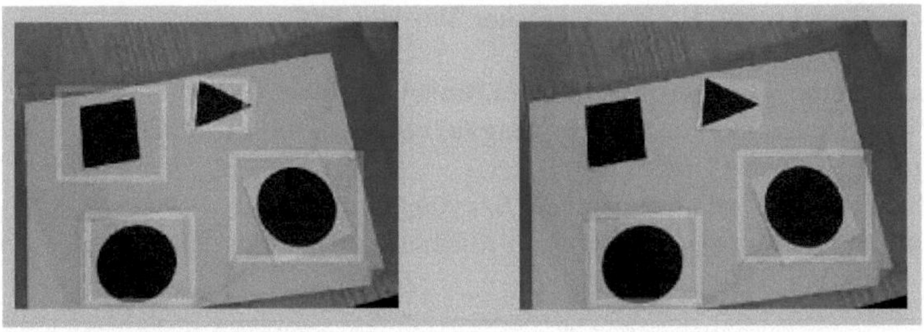

a. b.

Fig. 6. (a) Improved CAMSHIFT without shape recognition and (b) improved CAMSHIFT with shape recognition

5 Conclusion and Future work

The improved CAMSHIFT in this paper still uses the same concept of the traditional CAMSHIFT that tracking is based on color. We added a multiple objects tracking function and shape recognition for finding more applications. In addition, it has object identification for eliminating noise in the frame. One possible weakness of the proposal method is that the probabilities of a small object may be reduced in the noise elimination step.

As future work, we plan to introduce a more advanced shape recognition technique method in order to handle more complicated shapes. We will also work in speed-up the proposal method.

Acknowledgements. This research is financially supported by Center of Excellence in Biomedical Engineering, Sirindhorn International Institute of Technology (SIIT), Thammasat University (TU) and the National Research University Project, Thailand Office of Higher Education Commission.

References

1. Hidayatullah, P., Konik, H.: CAMSHIFT Improvement on Multi-Hue and Multi-Object Tracking. In: International Conference on Electrical Engineering and Informatics, Bandung, Indonesia (2011)
2. Gorry, B., Chen, Z., Hammond, K., Wallace, A., Michaelson, A.: Using Mean-Shift Tracking Algorithms for Real-Time Tracking of Moving Images on an Autonomous Vehicle Testbed Platform. World Academy of Science, Engineering and Technology (2007)
3. Amores, P.M.M., Gagwis, J., Jamora Marie, A.: PIC – Based Color And Shape Recognition: Using 3-Link Robotic Arm (2011)
4. Bradski, R.G.: Computer Vision Face Tracking For Use in a Perceptual User Interface. Microcomputer Research Lab. Santa Clara, CA. Intel Corporation (1998)
5. Exner, D., Bruns, E., Kurz, D., Grundhofer, A.: Fast and Robust CAMSHIFT Tracking. Bauhaus-University Weimar, Germany (2010)

6. Cognotics. How OpenCV's Face Tracker Works (2007), http://www.cognotics.com
7. Mathwork. Remove small objects from binary image (2012), http://www.mathworks.com/help/images/ref/bwareaopen.html
8. Pomplun, M.: Compactness (2007), http://www.cs.umb.edu/~marc/cs675/cv09-11.pdf
9. Zhongwan, L.: Compactness theorem. Mathematical Logic for Computer Science 2, 147–158 (1998)

Author Index

MIX
Papier aus verantwortungsvollen Quellen
Paper from responsible sources
FSC® C105338

If you have any concerns about our products,
you can contact us on
ProductSafety@springernature.com

In case Publisher is established outside the EU,
the EU authorized representative is:
Springer Nature Customer Service Center GmbH
Europaplatz 3, 69115 Heidelberg, Germany

Printed by Libri Plureos GmbH
in Hamburg, Germany